CRJ 900 Company Systems Manual

Printed in U.S.A.

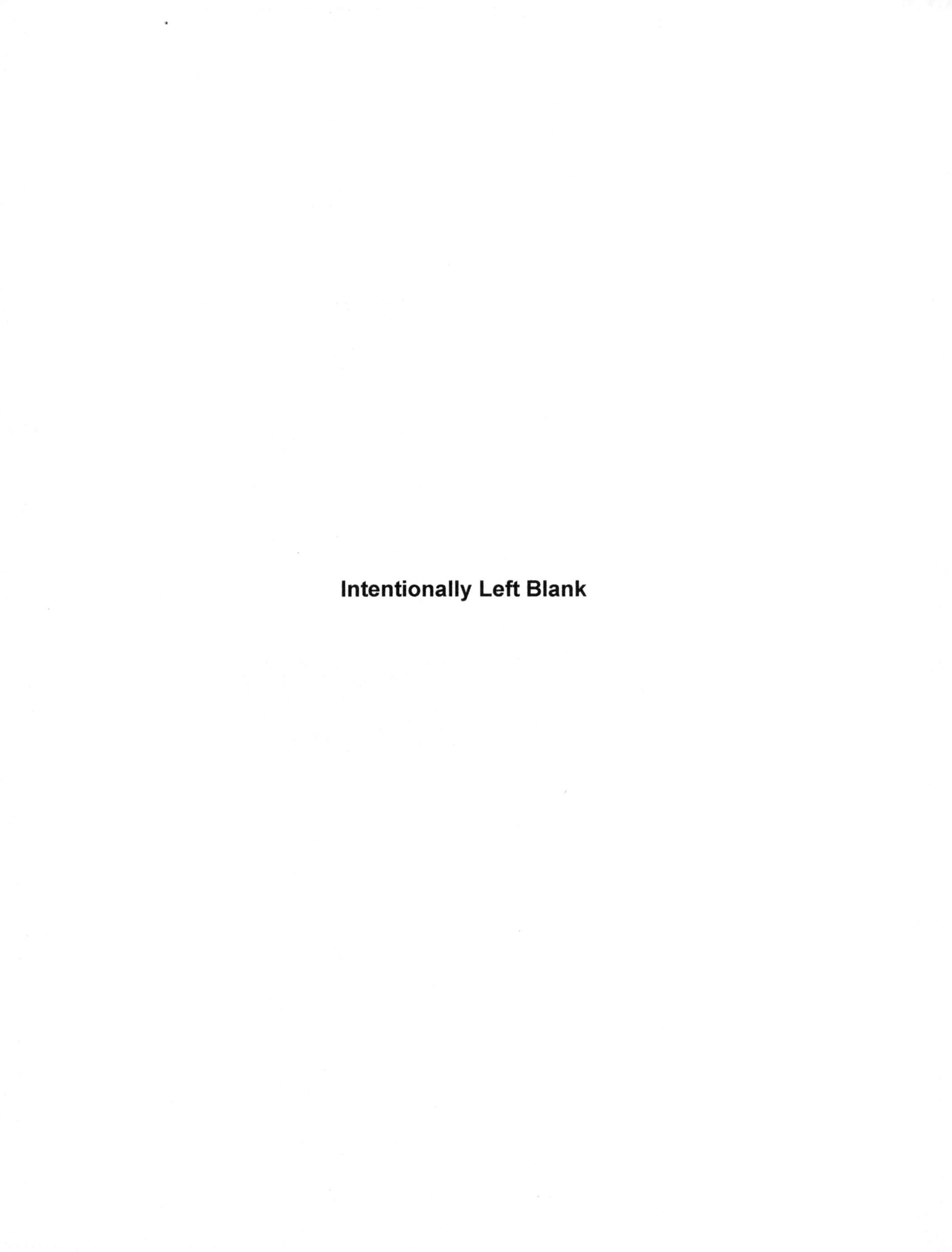

Intentionally Left Blank

Record of Revisions

Rev. No.	Effective Date	Rev. No.	Effective Date	Rev. No.	Effective Date
0	20 Jul 2017	16		32	
1		17		33	
2		18		34	
3		19		35	
4		20		36	
5		21		37	
6		22		38	
7		23		39	
8		24		40	
9		25		41	
10		26		42	
11		27		43	
12		28		44	
13		29		45	
14		30		46	
15		31		47	

Intentionally Left Blank

List of Effective Pages

Page No.	Revision No.	Effective Date
Record of Revisions		
i	0	20 Jul 2017
ii	0	20 Jul 2017
List of Effective Pages		
iii	0	20 Jul 2017
iv	0	20 Jul 2017
v	0	20 Jul 2017
vi	0	20 Jul 2017
vii	0	20 Jul 2017
viii	0	20 Jul 2017
ix	0	20 Jul 2017
x	0	20 Jul 2017
xi	0	20 Jul 2017
xii	0	20 Jul 2017
xiii	0	20 Jul 2017
xiv	0	20 Jul 2017
xv	0	20 Jul 2017
xvi	0	20 Jul 2017
xvii	0	20 Jul 2017
xviii	0	20 Jul 2017
Master Table of Contents		
xix	0	20 Jul 2017
xx	0	20 Jul 2017
xxi	0	20 Jul 2017
xxii	0	20 Jul 2017
xxiii	0	20 Jul 2017
xxiv	0	20 Jul 2017
xxv	0	20 Jul 2017
xxvi	0	20 Jul 2017
xxvii	0	20 Jul 2017
xxviii	0	20 Jul 2017
xxix	0	20 Jul 2017
xxx	0	20 Jul 2017
xxxi	0	20 Jul 2017
xxxii	0	20 Jul 2017
xxxiii	0	20 Jul 2017
xxxiv	0	20 Jul 2017
xxxv	0	20 Jul 2017
xxxvi	0	20 Jul 2017
xxxvii	0	20 Jul 2017
xxxviii	0	20 Jul 2017
xxxix	0	20 Jul 2017
xl	0	20 Jul 2017
xli	0	20 Jul 2017
xlii	0	20 Jul 2017
xliii	0	20 Jul 2017
xliv	0	20 Jul 2017
xlv	0	20 Jul 2017
xlvi	0	20 Jul 2017
Chapter 1 TOC		
i	0	20 Jul 2017
ii	0	20 Jul 2017
Chapter 1		
1	0	20 Jun 2017
2	0	20 Jun 2017
Chapter 2 TOC		
i	0	20 Jul 2017
ii	0	20 Jul 2017
Chapter 2		
1	0	20 Jul 2017
2	0	20 Jul 2017
3	0	20 Jul 2017
4	0	20 Jul 2017
5	0	20 Jul 2017
6	0	20 Jul 2017
7	0	20 Jul 2017
8	0	20 Jul 2017

Page No.	Revision No.	Effective Date
9	0	20 Jul 2017
10	0	20 Jul 2017
11	0	20 Jul 2017
12	0	20 Jul 2017
13	0	20 Jul 2017
14	0	20 Jul 2017
15	0	20 Jul 2017
16	0	20 Jul 2017
17	0	20 Jul 2017
18	0	20 Jul 2017
19	0	20 Jul 2017
20	0	20 Jul 2017
21	0	20 Jul 2017
22	0	20 Jul 2017
23	0	20 Jul 2017
24	0	20 Jul 2017
25	0	20 Jul 2017
26	0	20 Jul 2017
27	0	20 Jul 2017
28	0	20 Jul 2017
29	0	20 Jul 2017
30	0	20 Jul 2017
31	0	20 Jul 2017
32	0	20 Jul 2017
33	0	20 Jul 2017
34	0	20 Jul 2017
35	0	20 Jul 2017
36	0	20 Jul 2017
37	0	20 Jul 2017
38	0	20 Jul 2017
39	0	20 Jul 2017
40	0	20 Jul 2017
41	0	20 Jul 2017
42	0	20 Jul 2017
43	0	20 Jul 2017
44	0	20 Jul 2017
45	0	20 Jul 2017
46	0	20 Jul 2017
47	0	20 Jul 2017
48	0	20 Jul 2017
49	0	20 Jul 2017
50	0	20 Jul 2017
51	0	20 Jul 2017
52	0	20 Jul 2017
53	0	20 Jul 2017
54	0	20 Jul 2017
55	0	20 Jul 2017
56	0	20 Jul 2017
57	0	20 Jul 2017
58	0	20 Jul 2017

Chapter 3 TOC

Page No.	Revision No.	Effective Date
i	0	20 Jul 2017
ii	0	20 Jul 2017

Chapter 3

Page No.	Revision No.	Effective Date
1	0	20 Jul 2017
2	0	20 Jul 2017
3	0	20 Jul 2017
4	0	20 Jul 2017
5	0	20 Jul 2017
6	0	20 Jul 2017
7	0	20 Jul 2017
8	0	20 Jul 2017
9	0	20 Jul 2017
10	0	20 Jul 2017
11	0	20 Jul 2017
12	0	20 Jul 2017
13	0	20 Jul 2017
14	0	20 Jul 2017
15	0	20 Jul 2017
16	0	20 Jul 2017
17	0	20 Jul 2017

Page No.	Revision No.	Effective Date
18	0	20 Jul 2017
19	0	20 Jul 2017
20	0	20 Jul 2017
21	0	20 Jul 2017
22	0	20 Jul 2017
23	0	20 Jul 2017
24	0	20 Jul 2017
25	0	20 Jul 2017
26	0	20 Jul 2017
27	0	20 Jul 2017
28	0	20 Jul 2017
29	0	20 Jul 2017
30	0	20 Jul 2017
31	0	20 Jul 2017
32	0	20 Jul 2017
33	0	20 Jul 2017
34	0	20 Jul 2017
35	0	20 Jul 2017
36	0	20 Jul 2017

Chapter 4 TOC

Page No.	Revision No.	Effective Date
i	0	20 Jul 2017
ii	0	20 Jul 2017

Chapter 4

Page No.	Revision No.	Effective Date
1	0	20 Jul 2017
2	0	20 Jul 2017
3	0	20 Jul 2017
4	0	20 Jul 2017
5	0	20 Jul 2017
6	0	20 Jul 2017
7	0	20 Jul 2017
8	0	20 Jul 2017
9	0	20 Jul 2017
10	0	20 Jul 2017
11	0	20 Jul 2017
12	0	20 Jul 2017
13	0	20 Jul 2017
14	0	20 Jul 2017
15	0	20 Jul 2017
16	0	20 Jul 2017
17	0	20 Jul 2017
18	0	20 Jul 2017
19	0	20 Jul 2017
20	0	20 Jul 2017
21	0	20 Jul 2017
22	0	20 Jul 2017
23	0	20 Jul 2017
24	0	20 Jul 2017
25	0	20 Jul 2017
26	0	20 Jul 2017
27	0	20 Jul 2017
28	0	20 Jul 2017
29	0	20 Jul 2017
30	0	20 Jul 2017
31	0	20 Jul 2017
32	0	20 Jul 2017
33	0	20 Jul 2017
34	0	20 Jul 2017
35	0	20 Jul 2017
36	0	20 Jul 2017
37	0	20 Jul 2017
38	0	20 Jul 2017
39	0	20 Jul 2017
40	0	20 Jul 2017
41	0	20 Jul 2017
42	0	20 Jul 2017
43	0	20 Jul 2017
44	0	20 Jul 2017
45	0	20 Jul 2017
46	0	20 Jul 2017
47	0	20 Jul 2017
48	0	20 Jul 2017

Page No.	Revision No.	Effective Date
49	0	20 Jul 2017
50	0	20 Jul 2017

Chapter 5 TOC

Page No.	Revision No.	Effective Date
i	0	20 Jul 2017
ii	0	20 Jul 2017
iii	0	20 Jul 2017
iv	0	20 Jul 2017

Chapter 5

Page No.	Revision No.	Effective Date
1	0	20 Jul 2017
2	0	20 Jul 2017
3	0	20 Jul 2017
4	0	20 Jul 2017
5	0	20 Jul 2017
6	0	20 Jul 2017
7	0	20 Jul 2017
8	0	20 Jul 2017
9	0	20 Jul 2017
10	0	20 Jul 2017
11	0	20 Jul 2017
12	0	20 Jul 2017
13	0	20 Jul 2017
14	0	20 Jul 2017
15	0	20 Jul 2017
16	0	20 Jul 2017
17	0	20 Jul 2017
18	0	20 Jul 2017
19	0	20 Jul 2017
20	0	20 Jul 2017
21	0	20 Jul 2017
22	0	20 Jul 2017
23	0	20 Jul 2017
24	0	20 Jul 2017
25	0	20 Jul 2017
26	0	20 Jul 2017
27	0	20 Jul 2017
28	0	20 Jul 2017
29	0	20 Jul 2017
30	0	20 Jul 2017
31	0	20 Jul 2017
32	0	20 Jul 2017
33	0	20 Jul 2017
34	0	20 Jul 2017
35	0	20 Jul 2017
36	0	20 Jul 2017
37	0	20 Jul 2017
38	0	20 Jul 2017
39	0	20 Jul 2017
40	0	20 Jul 2017
41	0	20 Jul 2017
42	0	20 Jul 2017
43	0	20 Jul 2017
44	0	20 Jul 2017
45	0	20 Jul 2017
46	0	20 Jul 2017
47	0	20 Jul 2017
48	0	20 Jul 2017

Chapter 6 TOC

Page No.	Revision No.	Effective Date
i	0	20 Jul 2017
ii	0	20 Jul 2017

Chapter 6

Page No.	Revision No.	Effective Date
1	0	20 Jul 2017
2	0	20 Jul 2017
3	0	20 Jul 2017
4	0	20 Jul 2017
5	0	20 Jul 2017
6	0	20 Jul 2017
7	0	20 Jul 2017
8	0	20 Jul 2017
9	0	20 Jul 2017
10	0	20 Jul 2017

Page No.	Revision No.	Effective Date
11	0	20 Jul 2017
12	0	20 Jul 2017
13	0	20 Jul 2017
14	0	20 Jul 2017
15	0	20 Jul 2017
16	0	20 Jul 2017
17	0	20 Jul 2017
18	0	20 Jul 2017
19	0	20 Jul 2017
20	0	20 Jul 2017
21	0	20 Jul 2017
22	0	20 Jul 2017
23	0	20 Jul 2017
24	0	20 Jul 2017
25	0	20 Jul 2017
26	0	20 Jul 2017
27	0	20 Jul 2017
28	0	20 Jul 2017
29	0	20 Jul 2017
30	0	20 Jul 2017
31	0	20 Jul 2017
32	0	20 Jul 2017
33	0	20 Jul 2017
34	0	20 Jul 2017
35	0	20 Jul 2017
36	0	20 Jul 2017
37	0	20 Jul 2017
38	0	20 Jul 2017
39	0	20 Jul 2017
40	0	20 Jul 2017
41	0	20 Jul 2017
42	0	20 Jul 2017
43	0	20 Jul 2017
44	0	20 Jul 2017
45	0	20 Jul 2017
46	0	20 Jul 2017
47	0	20 Jul 2017
48	0	20 Jul 2017
49	0	20 Jul 2017
50	0	20 Jul 2017
51	0	20 Jul 2017
52	0	20 Jul 2017

Chapter 7 TOC

Page No.	Revision No.	Effective Date
i	0	20 Jul 2017
ii	0	20 Jul 2017

Chapter 7

Page No.	Revision No.	Effective Date
1	0	20 Jul 2017
2	0	20 Jul 2017
3	0	20 Jul 2017
4	0	20 Jul 2017
5	0	20 Jul 2017
6	0	20 Jul 2017
7	0	20 Jul 2017
8	0	20 Jul 2017
9	0	20 Jul 2017
10	0	20 Jul 2017
11	0	20 Jul 2017
12	0	20 Jul 2017
13	0	20 Jul 2017
14	0	20 Jul 2017
15	0	20 Jul 2017
16	0	20 Jul 2017
17	0	20 Jul 2017
18	0	20 Jul 2017
19	0	20 Jul 2017
20	0	20 Jul 2017
21	0	20 Jul 2017
22	0	20 Jul 2017

Chapter 8 TOC

Page No.	Revision No.	Effective Date
i	0	20 Jul 2017

Page No.	Revision No.	Effective Date
ii	0	20 Jul 2017

Chapter 8

Page No.	Revision No.	Effective Date
1	0	20 Jul 2017
2	0	20 Jul 2017
3	0	20 Jul 2017
4	0	20 Jul 2017
5	0	20 Jul 2017
6	0	20 Jul 2017
7	0	20 Jul 2017
8	0	20 Jul 2017
9	0	20 Jul 2017
10	0	20 Jul 2017
11	0	20 Jul 2017
12	0	20 Jul 2017
13	0	20 Jul 2017
14	0	20 Jul 2017
15	0	20 Jul 2017
16	0	20 Jul 2017
17	0	20 Jul 2017
18	0	20 Jul 2017
19	0	20 Jul 2017
20	0	20 Jul 2017
21	0	20 Jul 2017
22	0	20 Jul 2017
23	0	20 Jul 2017
24	0	20 Jul 2017
25	0	20 Jul 2017
26	0	20 Jul 2017
27	0	20 Jul 2017
28	0	20 Jul 2017
29	0	20 Jul 2017
30	0	20 Jul 2017
31	0	20 Jul 2017
32	0	20 Jul 2017

Chapter 9 TOC

Page No.	Revision No.	Effective Date
i	0	20 Jul 2017
ii	0	20 Jul 2017

Chapter 9

Page No.	Revision No.	Effective Date
1	0	20 Jul 2017
2	0	20 Jul 2017
3	0	20 Jul 2017
4	0	20 Jul 2017
5	0	20 Jul 2017
6	0	20 Jul 2017
7	0	20 Jul 2017
8	0	20 Jul 2017
9	0	20 Jul 2017
10	0	20 Jul 2017
11	0	20 Jul 2017
12	0	20 Jul 2017
13	0	20 Jul 2017
14	0	20 Jul 2017
15	0	20 Jul 2017
16	0	20 Jul 2017
17	0	20 Jul 2017
18	0	20 Jul 2017
19	0	20 Jul 2017
20	0	20 Jul 2017
21	0	20 Jul 2017
22	0	20 Jul 2017
23	0	20 Jul 2017
24	0	20 Jul 2017
25	0	20 Jul 2017
26	0	20 Jul 2017
27	0	20 Jul 2017
28	0	20 Jul 2017
29	0	20 Jul 2017
30	0	20 Jul 2017
31	0	20 Jul 2017
32	0	20 Jul 2017

Page No.	Revision No.	Effective Date
33	0	20 Jul 2017
34	0	20 Jul 2017
35	0	20 Jul 2017
36	0	20 Jul 2017
37	0	20 Jul 2017
38	0	20 Jul 2017
39	0	20 Jul 2017
40	0	20 Jul 2017
41	0	20 Jul 2017
42	0	20 Jul 2017
43	0	20 Jul 2017
44	0	20 Jul 2017
45	0	20 Jul 2017
46	0	20 Jul 2017
47	0	20 Jul 2017
48	0	20 Jul 2017
49	0	20 Jul 2017
50	0	20 Jul 2017
51	0	20 Jul 2017
52	0	20 Jul 2017
Chapter 10 TOC		
i	0	20 Jul 2017
ii	0	20 Jul 2017
Chapter 10		
1	0	20 Jul 2017
2	0	20 Jul 2017
3	0	20 Jul 2017
4	0	20 Jul 2017
5	0	20 Jul 2017
6	0	20 Jul 2017
7	0	20 Jul 2017
8	0	20 Jul 2017
9	0	20 Jul 2017
10	0	20 Jul 2017
11	0	20 Jul 2017
12	0	20 Jul 2017
13	0	20 Jul 2017
14	0	20 Jul 2017
15	0	20 Jul 2017
16	0	20 Jul 2017
17	0	20 Jul 2017
18	0	20 Jul 2017
19	0	20 Jul 2017
20	0	20 Jul 2017
21	0	20 Jul 2017
22	0	20 Jul 2017
23	0	20 Jul 2017
24	0	20 Jul 2017
25	0	20 Jul 2017
26	0	20 Jul 2017
27	0	20 Jul 2017
28	0	20 Jul 2017
29	0	20 Jul 2017
30	0	20 Jul 2017
Chapter 11 TOC		
i	0	20 Jul 2017
ii	0	20 Jul 2017
Chapter 11		
1	0	20 Jul 2017
2	0	20 Jul 2017
3	0	20 Jul 2017
4	0	20 Jul 2017
5	0	20 Jul 2017
6	0	20 Jul 2017
7	0	20 Jul 2017
8	0	20 Jul 2017
9	0	20 Jul 2017
10	0	20 Jul 2017
11	0	20 Jul 2017
12	0	20 Jul 2017

Page No.	Revision No.	Effective Date
13	0	20 Jul 2017
14	0	20 Jul 2017
15	0	20 Jul 2017
16	0	20 Jul 2017
17	0	20 Jul 2017
18	0	20 Jul 2017
19	0	20 Jul 2017
20	0	20 Jul 2017

Chapter 12 TOC

Page No.	Revision No.	Effective Date
i	0	20 Jul 2017
ii	0	20 Jul 2017

Chapter 12

Page No.	Revision No.	Effective Date
1	0	20 Jul 2017
2	0	20 Jul 2017
3	0	20 Jul 2017
4	0	20 Jul 2017
5	0	20 Jul 2017
6	0	20 Jul 2017
7	0	20 Jul 2017
8	0	20 Jul 2017
9	0	20 Jul 2017
10	0	20 Jul 2017
11	0	20 Jul 2017
12	0	20 Jul 2017
13	0	20 Jul 2017
14	0	20 Jul 2017
15	0	20 Jul 2017
16	0	20 Jul 2017
17	0	20 Jul 2017
18	0	20 Jul 2017
19	0	20 Jul 2017
20	0	20 Jul 2017
21	0	20 Jul 2017
22	0	20 Jul 2017
23	0	20 Jul 2017
24	0	20 Jul 2017
25	0	20 Jul 2017
26	0	20 Jul 2017

Chapter 13 TOC

Page No.	Revision No.	Effective Date
i	0	20 Jul 2017
ii	0	20 Jul 2017

Chapter 13

Page No.	Revision No.	Effective Date
1	0	20 Jul 2017
2	0	20 Jul 2017
3	0	20 Jul 2017
4	0	20 Jul 2017
5	0	20 Jul 2017
6	0	20 Jul 2017
7	0	20 Jul 2017
8	0	20 Jul 2017
9	0	20 Jul 2017
10	0	20 Jul 2017
11	0	20 Jul 2017
12	0	20 Jul 2017
13	0	20 Jul 2017
14	0	20 Jul 2017
15	0	20 Jul 2017
16	0	20 Jul 2017
17	0	20 Jul 2017
18	0	20 Jul 2017
19	0	20 Jul 2017
20	0	20 Jul 2017
21	0	20 Jul 2017
22	0	20 Jul 2017
23	0	20 Jul 2017
24	0	20 Jul 2017
25	0	20 Jul 2017
26	0	20 Jul 2017
27	0	20 Jul 2017
28	0	20 Jul 2017

Page No.	Revision No.	Effective Date
29	0	20 Jul 2017
30	0	20 Jul 2017
31	0	20 Jul 2017
32	0	20 Jul 2017
33	0	20 Jul 2017
34	0	20 Jul 2017
35	0	20 Jul 2017
36	0	20 Jul 2017
37	0	20 Jul 2017
38	0	20 Jul 2017
39	0	20 Jul 2017
40	0	20 Jul 2017
41	0	20 Jul 2017
42	0	20 Jul 2017
43	0	20 Jul 2017
44	0	20 Jul 2017
45	0	20 Jul 2017
46	0	20 Jul 2017
47	0	20 Jul 2017
48	0	20 Jul 2017
49	0	20 Jul 2017
50	0	20 Jul 2017
51	0	20 Jul 2017
52	0	20 Jul 2017
53	0	20 Jul 2017
54	0	20 Jul 2017
55	0	20 Jul 2017
56	0	20 Jul 2017
57	0	20 Jul 2017
58	0	20 Jul 2017
59	0	20 Jul 2017
60	0	20 Jul 2017
Chapter 14 TOC		
i	0	20 Jul 2017
ii	0	20 Jul 2017
Chapter 14		
1	0	20 Jul 2017
2	0	20 Jul 2017
3	0	20 Jul 2017
4	0	20 Jul 2017
5	0	20 Jul 2017
6	0	20 Jul 2017
7	0	20 Jul 2017
8	0	20 Jul 2017
9	0	20 Jul 2017
10	0	20 Jul 2017
11	0	20 Jul 2017
12	0	20 Jul 2017
13	0	20 Jul 2017
14	0	20 Jul 2017
15	0	20 Jul 2017
16	0	20 Jul 2017
17	0	20 Jul 2017
18	0	20 Jul 2017
19	0	20 Jul 2017
20	0	20 Jul 2017
21	0	20 Jul 2017
22	0	20 Jul 2017
23	0	20 Jul 2017
24	0	20 Jul 2017
25	0	20 Jul 2017
26	0	20 Jul 2017
27	0	20 Jul 2017
28	0	20 Jul 2017
29	0	20 Jul 2017
30	0	20 Jul 2017
31	0	20 Jul 2017
32	0	20 Jul 2017
33	0	20 Jul 2017
34	0	20 Jul 2017
35	0	20 Jul 2017

Page No.	Revision No.	Effective Date
36	0	20 Jul 2017
37	0	20 Jul 2017
38	0	20 Jul 2017
39	0	20 Jul 2017
40	0	20 Jul 2017
41	0	20 Jul 2017
42	0	20 Jul 2017
43	0	20 Jul 2017
44	0	20 Jul 2017

Chapter 15 TOC

Page No.	Revision No.	Effective Date
i	0	20 Jul 2017
ii	0	20 Jul 2017

Chapter 15

Page No.	Revision No.	Effective Date
1	0	20 Jul 2017
2	0	20 Jul 2017
3	0	20 Jul 2017
4	0	20 Jul 2017
5	0	20 Jul 2017
6	0	20 Jul 2017
7	0	20 Jul 2017
8	0	20 Jul 2017
9	0	20 Jul 2017
10	0	20 Jul 2017
11	0	20 Jul 2017
12	0	20 Jul 2017
13	0	20 Jul 2017
14	0	20 Jul 2017
15	0	20 Jul 2017
16	0	20 Jul 2017
17	0	20 Jul 2017
18	0	20 Jul 2017
19	0	20 Jul 2017
20	0	20 Jul 2017
21	0	20 Jul 2017
22	0	20 Jul 2017
23	0	20 Jul 2017
24	0	20 Jul 2017
25	0	20 Jul 2017
26	0	20 Jul 2017

Chapter 16 TOC

Page No.	Revision No.	Effective Date
i	0	20 Jul 2017
ii	0	20 Jul 2017
iii	0	20 Jul 2017
iv	0	20 Jul 2017

Chapter 16

Page No.	Revision No.	Effective Date
1	0	20 Jul 2017
2	0	20 Jul 2017
3	0	20 Jul 2017
4	0	20 Jul 2017
5	0	20 Jul 2017
6	0	20 Jul 2017
7	0	20 Jul 2017
8	0	20 Jul 2017
9	0	20 Jul 2017
10	0	20 Jul 2017
11	0	20 Jul 2017
12	0	20 Jul 2017
13	0	20 Jul 2017
14	0	20 Jul 2017
15	0	20 Jul 2017
16	0	20 Jul 2017
17	0	20 Jul 2017
18	0	20 Jul 2017
19	0	20 Jul 2017
20	0	20 Jul 2017
21	0	20 Jul 2017
22	0	20 Jul 2017
23	0	20 Jul 2017
24	0	20 Jul 2017
25	0	20 Jul 2017

Page No.	Revision No.	Effective Date
26	0	20 Jul 2017
27	0	20 Jul 2017
28	0	20 Jul 2017
29	0	20 Jul 2017
30	0	20 Jul 2017
31	0	20 Jul 2017
32	0	20 Jul 2017
33	0	20 Jul 2017
34	0	20 Jul 2017
35	0	20 Jul 2017
36	0	20 Jul 2017
37	0	20 Jul 2017
38	0	20 Jul 2017
39	0	20 Jul 2017
40	0	20 Jul 2017
41	0	20 Jul 2017
42	0	20 Jul 2017
43	0	20 Jul 2017
44	0	20 Jul 2017
45	0	20 Jul 2017
46	0	20 Jul 2017
47	0	20 Jul 2017
48	0	20 Jul 2017
49	0	20 Jul 2017
50	0	20 Jul 2017
51	0	20 Jul 2017
52	0	20 Jul 2017
53	0	20 Jul 2017
54	0	20 Jul 2017
55	0	20 Jul 2017
56	0	20 Jul 2017
57	0	20 Jul 2017
58	0	20 Jul 2017
59	0	20 Jul 2017
60	0	20 Jul 2017
61	0	20 Jul 2017
62	0	20 Jul 2017
63	0	20 Jul 2017
64	0	20 Jul 2017
65	0	20 Jul 2017
66	0	20 Jul 2017
67	0	20 Jul 2017
68	0	20 Jul 2017
69	0	20 Jul 2017
70	0	20 Jul 2017
71	0	20 Jul 2017
72	0	20 Jul 2017
73	0	20 Jul 2017
74	0	20 Jul 2017
75	0	20 Jul 2017
76	0	20 Jul 2017
77	0	20 Jul 2017
78	0	20 Jul 2017
79	0	20 Jul 2017
80	0	20 Jul 2017
81	0	20 Jul 2017
82	0	20 Jul 2017
83	0	20 Jul 2017
84	0	20 Jul 2017
85	0	20 Jul 2017
86	0	20 Jul 2017
87	0	20 Jul 2017
88	0	20 Jul 2017
89	0	20 Jul 2017
90	0	20 Jul 2017
91	0	20 Jul 2017
92	0	20 Jul 2017

Chapter 17 TOC

Page No.	Revision No.	Effective Date
i	0	20 Jul 2017
ii	0	20 Jul 2017

Page No.	Revision No.	Effective Date
Chapter 17		
1	0	20 Jul 2017
2	0	20 Jul 2017
3	0	20 Jul 2017
4	0	20 Jul 2017
5	0	20 Jul 2017
6	0	20 Jul 2017
7	0	20 Jul 2017
8	0	20 Jul 2017
9	0	20 Jul 2017
10	0	20 Jul 2017
11	0	20 Jul 2017
12	0	20 Jul 2017
13	0	20 Jul 2017
14	0	20 Jul 2017
15	0	20 Jul 2017
16	0	20 Jul 2017
17	0	20 Jul 2017
18	0	20 Jul 2017
19	0	20 Jul 2017
20	0	20 Jul 2017
21	0	20 Jul 2017
22	0	20 Jul 2017
23	0	20 Jul 2017
24	0	20 Jul 2017
25	0	20 Jul 2017
26	0	20 Jul 2017
Chapter 18 TOC		
i	0	20 Jul 2017
ii	0	20 Jul 2017
Chapter 18		
1	0	20 Jul 2017
2	0	20 Jul 2017
3	0	20 Jul 2017
4	0	20 Jul 2017
5	0	20 Jul 2017
6	0	20 Jul 2017
7	0	20 Jul 2017
8	0	20 Jul 2017
9	0	20 Jul 2017
10	0	20 Jul 2017
11	0	20 Jul 2017
12	0	20 Jul 2017
13	0	20 Jul 2017
14	0	20 Jul 2017
15	0	20 Jul 2017
16	0	20 Jul 2017
17	0	20 Jul 2017
18	0	20 Jul 2017
Chapter 19 TOC		
i	0	20 Jul 2017
ii	0	20 Jul 2017
Chapter 19		
1	0	20 Jul 2017
2	0	20 Jul 2017
3	0	20 Jul 2017
4	0	20 Jul 2017
5	0	20 Jul 2017
6	0	20 Jul 2017
7	0	20 Jul 2017
8	0	20 Jul 2017
9	0	20 Jul 2017
10	0	20 Jul 2017
11	0	20 Jul 2017
12	0	20 Jul 2017
Chapter 20 TOC		
i	0	20 Jul 2017
ii	0	20 Jul 2017

Page No.	Revision No.	Effective Date
Chapter 20		
1	0	20 Jul 2017
2	0	20 Jul 2017
3	0	20 Jul 2017
4	0	20 Jul 2017
5	0	20 Jul 2017
6	0	20 Jul 2017
7	0	20 Jul 2017
8	0	20 Jul 2017
9	0	20 Jul 2017
10	0	20 Jul 2017
11	0	20 Jul 2017
12	0	20 Jul 2017
13	0	20 Jul 2017
14	0	20 Jul 2017
15	0	20 Jul 2017
16	0	20 Jul 2017
17	0	20 Jul 2017
18	0	20 Jul 2017
19	0	20 Jul 2017
20	0	20 Jul 2017
21	0	20 Jul 2017
22	0	20 Jul 2017
23	0	20 Jul 2017
24	0	20 Jul 2017
25	0	20 Jul 2017
26	0	20 Jul 2017
27	0	20 Jul 2017
28	0	20 Jul 2017
29	0	20 Jul 2017
30	0	20 Jul 2017
31	0	20 Jul 2017
32	0	20 Jul 2017
33	0	20 Jul 2017
34	0	20 Jul 2017
35	0	20 Jul 2017
36	0	20 Jul 2017
Chapter 21 TOC		
i	0	20 Jul 2017
ii	0	20 Jul 2017
Chapter 21		
1	0	20 Jun 2017
2	0	20 Jun 2017
3	0	20 Jun 2017
4	0	20 Jun 2017
5	0	20 Jun 2017
6	0	20 Jun 2017
7	0	20 Jun 2017
8	0	20 Jun 2017
9	0	20 Jun 2017
10	0	20 Jun 2017
11	0	20 Jun 2017
12	0	20 Jun 2017
13	0	20 Jun 2017
14	0	20 Jun 2017
15	0	20 Jun 2017
16	0	20 Jun 2017
17	0	20 Jun 2017
18	0	20 Jun 2017
19	0	20 Jun 2017
20	0	20 Jun 2017
21	0	20 Jun 2017
22	0	20 Jun 2017
23	0	20 Jun 2017
24	0	20 Jun 2017
25	0	20 Jun 2017
26	0	20 Jul 2017
Chapter 22 TOC		
i	0	20 Jul 2017
ii	0	20 Jul 2017

Page No.	Revision No.	Effective Date
iii	0	20 Jul 2017
iv	0	20 Jul 2017

Chapter 22

Page No.	Revision No.	Effective Date
1	0	20 Jul 2017
2	0	20 Jul 2017
3	0	20 Jul 2017
4	0	20 Jul 2017
5	0	20 Jul 2017
6	0	20 Jul 2017
7	0	20 Jul 2017
8	0	20 Jul 2017
9	0	20 Jul 2017
10	0	20 Jul 2017
11	0	20 Jul 2017
12	0	20 Jul 2017
13	0	20 Jul 2017
14	0	20 Jul 2017
15	0	20 Jul 2017
16	0	20 Jul 2017
17	0	20 Jul 2017
18	0	20 Jul 2017
19	0	20 Jul 2017
20	0	20 Jul 2017
21	0	20 Jul 2017
22	0	20 Jul 2017
23	0	20 Jul 2017
24	0	20 Jul 2017
25	0	20 Jul 2017
26	0	20 Jul 2017
27	0	20 Jul 2017
28	0	20 Jul 2017
29	0	20 Jul 2017
30	0	20 Jul 2017

Chapter 23 TOC

Page No.	Revision No.	Effective Date
i	0	20 Jul 2017
ii	0	20 Jul 2017
iii	0	20 Jul 2017
iv	0	20 Jul 2017
v	0	20 Jul 2017
vi	0	20 Jul 2017

Chapter 23

Page No.	Revision No.	Effective Date
1	0	20 Jul 2017
2	0	20 Jul 2017
3	0	20 Jul 2017
4	0	20 Jul 2017
5	0	20 Jul 2017
6	0	20 Jul 2017
7	0	20 Jul 2017
8	0	20 Jul 2017
9	0	20 Jul 2017
10	0	20 Jul 2017
11	0	20 Jul 2017
12	0	20 Jul 2017
13	0	20 Jul 2017
14	0	20 Jul 2017
15	0	20 Jul 2017
16	0	20 Jul 2017
17	0	20 Jul 2017
18	0	20 Jul 2017
19	0	20 Jul 2017
20	0	20 Jul 2017
21	0	20 Jul 2017
22	0	20 Jul 2017
23	0	20 Jul 2017
24	0	20 Jul 2017
25	0	20 Jul 2017
26	0	20 Jul 2017
27	0	20 Jul 2017
28	0	20 Jul 2017
29	0	20 Jul 2017

Page No.	Revision No.	Effective Date	Page No.	Revision No.	Effective Date
30	0	20 Jul 2017	66	0	20 Jul 2017
31	0	20 Jul 2017	67	0	20 Jul 2017
32	0	20 Jul 2017	68	0	20 Jul 2017
33	0	20 Jul 2017	69	0	20 Jul 2017
34	0	20 Jul 2017	70	0	20 Jul 2017
35	0	20 Jul 2017	71	0	20 Jul 2017
36	0	20 Jul 2017	72	0	20 Jul 2017
37	0	20 Jul 2017	73	0	20 Jul 2017
38	0	20 Jul 2017	74	0	20 Jul 2017
39	0	20 Jul 2017	75	0	20 Jul 2017
40	0	20 Jul 2017	76	0	20 Jul 2017
41	0	20 Jul 2017	77	0	20 Jul 2017
42	0	20 Jul 2017	78	0	20 Jul 2017
43	0	20 Jul 2017			
44	0	20 Jul 2017			
45	0	20 Jul 2017			
46	0	20 Jul 2017			
47	0	20 Jul 2017			
48	0	20 Jul 2017			
49	0	20 Jul 2017			
50	0	20 Jul 2017			
51	0	20 Jul 2017			
52	0	20 Jul 2017			
53	0	20 Jul 2017			
54	0	20 Jul 2017			
55	0	20 Jul 2017			
56	0	20 Jul 2017			
57	0	20 Jul 2017			
58	0	20 Jul 2017			
59	0	20 Jul 2017			
60	0	20 Jul 2017			
61	0	20 Jul 2017			
62	0	20 Jul 2017			
63	0	20 Jul 2017			
64	0	20 Jul 2017			
65	0	20 Jul 2017			

Page No.	Revision No.	Effective Date	Page No.	Revision No.	Effective Date

Intentionally Left Blank

Master Table of Contents

Chapter 1: Reserved

Chapter 2: Aircraft General

Chapter 3: Air Conditioning and Pressurization

Chapter 4: Auto Flight

Chapter 5: Communication

Chapter 6: Electrics

Chapter 7: Equipment and Furnishings

Chapter 8: Fire and Overheat Protection

Chapter 9: Flight Controls

Chapter 10: Fuel System

Chapter 11: Hydraulics

Chapter 12: Ice and Rain Protection

Chapter 13: Indicating and Recording Systems

Chapter 14: Landing Gear

Chapter 15: Aircraft Lighting

Chapter 16: Navigation

Chapter 17: Oxygen and Emergency Equipment

Chapter 18: Pneumatics

Chapter 19: Water and Waste

Chapter 20: Airborne Auxiliary Power

Chapter 21: Doors

Chapter 22: Power Plant

Chapter 23: Flight Instruments

Chapter 13: Indicating and Recording Systems

Chapter 13: Indicating and Recording Systems

General

The indicating and recording system consists of components that display or record critical aircraft information and system operation.

The engine indication and crew alerting system (EICAS) provides:

- Automatic system monitoring
- Integration of system information
- Color logic for presentation of minimum/maximum limits
- Aural alert warning messages

Information is provided on two cathode ray tubes (CRTs) located on the center instrument panel. Master warning and caution lights on the glareshield enhance the indication system Aural signals are generated within the data concentrator units (DCUs) and are heard through the flight deck speakers and headsets.

The flight data recording system (FDR) receives and stores aircraft flight data information for later analysis.

The maintenance diagnostic computer (MDC) is used to record engine and line replaceable unit (LRU) status and parameters. MDC data can be downloaded for system diagnosis and fault identification.

The aircraft clocks are connected to the air radio incorporated (ARING) bus and provide time data for specific aircraft systems.

Included in this chapter are TCAS and EGPWS related voice messages. These messages are totally independent from the EICAS system. They are generated respectively from the TCAS the EGPWS computers. TCAS related voice messages are shown in Navigation.

GPS Clocks

A digital electronic clock is installed on the pilot and copilot side panels. The clocks have the capability of being synchronized with the Global Positioning System (GPS). Each clock is capable of displaying date [GPS or internal Universal Time Coordinated (UTC)], current time (GPS, internal UTC, or local), chronometer (CHR), as well as elapsed time (ET) functions.

The clocks are synchronized to the GPS input as soon as valid GPS information is received. In the case of invalid GPS data or signal loss, the clocks will operate in internal (INT) mode using the integrated time base of each clock. If there is a valid GPS signal, the clocks do not need to be set, as this will be done automatically at power up. The flight crew can disable the GPS signal by entering the time setting mode. The clocks will then ignore the GPS signal until the next primary power reset.

The MODE, ET SEL and ET RST buttons are used to set the time and date. To set the clock, push the MODE button for two seconds, then push the MODE button again to toggle between UTC hours and minutes (when the INT is lit), year, month, and day, (when the DT is lit), and local time hours and minutes (when the LT is lit). In any of these modes, the ET SEL button is used to decrease the data and the ET RST button is used to increase the data. Data changes are in increments of one digit for each press of the ET SEL or ET RST button. At any time during the time setting process, pressing the MODE button for a minimum of two seconds will exit the time setting mode and restart the clock operation.

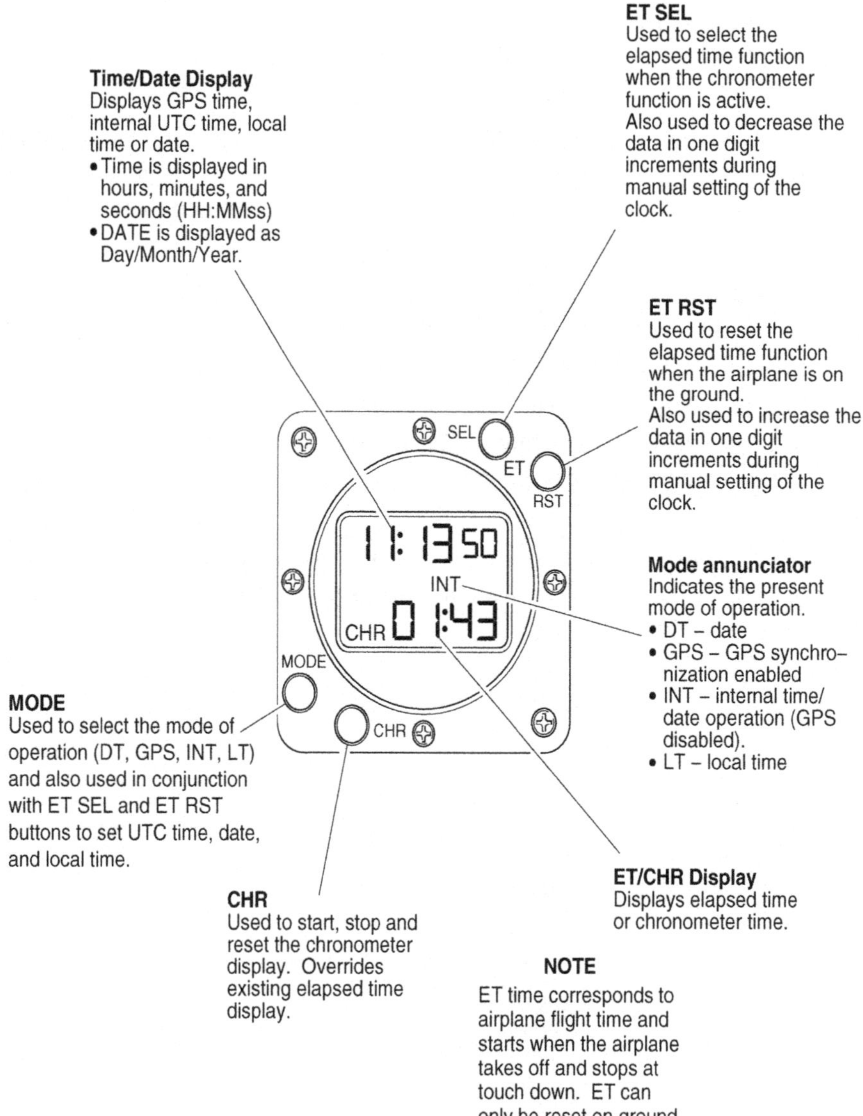

Figure 13 - 1: GPS Clock

Engine Indicating Crew Alerting System (EICAS)

The function of the EICAS is to display the engine instruments, visual and aural crew alert messages and real-time interpretation of aircraft systems operation.

Two EICAS displays (ED 1 and ED 2) present the data on different selectable display pages. Some display pages are shown by default, others are available through selection. Control of displayed information is provided by the EICAS control panel (ECP) located on the center pedestal.

Engine indications are provided on the EICAS primary page. Color is used to depict normal and non-normal ranges of operation.

The crew alerting system (CAS) provides visual and aural alerts when the DCUs detect a malfunction. The CAS prioritizes messages by order of occurrence and order of importance. The order of occurrence is easily defined. The most recent message always appears on the top of its associated list.

The order of importance is accomplished by the use of color and aural alerts. There are four levels of CAS message importance: warning, caution, advisory and status.

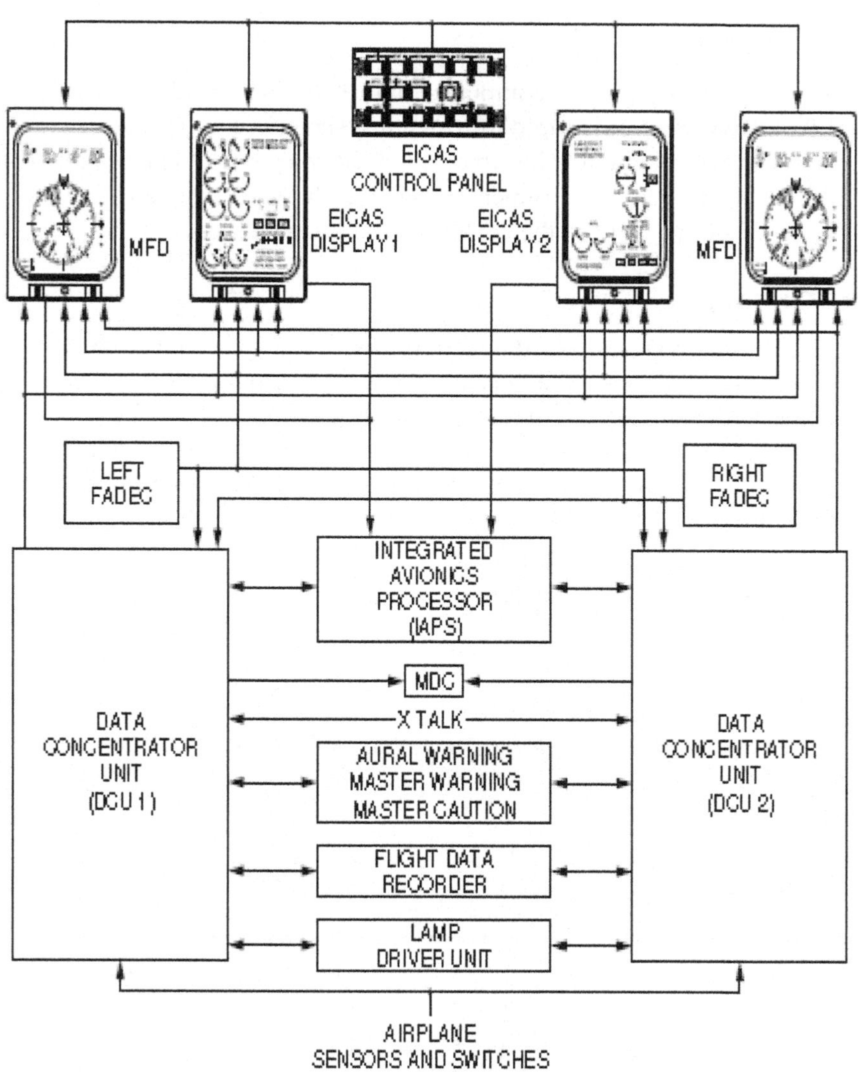

Figure 13 - 2: EICAS Schematic

Components and Operation

EICAS Displays

The EICAS displays are computer-controlled video displays. All six EICAS and electronic flight instrument system (EFIS) cathode ray tubes are identical and interchangeable. EICAS display 1 (ED 1) is installed on the left of the center instrument panel and EICAS display 2 (ED 2) is installed on the right of the center instrument panel.

CRT Cooling

The CRTs are DC-powered but cooled by AC-powered fans. To provide redundancy, two separate fans are installed. Normal fan operation is controlled automatically by the proximity sensing system (PSS). If a cooling fan fails, the DSPLY FAN rotary switch is used to select a backup fan. In the event that both fans fail, cooling is accomplished by the conditioned air.

DSPLY and AVIONICS FAN
GND ALTN - Flight fan is activated during ground ops
NORM - Appropriate fan is activated for either flight or ground operations
FLT ALTN - Ground fan is activated during flight.

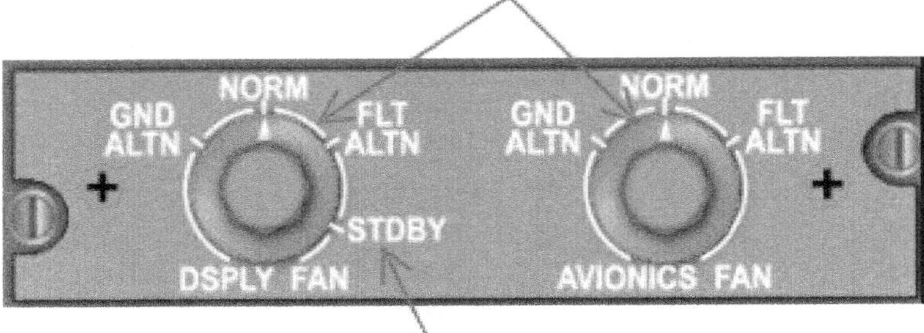

STDBY - Both fans deactivated, air is supplied from flight deck air conditioning duct.

Figure 13 - 3: DSPLY/AVIONICS FAN Control Panel

CRT Lighting Adjustment

Two separate control knobs are used to adjust CRT lighting intensity. In the upper left corner of the display unit, a brightness (BRT) adjustment knob is used to set the minimum lighting intensity for the associated screen. After adjusting the BRT knob to a minimum visible level, the pilot can select the desired level of lighting for both displays by using the display (OSPL) knob located on the associated LIGHTING control panel.

The EICAS displays present system information on the primary, status, synoptic and menu pages. ED 1 displays the primary page by default. ED 2 is defaulted to the status page. Page selection is accomplished via the EICAS control panel (ECP).

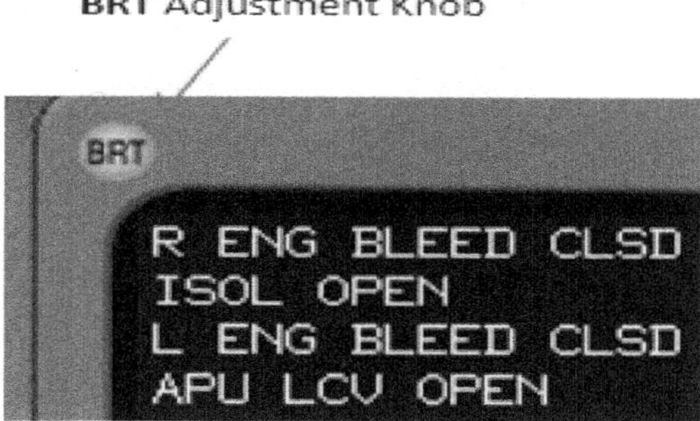

Figure 13 - 4: CRT Lighting Adjustment

Data Concentrator Units

The data concentrator units (DCUs) are the heart of the engine indication and crew alerting system (EICAS). The DCU collects data from various aircraft systems, processes the information and relays it to the proper component or display. There are two DCUs installed in the aircraft. They are designated DCU 1 and DCU 2.

All DCUs share in providing information to the EICAS displays. Internal switching logic determines how the information is provided by the DCUs. If a partial or complete failure of a DCU occurs, system redundancy ensures DCU that no displayed data is lost.

Each of the DCUs receives identical analog or digital information via an aeronautical radio incorporated (ARINC) bus from the following sources:

- Aircraft systems

- Integrated avionics processor system (IAPS)
- Stall protection computer
- Enhanced ground proximity warning system (EGPWS)
- Clocks
- Attitude heading and reference system (AHRS)

The DCUs process and format this information then transmit the data on the ARINC bus to the:

- EICAS displays
- Lamp driver unit (LOU)
- Maintenance diagnostic computer (MDC)
- Flight data recorder (FDR)

Lamp Driver Unit

The two-channel lamp driver unit (LDU) receives information from the DCUs and controls the panel and glareshield switch-light illumination. When the DCU generates an EICAS message the data is sent to the LDU.

The LDU interprets the information and then illuminates the switch-light on the appropriate control panel.

Should one channel of the LDU fail, the remaining channel continues to operate. Testing of the LDU and panel lamps is accomplished by using the test switch located on the miscellaneous test panel. The three-position LAMP TEST switch is spring-loaded to the center OFF position. Holding the switch to either position 1 or 2 tests the associated LDU channel and lamps.

The IND LTS switch on the miscellaneous test panel is used to set the switch-light lamp intensity to bright or dim.

LAMP TEST 1 & 2 Switch

Tests respective lamp driver unit

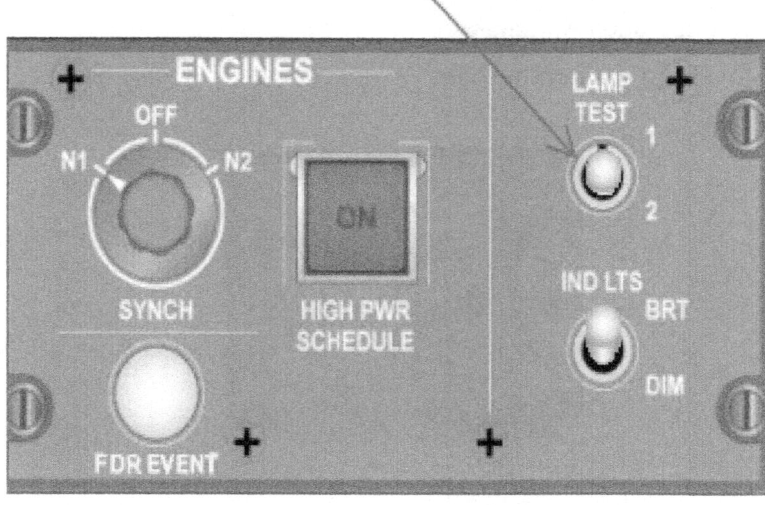

Figure 13 - 5: LAMP TEST Switch

MASTER WARNING Switch-lights

The two red MASTER WARNING switch-lights are located on the left and right glareshield panels. Both switch-lights come on flashing when any warning occurs. A triple chime always accompanies a MASTER WARNING. Pushing either MASTER WARNING switch-light will extinguish both MASTER WARNING lights for the duration of that warning and resets the lights for future warnings. The warning message on the EICAS primary page will remain displayed as long as the warning exists. Pushing the MASTER WARNING switch-light also silences the aural warnings except for the following:

- Stall warbler
- EGPWS/TCAS (voice and aural)
- Over-speed clacker
- Flap clacker
- Stabilizer trim clacker
- AP disconnect cavalry charge
- Configuration warnings
- Landing gear warning horn

MASTER CAUTION Switch-lights

The two amber MASTER CAUTION switch-lights are located on the left and right glareshield panels. Both switch-lights come on flashing when any caution occurs. A single chime always accompanies a MASTER CAUTION. Pushing either MASTER CAUTION switch-light will extinguish both MASTER CAUTION lights for the duration of that caution and resets the lights for future cautions. The caution message on the EICAS primary page will remain displayed as long as the caution exists. Pushing the MASTER CAUTION switch-light will not silence the following:

- GPWS and TCAS voice alerts
- Altitude alert (C-chord) aural

Audio Integrating System

The audio integrating system receives signals from the DCUs and EGPWS and generates the aural alerts. When multiple aural alerts are generated at the same time, the audio integration system determines the priority in which the aural alerts are heard.

There are two separate categories of aural alerts, warning aurals and voice messages. Voice messages are related to three separate systems. EGPWS voice messages are generated independently of the DCUs.

Table 13 - 1: Aural Warning Types

Sounds	Indications
Warbler	Stall
Siren	Wind shear
"Whoop whoop"	GPWS mode 1 or 2 (excessive descent or closure rate)
Fire bell	Fire Warning
Voice	Voice Aural Warning
Clacker	• Excessive stabilizer trim movement • VMO/MMO exceedance • Airspeed too high for current flap setting
Horn	Gear not down
Cavalry charge	Autopilot disconnected
Warning (triple chime)	Warning tone preceding a voice advisory
C-cord	Altitude Alert

Table 13 - 2: EICAS VOICE AND WARNING MESSAGES

Voice Message	Display Message
"Anti-ice duct"	ANTI-ICE DUCT, L(R) COWL A/I DUCT
"APU"	APU OVERSPEED, APU OVERTEMP
"Bleed-Air Duct"	L/R 10TH DUCT L/R 14TH DUCT
"Brakes"	BRAKE OVHT
"Cabin Pressure"	DIFF PRESS, CABIN ALT
"Engine Oil"	L (or R) ENG OIL PRESS
"Gear Bay Overheat"	MLG BAY OVHT
"Gear Disagree"	GEAR DISAGREE
"Nose Door"	NOSE DOOR OPEN
"Smoke"	SMOKE AFT CARGO, SMOKE FWD CARGO
"Wing Overheat"	WING OVHT
"Door"	PASSENGER DOOR
"Config Trim"	CONFIG AILERON, CONFIG RUDDER CONFIG STAB
"Config Autopilot"	CONFIG AP
"Config Brakes"	PARKING BRAKE
"Config Flaps"	CONFIG FLAPS
"Config Spoiler"	CONFIG SPLRS

Table 13 - 3: EGPWS RELATED VOICE MESSAGES

Voice Message	Display Message
"Bank Angle"	N/A
"Don't Sink"	Amber GND PROX switch-light
"Glide-slope"	Amber GND PROX switch-light
"Minimums"	N/A
"Decision Height"	DH (on PFD)
"Pull up"	Red PULL UP switch-light
"Sink Rate"	Amber GND PROX switch-light
"Terrain, Terrain"	Amber GND PROX switch-light
"Terrain, Terrain Whoop Whoop Pull Up"	Terrain awareness warning on MFDs, Red PULL UP switch-light
"Too Low Flap"	Amber GND PROX switch-light
"Too Low Gear"	Amber GND PROX switch-light
"Too Low Terrain"	Amber GND PROX switch-light
"Windshear"	WINDSHEAR (on PFD)
"Obstacle, Obstacle, Whoop Whoop, Pull UP"	Obstacle awareness warning on MFDs Red PULL UP switch-light
"Caution Obstacle"	Obstacle awareness caution display on MFDs Amber GND PROX switch-light
"Caution Terrain"	Terrain awareness caution display on MFDs Amber GND PROX switch-light

"10, 20, 30,40, 50, 100, 500, 1000"(altitude call outs)	N/A
700, 960, 1400, and 2,800 Hz tones	N/A

Table 13 - 4: TCAS Voice Messages

Voice Message	Display Message
"Clear of Conflict"	N/A
"Climb, Climb, Climb"	TCAS resolution advisory on PFD or MFD
"Climb, Climb Now"	TCAS resolution advisory on PFD or MFD
"Climb, Cross Climb"	TCAS resolution advisory on PFD or MFD
"Descend, Crossing, Descend"	TCAS resolution advisory on PFD or MFD
"Descend, Descend, Descend"	TCAS resolution advisory on PFD or MFD
"Descend, Descend, Now"	TCAS resolution advisory on PFD or MFD
"Increase Climb"	TCAS resolution advisory on PFD or MFD
"Increase Descent"	TCAS resolution advisory on PFD or MFD
"Monitor Vertical Speed"	TCAS resolution advisory on PFD or MFD
"Reduce Climb"	TCAS resolution advisory on PFD or MFD
"Reduce Descent"	TCAS resolution advisory on PFD or MFD
"Traffic, Traffic"	TCAS traffic advisory on PFD or MFD

Aural Alert Disable

Two AUDIO WARNING DISABLE switch-lights located on the AUDIO WARNING panel on the copilot's side console are used to disable and silence the aural warnings of a malfunctioning DCU. When DCU 1 is selected to DISABLE, DCU 2 will provide the aural warning function.

Selecting all DCU switch-lights to DISABLE will disable all EICAS aural warnings. EGPWS and TCAS aural warnings are not disabled by selection at this panel.

EICAS Control Panel

The EICAS control panel (ECP) is located on the center pedestal. A single channel microprocessor control unit, it has dedicated momentary contact buttons for control of the information displayed on EICAS. Four of the buttons are mechanical which allows them to operate should the ECP fail.

The four mechanical buttons are the PRI, STAT, CAS, and STEP buttons.

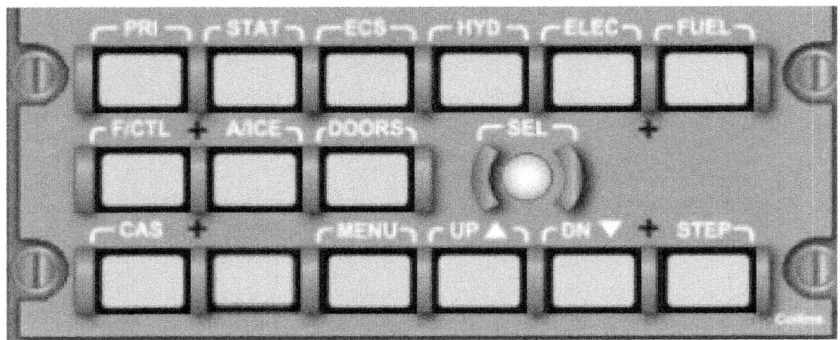

Figure 13 - 6: EICAS Control Panel

PRI

Selecting PRI causes the primary page to be displayed on ED 2.

STAT

Selecting STAT causes the status page to be displayed on ED 2. If the status page is already displayed, the button will remove the status messages and display a white MSGS icon. Pressing the ST AT button a second time will cause the status messages to reappear. DCU generation of a new status message will remove the MSGS icon and the new status message will be displayed on the top of the status message stack.

CAS

When both engines are stabilized at idle, it is possible to remove the displayed caution messages from the primary page. Selecting the CAS button will remove the caution messages and display a white MSGS icon.

Pressing the CAS button a second time will cause the caution messages to reappear. If a new caution message is generated it appears above the MSGS icon. Pressing the CAS button again combines the new message with the already boxed messages.

STEP

The STEP button will sequentially step through all available pages. starting at the page currently selected and moves left to right as labeled on the ECP. There are seven synoptic pages plus a menu page. They can be accessed through buttons on the ECP.

ECS

Selecting ECS displays the ECS synoptic page.

HYD

Selecting HYD displays the HYDRAULIC synoptic page.

ELEC

Selecting ELEC once displays the AC ELECTRICAL synoptic page.

Selecting ELEC a second time displays the DC ELECTRICAL synoptic page.

FUEL

Selecting FUEL displays the FUEL synoptic page.

F/CTL

Selecting F/CTL displays the FLIGHT CONTROLS synoptic page.

A/ICE

Selecting A/ICE displays the ANTI-ICE synoptic page.

DOORS

Selecting DOORS displays the DOORS synoptic page.

MENU

The MENU page is used to reset the FUEL USED value present on the FUEL synoptic page. The left and right engine oil tank quantities are displayed on the MENU page. The buttons labeled SEL, UP, and ON will only function when the MENU page is selected.

Primary Page

EICAS display 1 (ED 1) is defaulted to display the primary page. Engine Indications Engine indications are presented in both the traditional analog gauge format and as digital readouts. Displayed engine parameters include:

- N1 speed
- Inter turbine temperature (ITT)
- N2 speed
- Fuel flow
- Oil temperature
- Oil pressure (digital)
- Oil pressure gauges
- N1 fan vibration gauges

Some engine indications are presented only on an "as required" basis. For example:

- After the engine start, when both engines are stabilized at idle and engine oil pressures are normal, the EICAS logic replaces the analog oil pressure gauges with N 1 fan vibration gauges. When on the ground and the engines are shut down, both oil pressure gauges return.
- The N1 gauges present APR (automatic performance reserve) and REV (reverse thrust) icons only when the systems are activated.
- The N2 gauge presents VIB icon only when the N2 fan vibration exceeds a predetermined value.
- The N2 gauge arc markings change from green to white (0 to 77%) when the wing anti-icing switch is selected on.

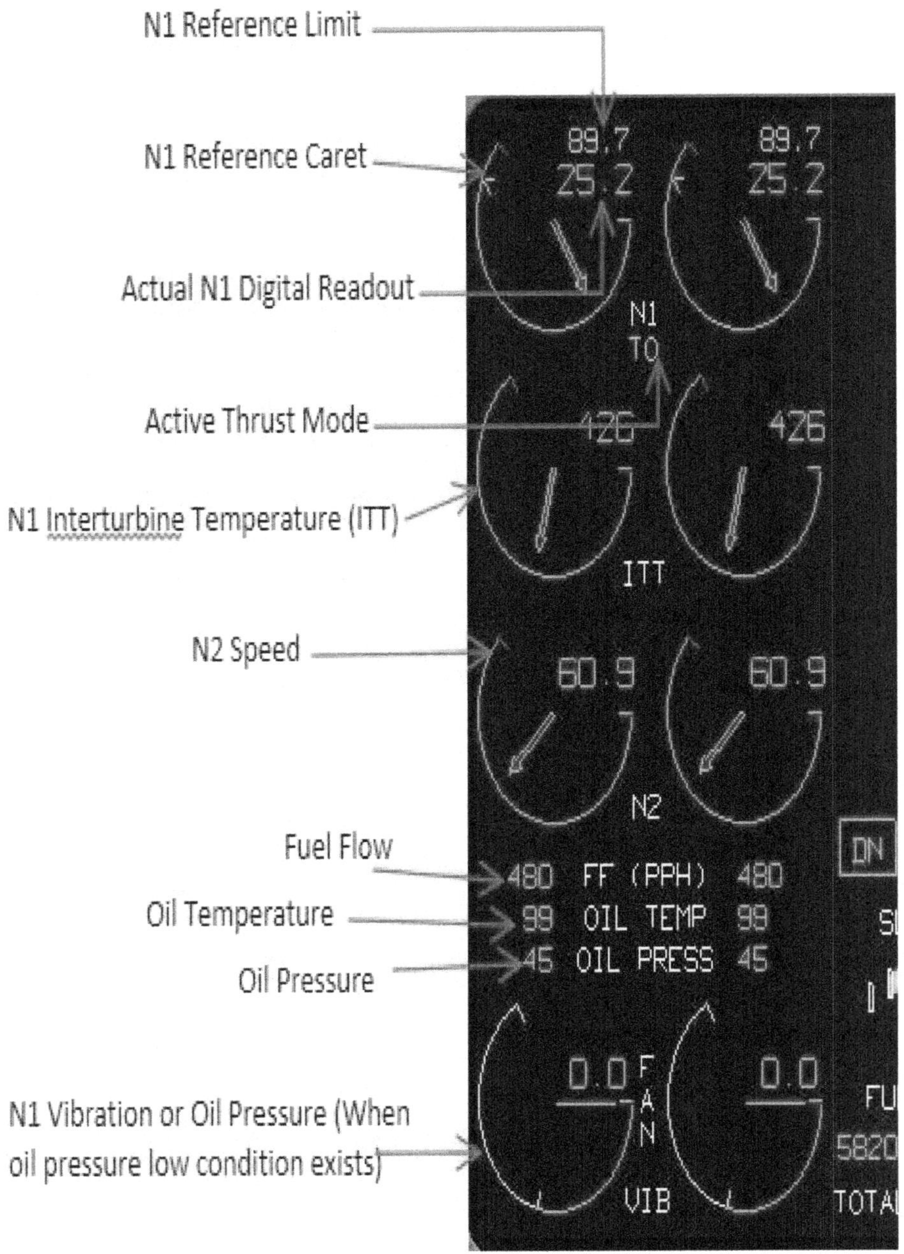

Figure 13 - 7: Primary Page Engine Indications

Crew Alerting System Messages

Warning and caution messages are only presented on the primary page.

Landing Gear and Slats/Flaps

Landing gear and slats/flaps information is presented both pictorially and in a digital format During flight, the landing gear and slats/flaps information is removed from view when the following conditions exist:

- Landing gear is up and locked
- Slats/flaps are up
- Wheel brake temperatures are normal

The information reappears during the approach when the gear or flaps are selected.

Fuel Quantity

In the bottom right hand corner of the primary page, EICAS provides the pilot with a summary of the current fuel quantities. The information presented includes individual tank and total fuel quantities.

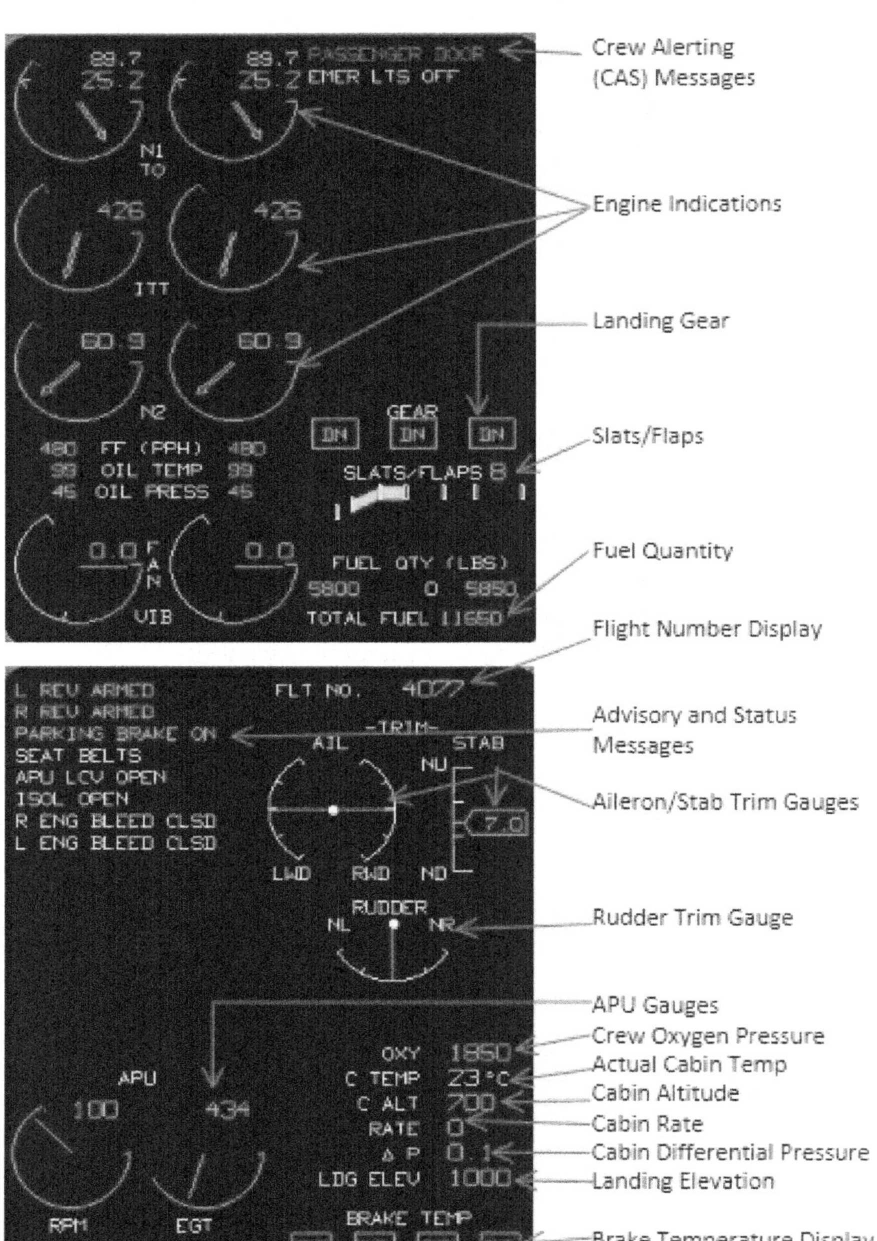

Figure 13 - 8: Primary and Status Page Indications

Status Page

The status page is the default page and is normally in view on ED 2.

Crew Alerting System Messages

Advisory and status messages are only presented on the status page.

Auxiliary Power Unit

The APU analog and digital RPM gauge and the EGT gauge are only presented on the status page when the APU PWR/FUEL switch is selected.

The APU door position is presented continuously.

FLT NO

The flight number is inserted via the flight management system (FMS).

Trim Settings

Aileron, horizontal stabilizer and rudder trim settings are continuously displayed on the status page.

The aileron and rudder trim indicators are green when the trims are properly set for takeoff. In flight, the indicators are white.

The horizontal stabilizer trim indicator should be in the green range for takeoff.

Crew Oxygen Pressure

The crew gaseous oxygen pressure is continuously displayed on the status page. Green indicates normal quantity and amber identifies low quantity.

Cabin Pressurization and Temperature

Cabin pressurization and temperature information are continuously displayed on the status page. The information includes:

- Actual cabin temperature
- Cabin altitude
- Rate of climb

- Differential pressure
- Landing elevation (pilot selectable)

Brake Temperature

Each main wheel brake temperature is monitored. During flight, the brake temperature information is removed from the status page if:

- All brake temperatures are normal
- Slats/flaps are up
- Landing gear is up and locked

Synoptic Pages

The synoptic pages can be selected for presentation on ED 2 by using the individual buttons on the ECP. The synoptic pages provide the pilot with real-time interpretation of aircraft system operation. Each page presents the aircraft systems as schematic diagrams.

ECS

The ECS synoptic page presents the:

- Bleed air system
- Air conditioning system
- Pressurization system

HYDRAULIC

All three hydraulic systems are presented. At the bottom of the page, a list of systems and components that are serviced by the hydraulic systems is provided

AC ELECTRICAL

The AC ELECTRICAL synoptic page provides the AC system graphically.

DC ELECTRICAL

The DC ELECTRICAL synoptic page provides the DC system graphically.

FUEL

The FUEL synoptic page depicts fuel quantities, fuel temperature and fuel system operation.

FLIGHT CONTROLS

The FLIGHT CONTROLS synoptic page displays all primary and secondary flight controls including control surface movement.

ANTI-ICE

The ANTI-ICE synoptic page depicts the operation of the:

- Wing anti-icing system
- Cowl anti-icing system

DOORS

The DOORS synoptic page provides the actual position of all doors in the pressurized portion of the fuselage. There is one exception. The flight deck escape hatch is not presented on the synoptic page.

MENU

When MENU is selected on the ECP, the MENU page appears on ED 2.

The MENU page allows the pilot to reset the FUEL USED indication on the FUEL synoptic page and monitor engine oil tank quantities.

EICAS Color Logic and Synoptic Page Presentation

The Bombardier CRJ uses a 'dark cockpit' philosophy for system indications. In flight if the aircraft is properly configured, the EICAS screens present normal system indications and the switch-lights are not illuminated.

Color Logic

Generally, the following color logic is used for EICAS presentation.

COLOR	MEANING
Red	Warning or exceedance
Amber	Caution or invalid data
Green	Normal Operations
Blue	Structure
Cyan	Labeling
Cyan 1/2 Intensity	Component is operational
Magenta	FLX Takeoff N1 Setting
Magenta 1/2 Intensity	Insufficient data to determine proper color coding
White	Everything else no covered above

Synoptic Page Presentation

Color logic is used to depict normal, precautionary and maximum system limits:

- Normal range, indication is green
- Precautionary range, the system indicator turns amber
- Exceeds the limits, the indicator turns red

Flow lines are used on a number of synoptic pages. Although not all pages can display the four colors, generally the color logic remains the same for all.

COLOR	MEANING
Red	Exceeds normal values
Amber	Below Normal values
Green	Normal value
Black	Structure
White/Cyan	Labeling

The synoptic page valves graphically represent the actual valve operation. Color is used to identify the serviceability of the valve. Operative valves are white and inoperative valves are amber.

Fuel or hydraulic pumps are white when off, green when operating and amber when failed.

EICAS Messages

The crew alerting system (CAS) provides visual and aural alerts when the DCU is alerted of a malfunction.

The CAS prioritizes messages by order of occurrence and order of importance.

The order of occurrence is easily defined. The most recent message always appears on the top of its associated list.

The order of importance is accomplished by the use of color and aural alerts. There are four levels of CAS message importance: warning, caution. advisory and status.

Warning Messages

The most urgent messages are called warnings and are red in color. Warning messages are presented on the top of the primary page and remain in view until the problem is resolved.

Warning messages require immediate action and are accompanied by:

- A triple chime

- Red flashing MASTER WARNING lights
- Red light on the faulted control switch or a red indication on EFIS or a synoptic page

Warning messages in addition to the above indications, can also generate one or both of the following:

- Aural warning tone
- Voice message

Pressing either MASTER WARNING switch-light:

- Silences the aural alerts
- Extinguishes the red flashing MASTER WARNING lights
- Resets the CAS, allowing it to annunciate other faults

Caution Messages

Caution messages are second in the order of importance. The amber caution messages are presented on the primary page and appear directly below any warning messages that may be displayed.

Caution messages require prompt action and are always accompanied by

- A single chime
- Amber flashing MASTER CAUTION lights
- Amber light on the faulted control switch or as an amber indication on EFIS or a synoptic page

Pressing either MASTER CAUTION switch-light:

- Extinguishes the amber flashing MASTER CAUTION lights
- Resets the CAS to allow it to annunciate another fault

Pagination

It is possible to have more than one page of caution messages. The CAS button on the ECP allows the pilot to page forward and backward to view the entire list of caution messages. A boxed amber PAGE 1/2 is displayed at the bottom of the message list to indicate that another page of messages exists. At the end of the second message list, PAGE 2/2 is displayed.

Clearing

When both engines are stabilized at idle, it is possible to remove from view the caution messages that are displayed.

Selecting the CAS button will remove the caution messages and display a white MSGS icon. Pressing the CAS button a second time will allow the caution messages to reappear.

DCU generation of a new caution message will display the new message above the MSGS icon. If the new messages extend to another page, a white boxed MORE appears at the bottom of the message list. At the end of the second page message list a boxed END is displayed.

To remove or box the new list, pressing the CAS button removes all the current amber messages and the MSGS icon is displayed

Advisory Messages

Green advisory messages are presented at the top of the status page.

Advisory messages are used to advise of:

- Aircraft configuration for a particular phase of flight
- Successful system test
- Confirmation of SOV closure
- SELCAL

Status Messages

White status messages are presented on the status page and appear directly below any advisory messages that may be present.

White status messages are used to:

- Provide status of a specific system that has been manually or automatically activated
- Identify a low-priority system failure

Pagination

It is possible to have more than one page of status messages. The STA T button on the ECP allows the pilot to page forward and backward to view the entire list of status messages.

Clearing

It is possible to remove from view the status messages that are displayed Selecting the STAT button will remove the status messages and display a white MSGS icon. Pressing the STAT button a second time will allow the status messages to reappear.

DCU generation of a new status message will remove the MSGS icon and the new message will be displayed on the top of the status message stack.

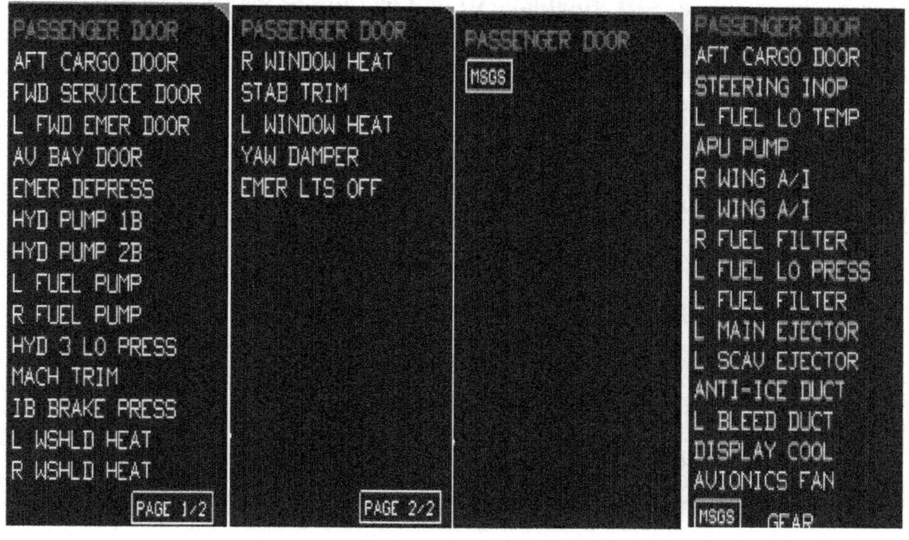

Messages appear on more than one page. CAS pressed.

Second page displays end of messages CAS is pressed

Messages are boxed.

New messages appear on top of boxed messages

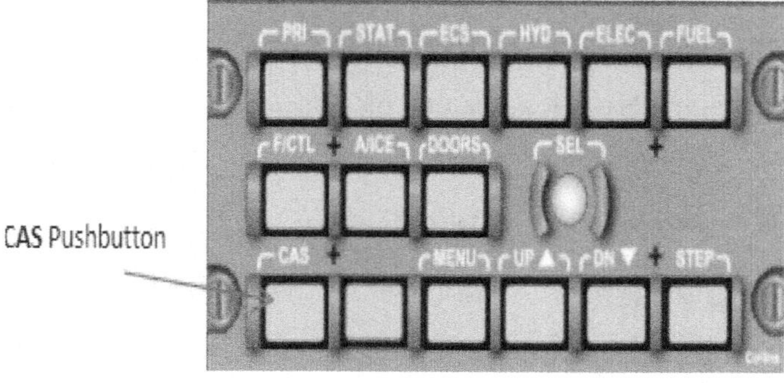

Figure 13 - 9: Primary Page - Caution Messages Clearing

EICAS Message Inhibit Logic

During the initial takeoff, final takeoff and landing phases, the DCUs will process inhibit logic to minimize spurious or distracting warning and/or caution messages. Condition 1 (initial takeoff) and Condition 2 (final takeoff) applies to the takeoff phases. Condition 3 applies to the landing phase. The message inhibit conditions are shown in the figure below.

Condition 1 Initial Take-off Phase

The initial take-off inhibits are enabled when:

- Left and right engine N1 is greater than 79%, and
- Weight-on-wheels, and airspeed is less than 100 knots.

The initial take-off inhibit is removed when:

- Left and right engine N1 is less than 67.6%, or
- Airplane is in the final take-off phase.

Condition 2 Final Take-off Phase

The final take-off inhibits are enabled when:

- Left and right engine N1 is greater than 79%, and
- Airspeed transitions to greater than 100 knots.

The final take-off inhibit is removed when:

- Left and right engine N1 is less than 67.6%, or
- Radio altitude is greater than 400 ft AGL, or
- 30 seconds after ground to air transition.

Condition 3 Landing Phase

Landing phase inhibits are enabled when:

- Radio altitude transitions to less than 400 ft AGL, and
- Landing gear down and locked.

The landing phase inhibit is removed when:

- 30 seconds after air to ground transition, or
- Radio altitude transitions from less than 400 ft to greater than 400 ft.

Table 13 - 5: Warning Inhibits

Warning Message	Initial Take-off Phase	Final Take-off Phase	Landing Phase
AFCS MSG FAIL		X	
ANTI-ICE DUCT			X
APU OVERTEMP		X	X
APU OVERSPEED		X	
CABIN ALT	X	X	X
CONFIG FLAPS			X
DIFF PRESS		X	X
GEAR DISAGREE	X	X	
L COWL A/I DUCT			X
L ENG OIL PRESS		X	X
NOSE DOOR OPEN	X	X	X
PASSENGER DOOR		X	X
R COWL A/I DUCT			X
R ENG OIL PRESS		X	X
WING OVHT			X

Table 13 - 6: Caution Inhibits

Caution Message	Initial Take-off Phase	Final Take-off Phase	Landing Phase
AC 1 AUTOXFER	X	X	X
AC 2 AUTOXFER	X	X	X
AC BUS 1		X	X
AC BUS 2		X	X
AC ESS BUS		X	X
AC SERV BUS	X	X	X
AFT CARGO DET	X	X	X
AFT CARGO DOOR	X	X	X
AFT CARGO OVHT	X	X	X
AFT CARGO SQB 1	X	X	X
AFT CARGO SQB 2	X	X	X

Caution Message	Initial Take-off Phase	Final Take-off Phase	Landing Phase
AFT SERV DOOR	X	X	X
ALT LIMITER	X	X	X
ANTI ICE LOOP	X	X	X
AP PITCH TRIM	X	X	
APR CMD SET	X	X	X
AP TRIM IS LWD	X	X	
AP TRIM IS ND	X	X	
AP TRIM IS NU	X	X	
AP TRIM IS RWD	X	X	
APU BATT OFF	X	X	X
APU BLEED ON	X	X	X
APU BTL LO	X	X	X
APU DOOR OPEN	X	X	X
APU ECU FAIL	X	X	X
APU FAULT	X	X	X
APU FIRE FAIL	X	X	X
APU GEN OFF		X	X
APU GEN OVLD	X	X	X
APU LCV CLSD	X	X	X
APU LCV OPEN	X	X	X
APU PUMP	X	X	X
APU SOV FAIL	X	X	X
APU SOV OPEN	X	X	X
APU SQB	X	X	X
AUTO PRESS	X	X	X
AV BAY DOOR	X	X	X
AVIONICS FAN	X	X	X
BATTERY BUS		X	X
BLEED MISCONFIG	X	X	X
BULK FUEL TEMP	X	X	X
CARGO BTL LO	X	X	X

Caution Message	Initial Take-off Phase	Final Take-off Phase	Landing Phase
CABIN ALT	X	X	X
CTR CARGO DOOR	X	X	X
DC BUS 1		X	X
DC EMER BUS		X	X
DC ESS BUS		X	X
DC SERV BUS	X	X	X
DISPLAY COOL	X	X	X
EFIS COMP INOP	X	X	X
ELT ON	X	X	X
EMER DEPRESS	X	X	
EMER LTS OFF	X	X	X
ENG BTL 1 LO	X	X	
ENG BTL 2 LO	X	X	
FLAPS FAIL		X	
FWD CARGO DET	X	X	X
FWD CARGO DOOR	X	X	X
FWD CARGO SQB 1	X	X	X
FWD CARGO SQB 2	X	X	X
FWD SERV DOOR	X	X	X
FUEL CH ½ FAIL	X	X	X
FUEL CTR TANK	X	X	X
FUEL IMBALANCE	X	X	X
FUEL TK MISMGMT	X	X	X
GEN 1 OFF		X	X
GEN 2 OFF		X	X
GEN 1 OVLD	X	X	X
GEN 2 OVLD	X	X	X
HYD EDP 1A	X	X	X
HYD EDP 2A	X	X	X
HYD 1 HI TEMP	X	X	X
HYD 2 HI TEMP	X	X	X

Caution Message	Initial Take-off Phase	Final Take-off Phase	Landing Phase
HYD 3 HI TEMP	X	X	X
HYD PUMP 1B	X	X	X
HYD PUMP 2B	X	X	X
HYD PUMP 3A	X	X	X
HYD PUMP 3B	X	X	X
HYD SOV 1 OPEN	X	X	
HYD SOV 2 OPEN	X	X	
IDG 1		X	X
IDG 2		X	X
ISOL FAIL	X	X	X
L AFT EMER DOOR	X	X	X
L AOA HEAT	X	X	X
L BLEED LOOP	X	X	X
L COWL A/I	X	X	
L COWL A/I OPEN	X	X	
L EMER DOOR	X	X	X
L ENG BLEED	X	X	X
L ENG DEGRADED	X	X	X
L ENG SOV CLSD	X	X	X
L ENG SOV FAIL	X	X	X
L ENG SOV OPEN	X	X	
L ENG SQB	X	X	X
L ENG SRG OPEN	X	X	
L ENG TAT HEAT	X	X	X
L FIRE FAIL	X	X	X
L FUEL FILTER	X	X	X
L FUEL LO PRESS		X	X
L FUEL PUMP	X	X	X
L FUEL LO TEMP	X	X	X
L FWD EMER DOOR	X	X	X
L MAIN EJECTOR		X	X

Caution Message	Initial Take-off Phase	Final Take-off Phase	Landing Phase
L PACK	X	X	X
L PACK AUTOFAIL	X	X	X
L PACK TEMP	X	X	X
L PITOT HEAT	X	X	X
L SCAV EJECTOR	X	X	X
L START ABORT	X	X	X
L START VALVE	X	X	X

Effectivity:

Airplanes 15260 and subsequent:

Caution Message	Initial Take-off Phase	Final Take-off Phase	Landing Phase
L STRT VLV OPEN (In flight)	X	X	X

Caution Message	Initial Take-off Phase	Final Take-off Phase	Landing Phase
L STATIC HEAT	X	X	X
L THROTTLE		X	
L WINDOW HEAT	X	X	X
L WSHLD HEAT	X	X	X
L XFER SOV	X	X	X
MACH TRIM	X	X	X
MAIN BATT OFF	X	X	X
MLG OVHT FAIL	X	X	X

Effectivity:

Airplanes 15001 thru 15259:

Caution Message	Initial Take-off Phase	Final Take-off Phase	Landing Phase
NO STRTR CUTOUT (In flight)	X	X	X

Caution Message	Initial Take-off Phase	Final Take-off Phase	Landing Phase
OVBD COOL	X	X	X
OXY LO PRESS	X	X	X
PASS OXY ON	X	X	X
PARK BRAKE SOV		X	
PAX DR LATCH	X	X	X
PAX DR OUT HNDL		X	X
PROX SYS CHAN		X	
R AFT EMER DOOR	X	X	X
R AOA HEAT	X	X	X
R BLEED LOOP	X	X	X
R COWL A/I	X	X	
R COWL A/I OPEN	X	X	
R EMER DOOR	X	X	X
R ENG BLEED	X	X	X
R ENG DEGRADED	X	X	X
R ENG SOV FAIL	X	X	X
R ENG SOV OPEN	X	X	
R ENG SQB	X	X	X
R ENG SRG OPEN	X	X	
R ENG TAT HEAT	X	X	X
R FIRE FAIL	X	X	X
R FUEL FILTER	X	X	X
R FUEL PUMP	X	X	X
R FUEL LO PRESS		X	X

Caution Message	Initial Take-off Phase	Final Take-off Phase	Landing Phase
R FUEL LO TEMP	X	X	X
R FWD EMER DOOR	X	X	X
R MAIN EJECTOR		X	X
R PACK	X	X	X
R PACK AUTOFAIL	X	X	X
R PACK TEMP	X	X	X
R PITOT HEAT	X	X	X
R SCAV EJECTOR	X	X	X
R STATIC HEAT	X	X	X
R START ABORT	X	X	X
R START VALVE	X	X	X

Effectivity:

Airplanes 15260 and subsequent:

Caution Message	Initial Take-off Phase	Final Take-off Phase	Landing Phase
R STRT VLV OPEN	X	X	X

Caution Message	Initial Take-off Phase	Final Take-off Phase	Landing Phase
R THROTTLE		X	
R WINDOW HEAT	X	X	X
R WSHLD HEAT	X	X	X
R XFER SOV	X	X	X
SERVICE DOOR	X	X	X
STALL FAIL	X	X	X
STEERING INOP		X	
STBY PITOT HEAT	X	X	X
TAT PROBE HEAT	X	X	X
WING A/I SNSR	X	X	X

Caution Message	Initial Take-off Phase	Final Take-off Phase	Landing Phase
WING XBLEED	X	X	X
XFLOW PUMP	X	X	X

Display Reversionary Control

EICAS Reversionary Mode

The EICAS reversionary mode provides an alternate method of displaying EICAS information should ED 1 or ED 2 fail.

Should ED 1 fail, the primary page is automatically transferred to ED 2.

This automatic feature ensures that engine indications and warning and caution messages are always available to the pilot. There is no automatic transfer of information should ED 2 fail.

Reversionary control of EICAS is accomplished by sharing displays and control panels that are used by other EICAS or EFIS modes. The list includes:

- EICAS control panel (ECP)
- EFIS multi-functional displays (MFDs)
- Display reversionary panels
- Source selection panel

Failure of ED 2 or automatic transfer of the primary page to ED 2 renders the ECP inactive and removes the status and synoptic pages from view. Recovery of these pages is performed with the display reversionary panels or the source selection panel.

Display Reversionary Panel

Display reversionary panels (DRPs) located on the pilot and copilot side consoles control the presentation on the associated multi-functional display (MFD). Each panel has a three-position rotary switch labeled PFD, NORM, and EICAS.

When the EICAS switch position is selected. the MFD is reconfigured to act as an EICAS display and the status page is presented as the default page. The ECP 1s now functional and can be used to view the synoptic or other EICAS pages.

PFD Reversionary Mode

When a primary flight display (PFD) fails. the PFD information can be transferred to the associated MFD. When the rotary knob on the display reversionary panel is selected to the PFD position, transfer occurs.

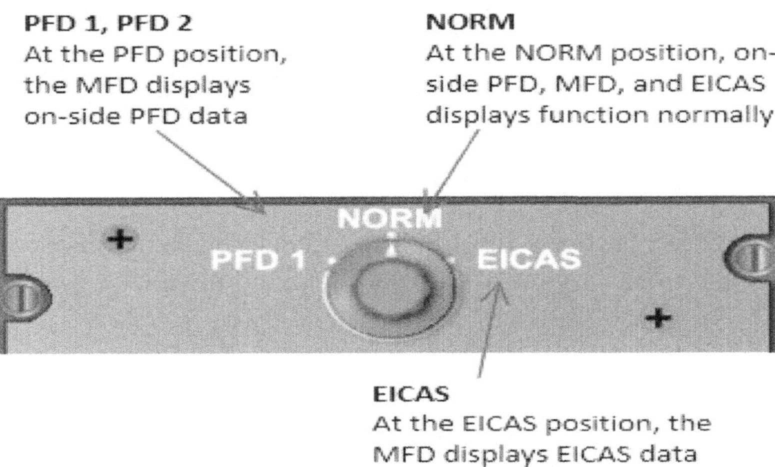

Figure 13 - 10: Display Reversionary Panel

Source Selection Panel

If either EICAS display fails, the operative ED can be used to present all EICAS information. This is accomplished by using the three-position EICAS selector knob on the source selection panel. The EICAS switch positions are designated ED 1, NORM, and ED 2.

If ED 1 fails, moving the EICAS selector knob to the ED 2 switch position allows all EICAS pages to be accessed through the reactivated ECP. Only one page can be viewed at a time. If ED 2 is inoperative, selecting the knob to ED 1 allows all EICAS information to be viewed on ED 1.

If the only operative ED is displaying a status. synoptic or menu page and the DCU generates a warning message, the ED automatically reverts back to the primary page. All aural alerts associated with the warning message sound.

MFD Reversionary Mode

There is no reversionary mode for the data normally presented on the multi-functional displays.

EICAS SELECTOR
NORM - Normal EICAS display
ED1 - Enable ECP to present all EICAS information EICAS Display 1
ED2 - Enable ECP to present all EICAS information EICAS Display 2

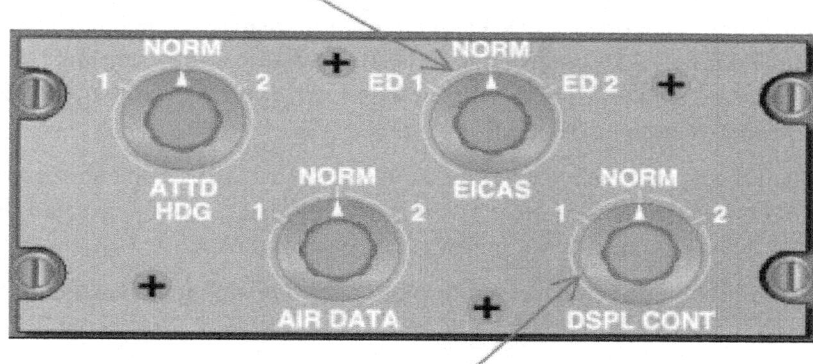

DSPL CONT
NORM - DCP1 controls left displays and DCP 2 controls right displays
1 - DCP 1 controls left and right displays
2 - DCP 2 controls left and right displays

Maintenance Diagnostic System

The maintenance diagnostic system is used by maintenance personnel to view current and historical information relating to specific aircraft systems health and operation.

Components and Operation

The system uses a maintenance diagnostic computer (MDC) to process and record avionics and aircraft systems data for future retrieval. A guarded three-position maintenance switch, located behind the pilot's seat on CBP-1, is used to enter the maintenance diagnostics mode. The multi-functional displays (MFDs) are used to display the maintenance data and the EICAS control panel (ECP) is used to control and select information on the MFD display. A data loader unit is used to upload or download data to or from a floppy disk.

When the maintenance switch is set to MFD 1 or MFD 2, the applicable MFD is configured to display maintenance related display pages and the ECP is configured as a maintenance page control panel.

The following can be accessed through the MDC maintenance menu page:

- Current faults - Displays fault(s) currently detected by the MDC and failure messages reported by the DCU.
- Current service messages - Displays maintenance messages received from the DCU.
- Aircraft history - Provides access to history displays for faults, service messages, engine excellence and engine trends. Also used to access life cycle data and flight leg summary.
- LRU testing - Used to initiate LRU test and display test results.
- LRU rigging - Used to initiate the LRU programing procedure.
- System parameters - Displays the aircraft system parameters.
- ATA index - Displays a list of ATA chapter numbers for all systems
- LRU index/operations - Displays a list of LRUs and is used to select any associated test or rigging procedure.
- MDC setup - Used to set aircraft identification and clock. Also used to load files.

- Configuration data - Used to access the configuration of the IAPS computers and to check the MDC version information.
- FCC diagnostic - Displays instructions to put flight control system into diagnostic mode.

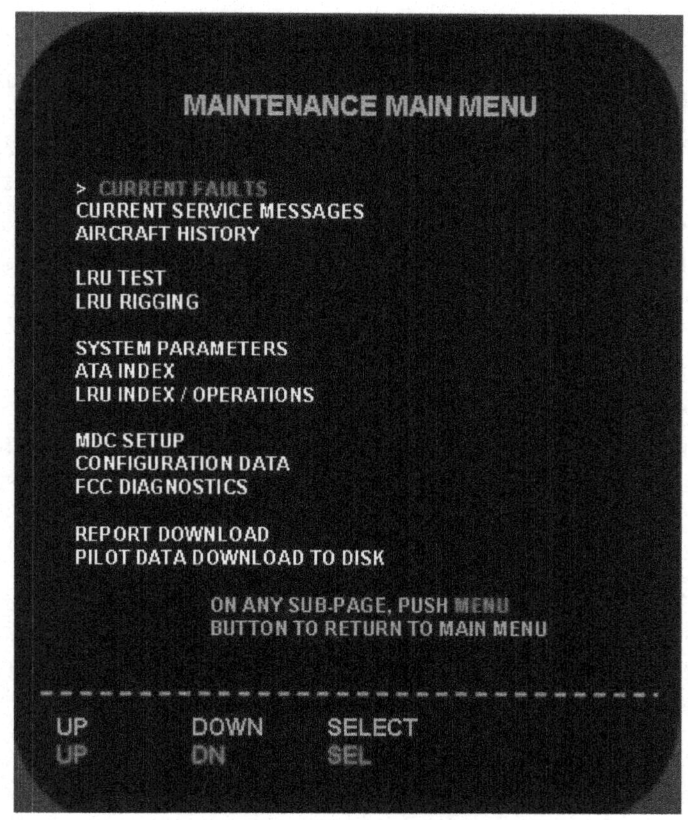

Flight Data Recorder

The flight data recorder (FDR) records aircraft flight parameters. The FDR includes a digital FDR, an underwater locater device (ULD), and a triaxial accelerometer.

Components and Operation

Flight Data Recorder

The FDR is located in the aft equipment bay and records the last 25 hours of flight data in a solid-state memory. The FDR is specially constructed to survive the shock of an aircraft impact.

The FDR starts to record when the BEACON or STROBE lights switch is selected ON. The system is also activated by the proximity sensing system's weight-off-wheels switches.

FDR EVENT Pushbutton

The FDR EVENT pushbutton is located on the ENGINES/miscellaneous test panel on the center pedestal When the pushbutton is selected, the tape is marked as a significant event for future ease in data analysis.

Underwater Locater Device

The ULD, mounted on the FDR front panel, is a battery-operated Underwater acoustic pulse generator. The ULD transmits when submerged after aircraft ditching.

Triaxial Accelerometer

The triaxial accelerometer is used for acceleration measurements. It contains three separate seismic sensors that measure vertical. lateral and longitudinal acceleration near the center-of-gravity of the aircraft. This data is sent to the DCUs where it is formatted for the FDR.

Controls and EICAS Indications

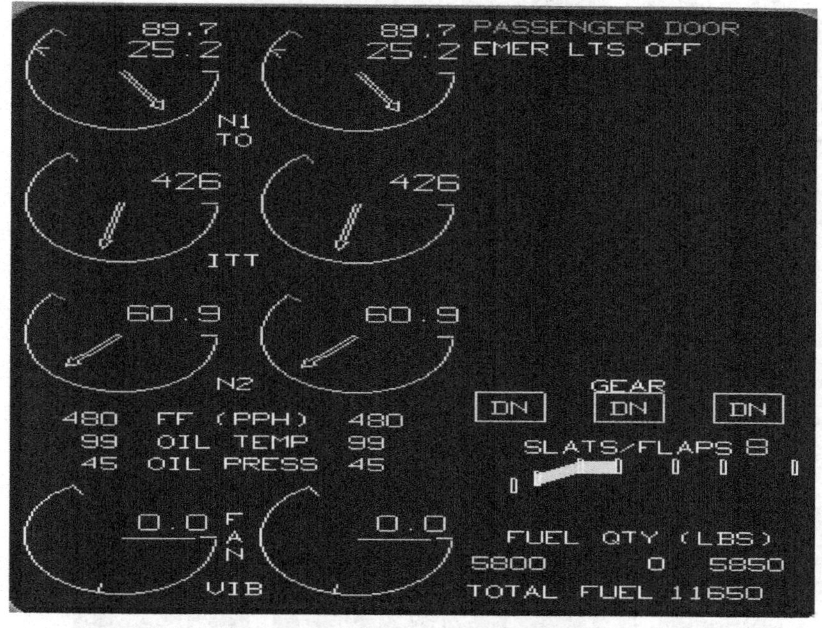

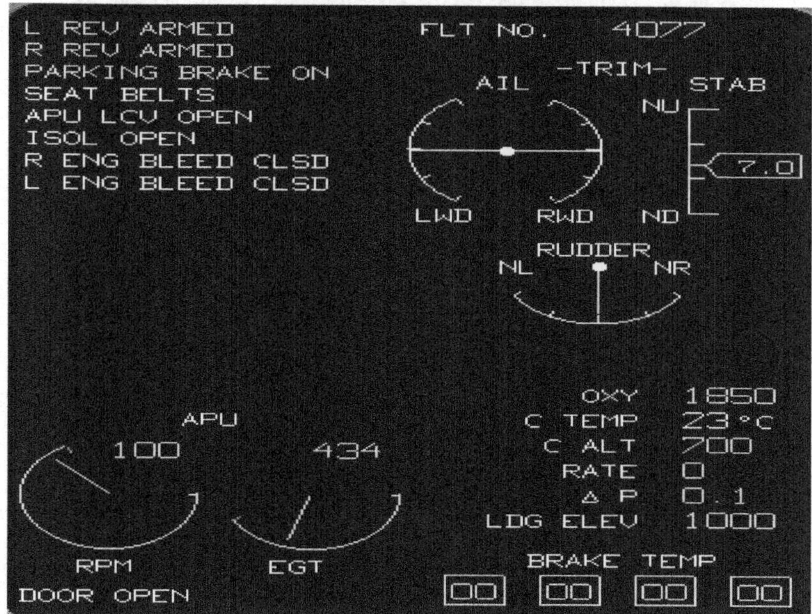

Figure 13 - 11: Primary and Status Pages

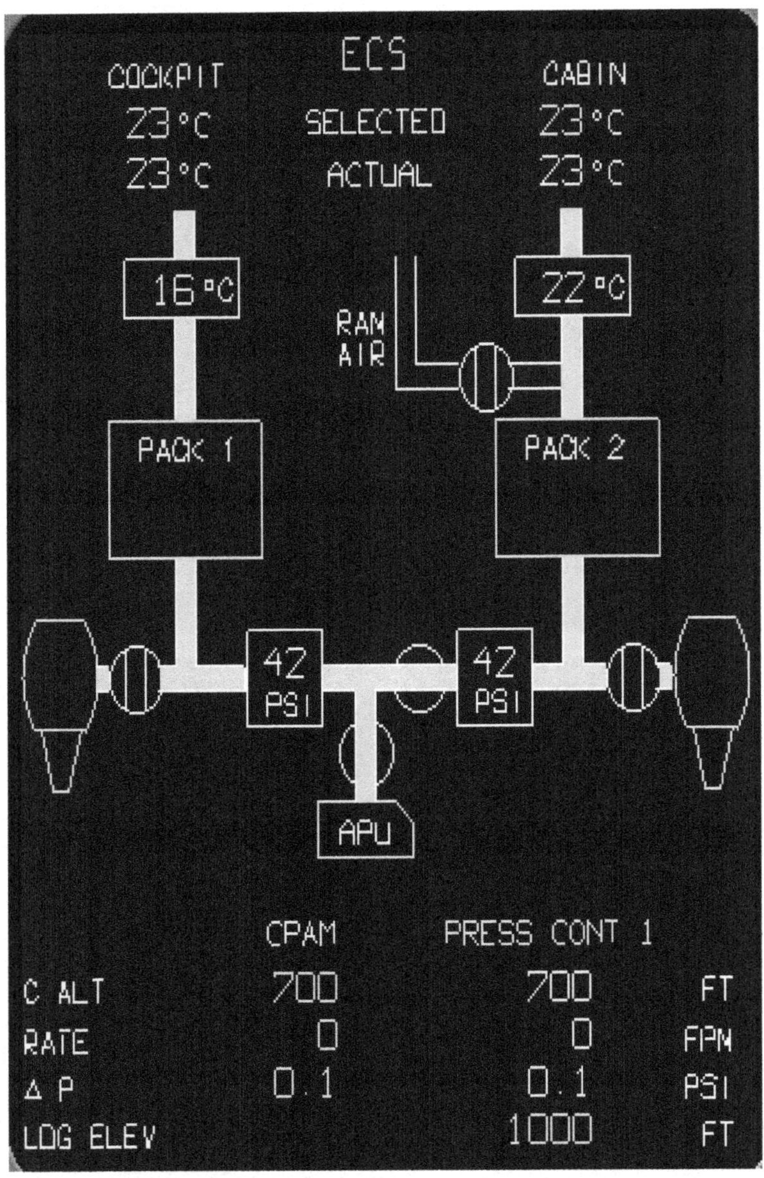

Figure 13 - 12: ECS Synoptic Page

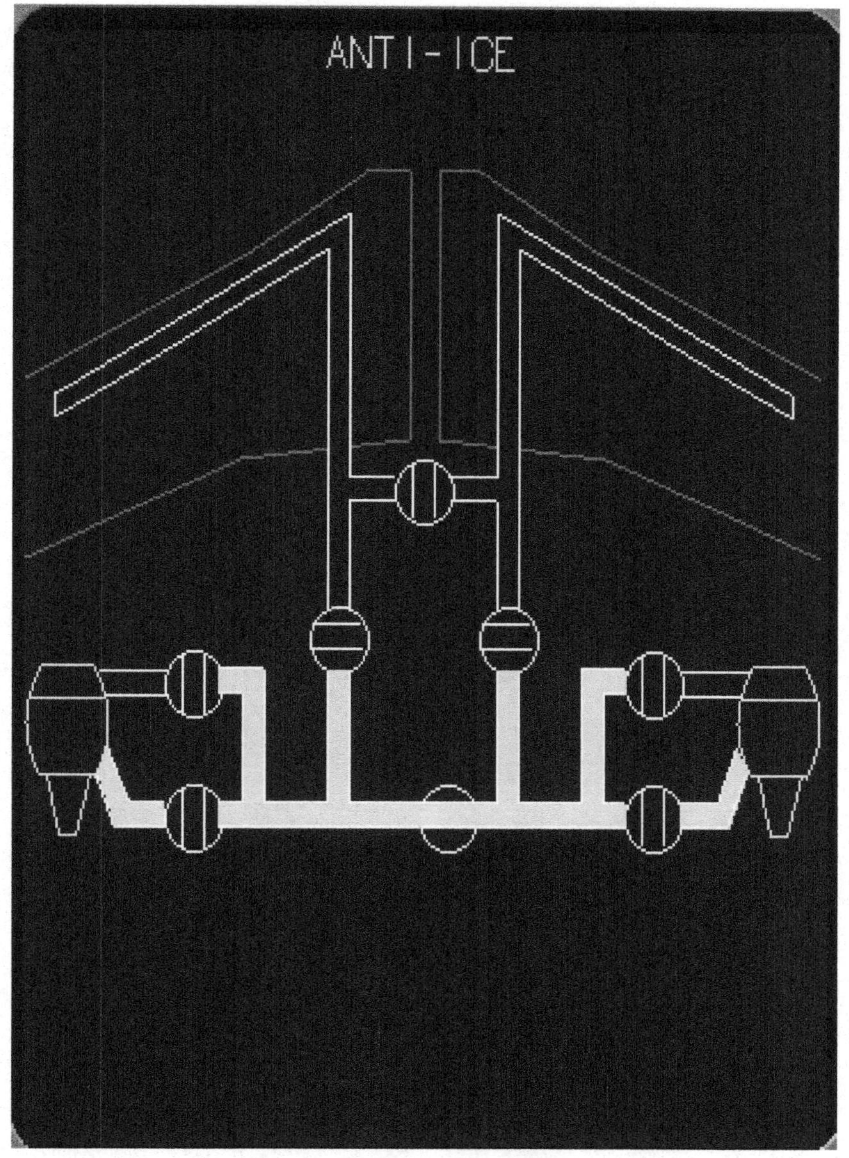

Figure 13 - 13: Anti-Ice Synoptic Page\

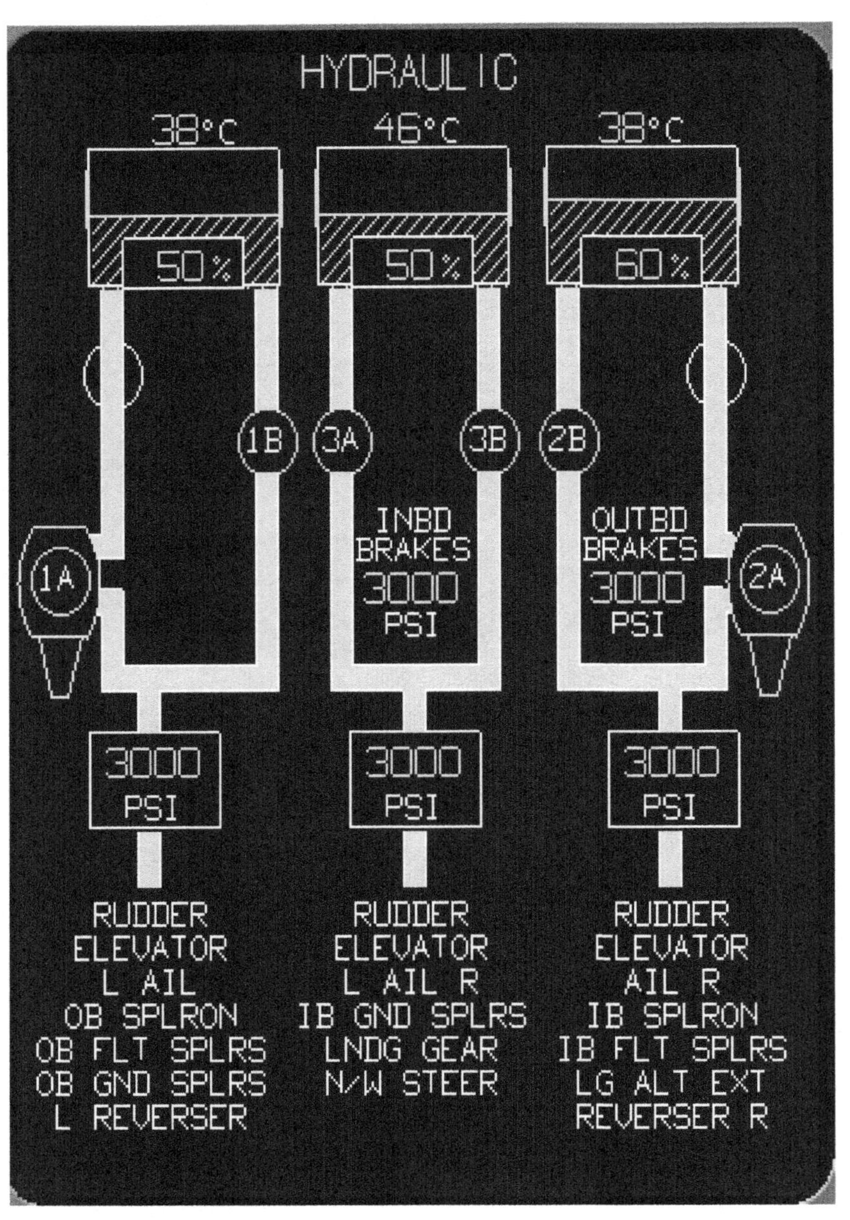

Figure 13 - 14: Hydraulic Synoptic Page

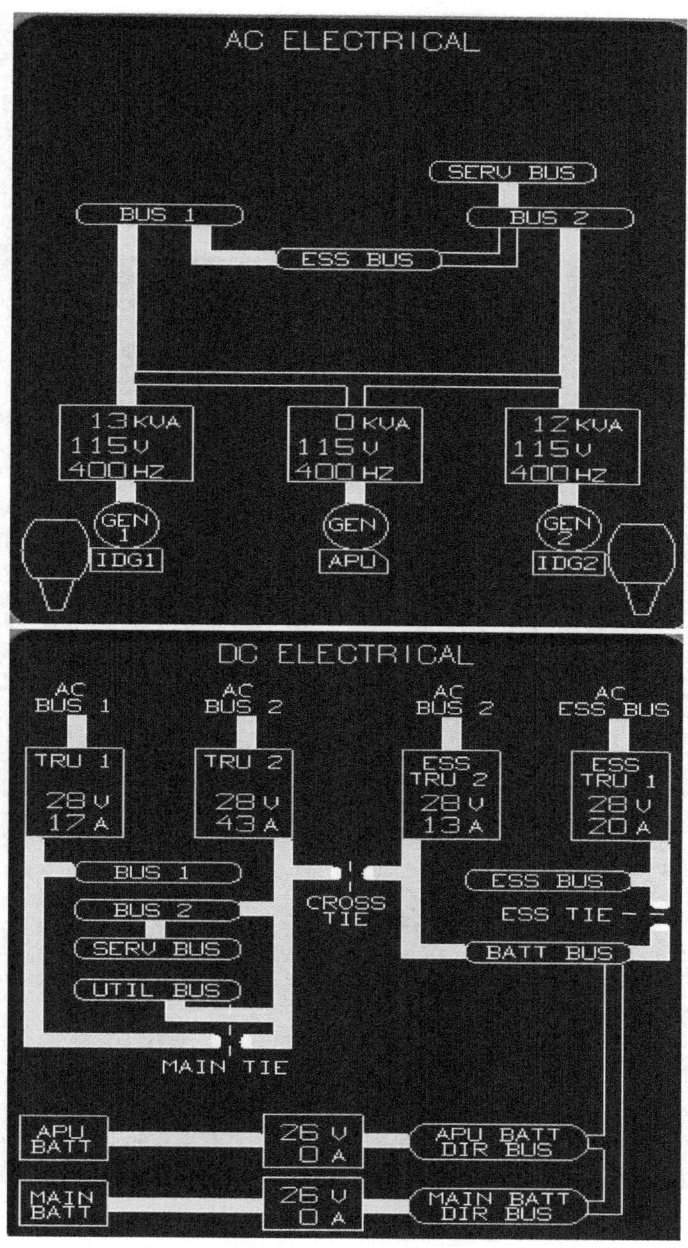

Figure 13 - 15: AC/DC Electrical Synoptic Pages

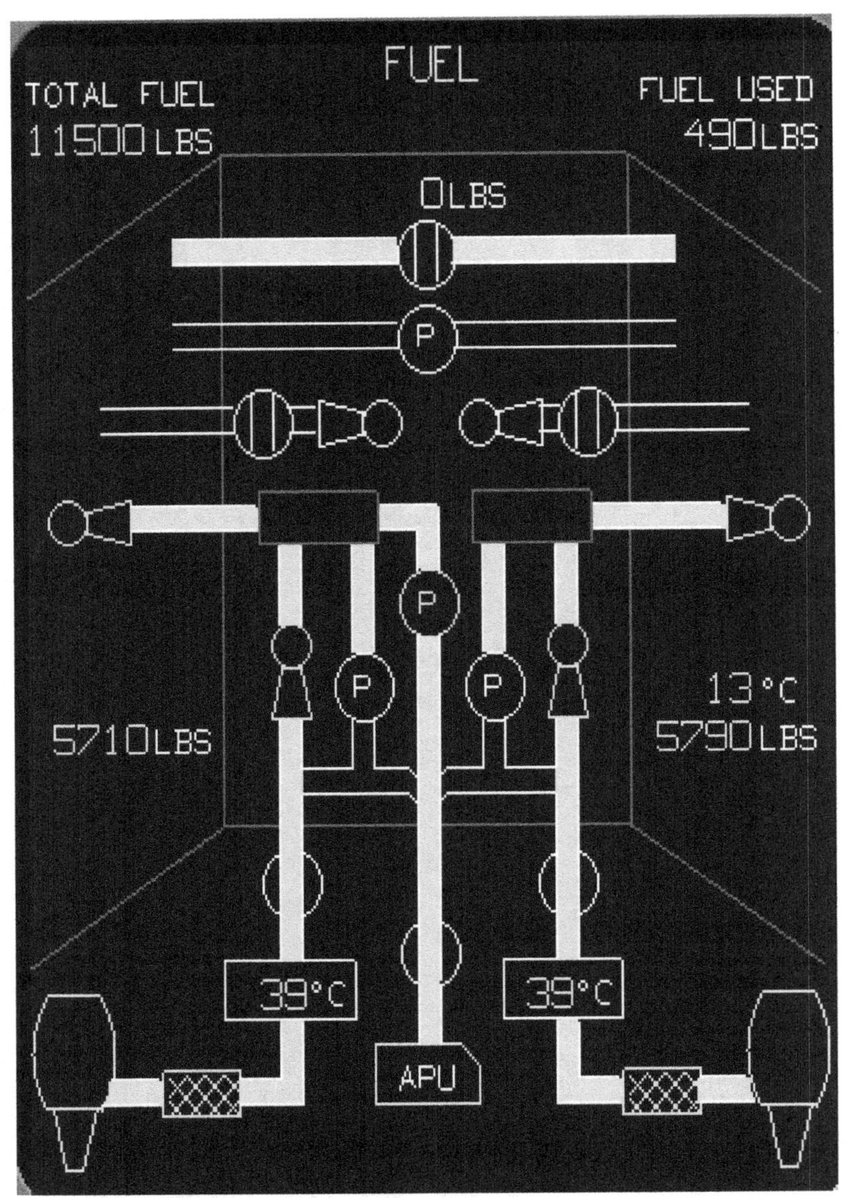

Figure 13 - 16: Fuel Synoptic Page

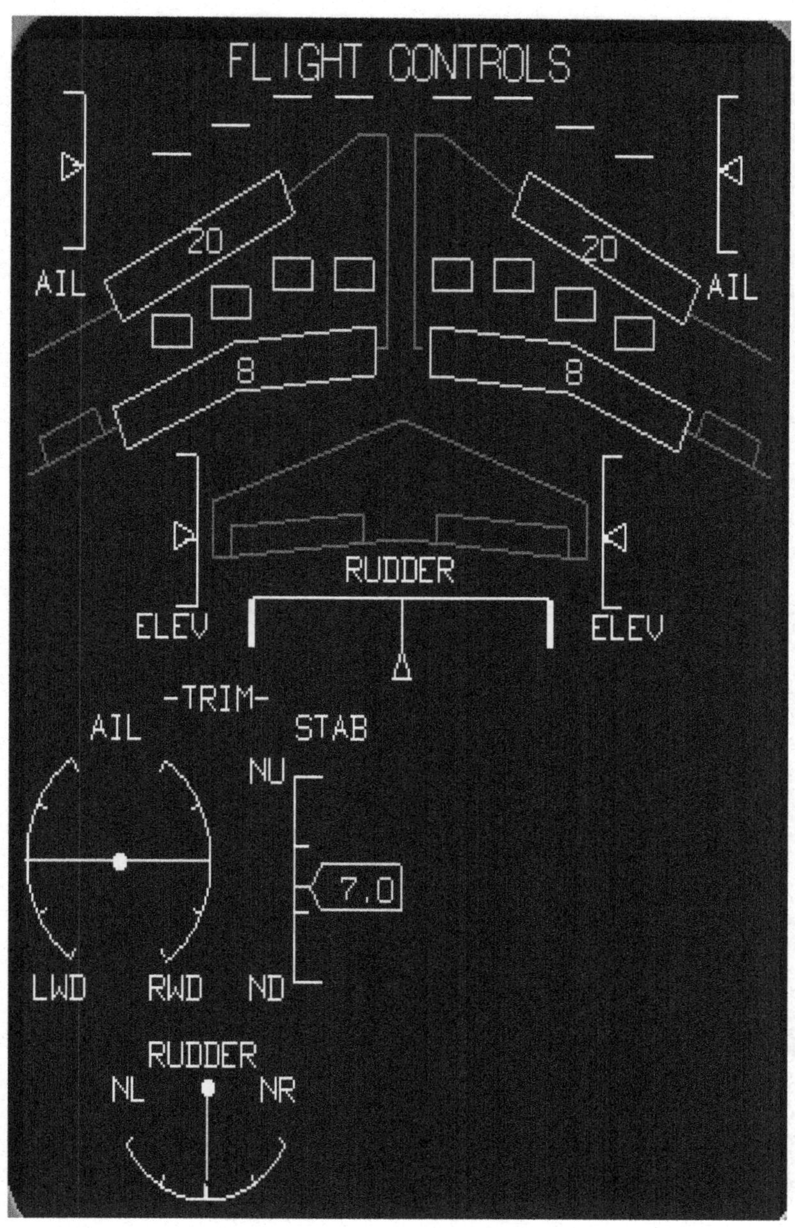

Figure 13 - 17: Flight Controls Synoptic Page

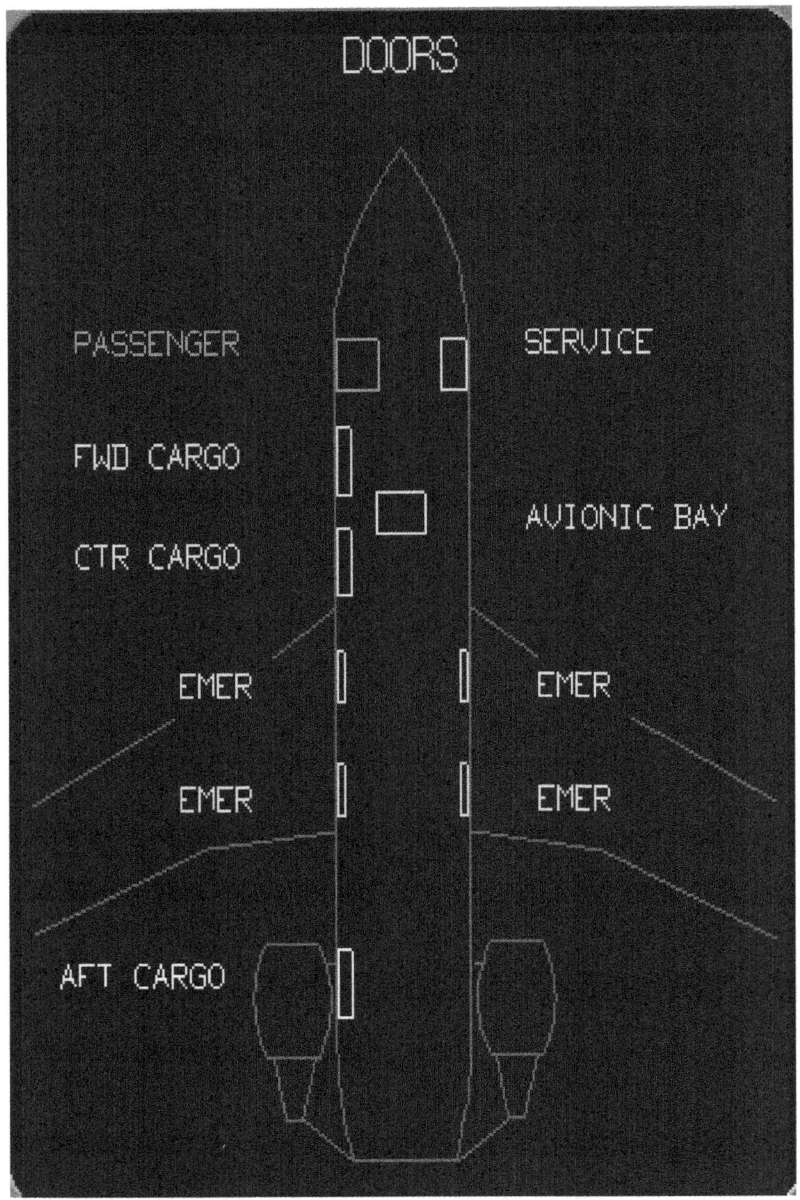

Figure 13 - 18: Doors Synoptic Page

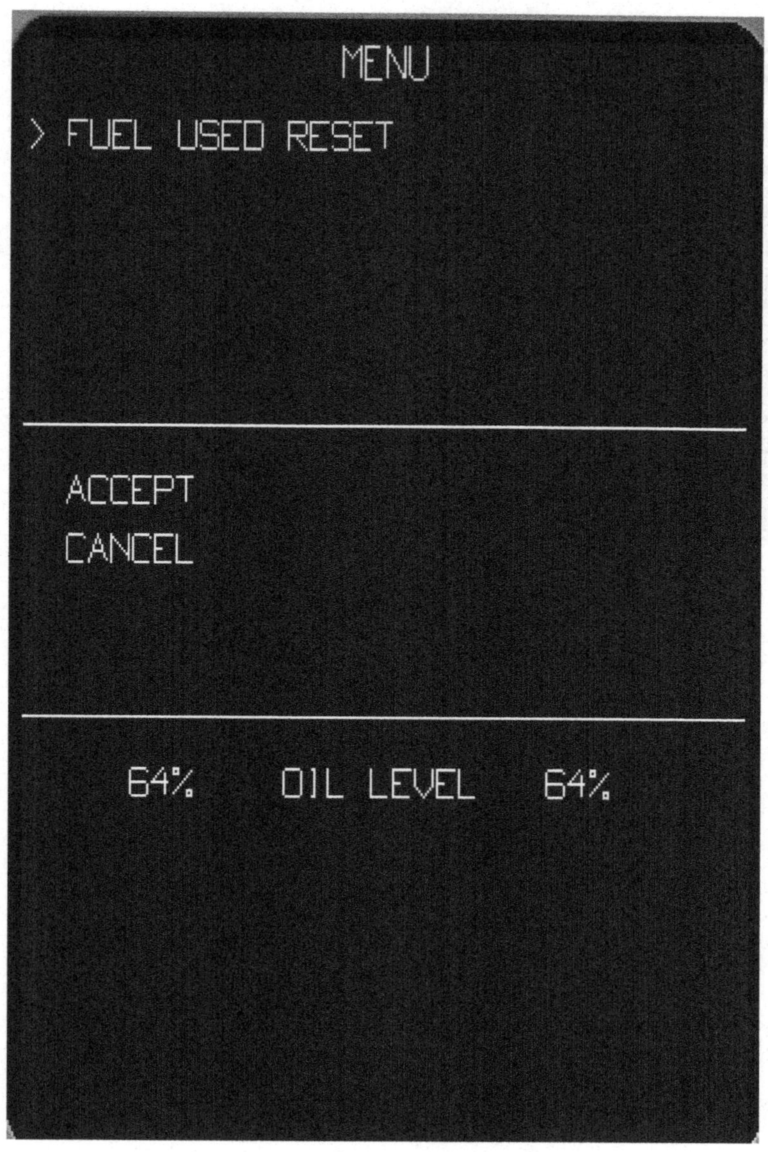

Figure 13 - 19: Menu Synoptic Page

Table 13 - 7: EICAS Messages

Status Message	Cause
CAS MISCOMP	Mis-comparison of detected warning, caution or aural alerts exists between DCUs
DCU 1 (2) AURAL INOP	Internal aural fault in respective data concentrator unit or indicates respective DCU aural output has been disabled
DCU 1 (2) INOP	Internal fault or crosstalk fault in respective data concentrator unit
FDR ACCEL FAIL	Triaxial accelerometer out of tolerance on ground with parking brake set and DC bus 1 on.
FDR FAIL	Difference between recorded data and data supplied by DCU
MDC FAULT	Maintenance diagnostic computer inoperative

System Schematic

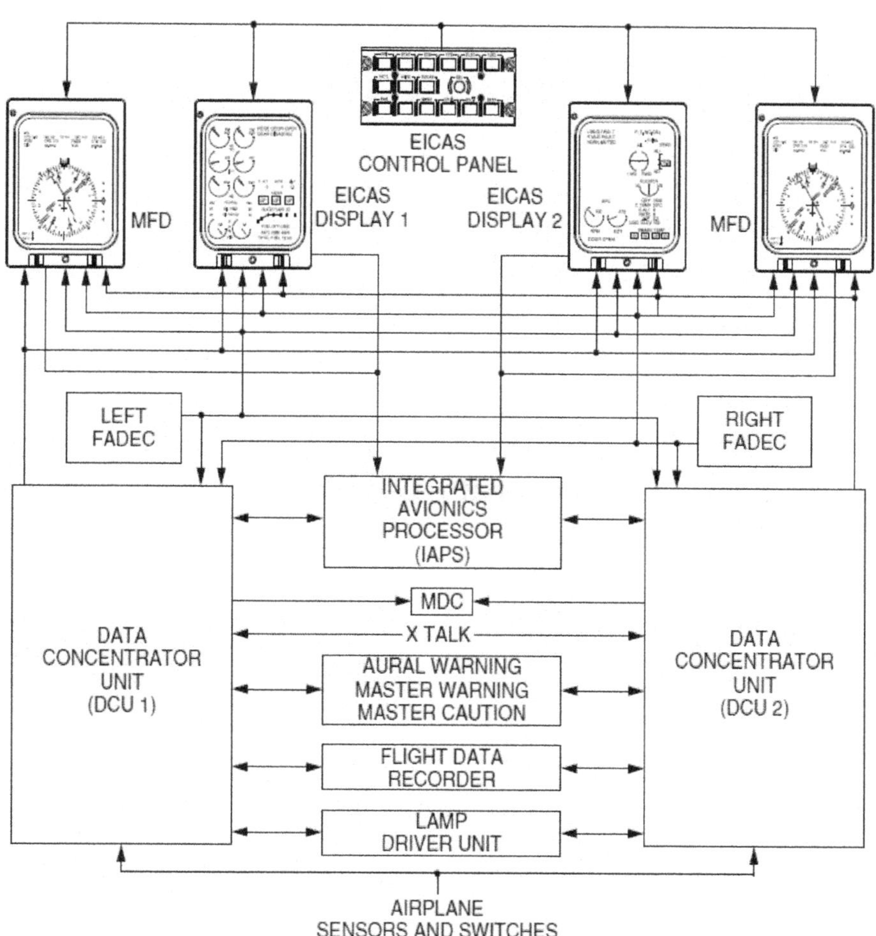

Figure 13 - 20: EICAS Schematic

Intentionally Left Blank

Chapter 14: Landing Gear

Chapter 14: Landing Gear

General

The aircraft landing gear is a retractable tricycle type consisting of two wing root mounted main landing gear (MLG) assemblies and a forward fuselage mounted steerable nose landing gear (NLG) assembly. Each gear assembly has two wheels. The MLG retracts inboard and the NLG retracts forward. Each landing gear has a shock strut to absorb and dissipate the shock loads encountered when the aircraft lands. Each MLG is fitted with steel multi-disc brakes.

Landing gear extension and retraction is electrically activated by the landing gear (LDG GEAR) selector lever and controlled by the proximity sensing electronic unit (PSEU). Sensors are mounted on the landing gear and landing gear doors to supply position and control information to the PSEU. The PSEU provides landing gear position indication to the DCUs for display on the EICAS primary page. In normal operation, the landing gear is hydraulically actuated by hydraulic system 3, but an alternate independent means of extending the landing gear is available should the normal extension system fail.

A tail bumper protects the airplane tail structure from tail strikes caused by over-rotation of the airplane on takeoff The tail bumper consists of a shock absorber, a skid assembly and a strike indicator.

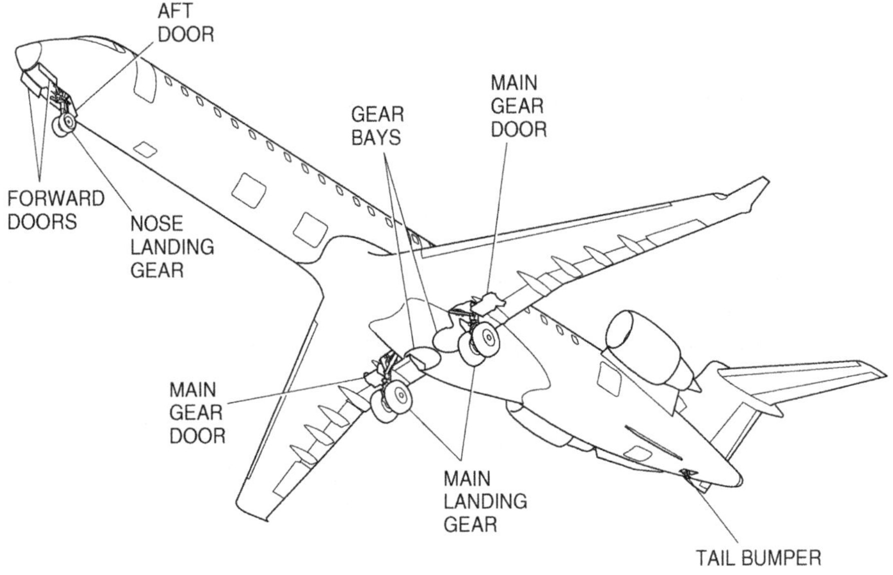

Figure 14 - 1: Landing Gear

Main Landing Gear and Doors

There are two main landing gear (MLG) assemblies. Each assembly is mounted on the underside of the wing center section. The purpose of the MLG is to provide stability and support when the airplane is on the ground and to protect the airplane from structural damage by absorbing and dissipating the shock loads that are generated when the airplane lands.

Components and Operation

Main Landing Gear

The MLG is a double-wheel retracting type that retracts inboard into the main wheel bays. Each MLG includes:

- Nitrogen-charged/oil-filled shock strut
- A folding side stay assembly
- Shimmy dampers
- Dual wheels and brake assemblies
- Hydraulic actuators
- Auxiliary actuators (S/N 15001-15211)

NOTE:
The MLG auxiliary actuators are not installed on new production aircraft (S/N 15212 and subsequent).

When retracted, the MLG is held in the up and locked position by a mechanical lock. The hydraulic actuators are used to unlock, extend and retract the MLG. The side stay assembly has an over-center lock link mechanism which locks the MLG in the extended position.

Airplanes With Auxiliary Actuators

In the event of an emergency, the landing gear can be extended by operating the LANDING GEAR MANUAL RELEASE T-handle in the flight deck and then the gear is assisted to the down position by the auxiliary actuator.

Airplanes Without Auxiliary Actuators

In the event of an emergency, the landing gear can be extended manually by operating the LANDING GEAR MANUAL RELEASE T-handle in the flight deck. The main and nose gear will then free-fall to the down position.

Main Landing Gear Doors

The MLG doors are hinged to the wing structure and mechanically attached to the MLG by an adjustable link. The doors remain open when the gear is extended and remain closed while the landing gear is retracted. When the MLG is retracted, the doors make a smooth aerodynamic surface with the airplane.

Figure 14 - 2: Main Landing Gear Actuators

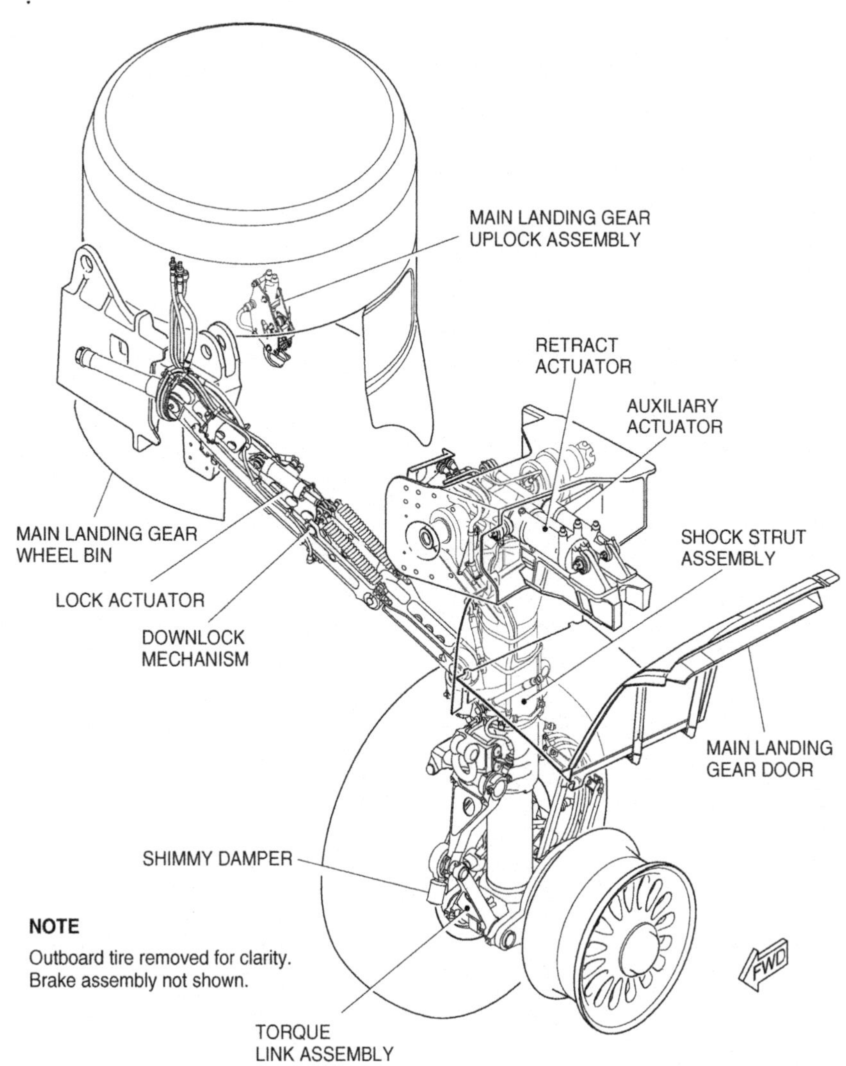

Figure 14 - 3: Main Landing Gear - Aircraft 15001-15211

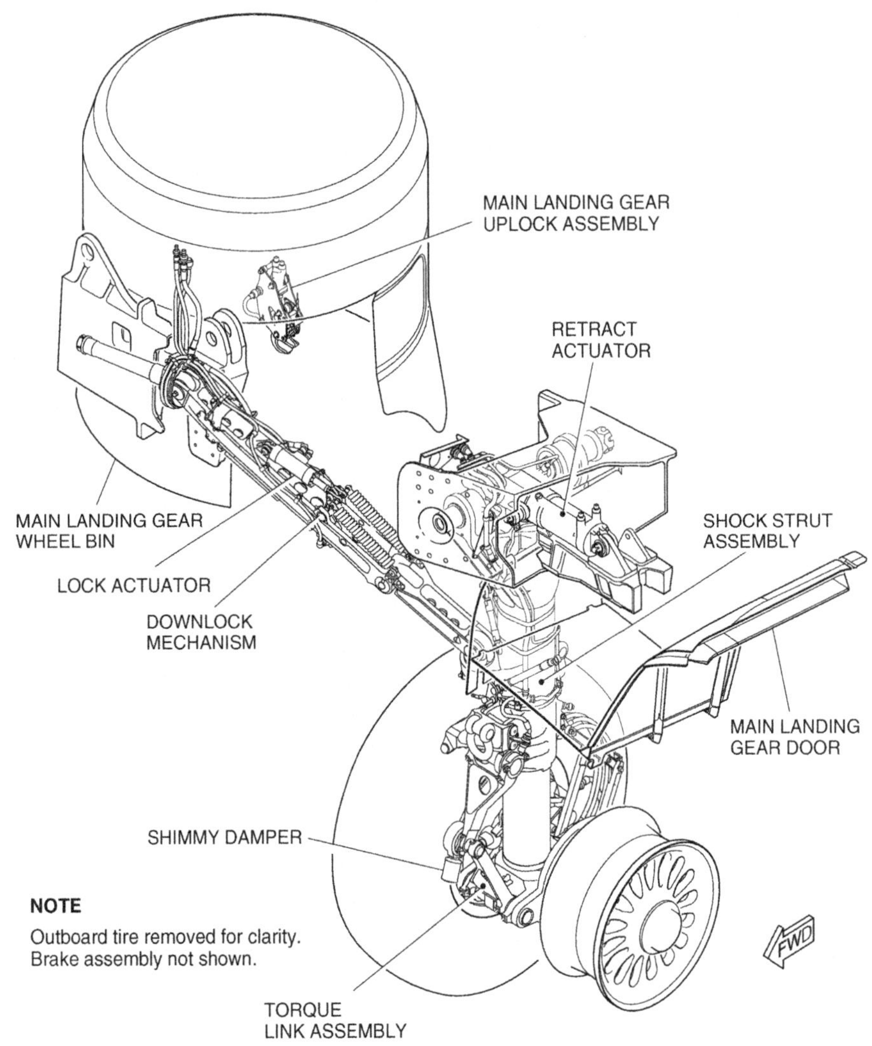

Figure 14 - 4: Main Landing Gear - Aircraft 15212 and Subsequent

Nose Landing Gear and Doors

The nose landing gear (NLG) supports the forward fuselage and also protects the forward fuselage from structural damage by absorbing and dissipating the shock loads generated during landing.

Components and Operation

Nose Landing Gear

The NLG is a double-wheel retracting type that retracts inboard into the nose wheel bay. Each NLG includes

- Nitrogen-charged/oil-filled shock strut
- A drag brace assembly
- Springs
- A lock actuator
- A retract actuator
- Dual wheel assemblies
- Nose wheel steering components

The NLG locks in both the extended and retracted positions with a spring loaded, over-center type locking mechanism. A lock actuator moves the locking mechanism out of the over-center condition at the beginning of each extend or retract cycle.

In the event of an emergency, the NLG can be extended by operating the LANDING GEAR MANUAL RELEASE T-handle in the flight deck. The gear then free-falls and is assisted to the down position by ram air.

Nose Landing Gear Doors

The NLG doors consist of two forward doors and an aft door. The doors open and close with nose gear motion through mechanical means. The doors are hinged to the fuselage and mechanically linked to the nose gear by control rod assemblies which force the doors to open or close as the NLG is extended or retracted.

Spin Down Assembly

The spin down assembly has been removed from company aircraft.

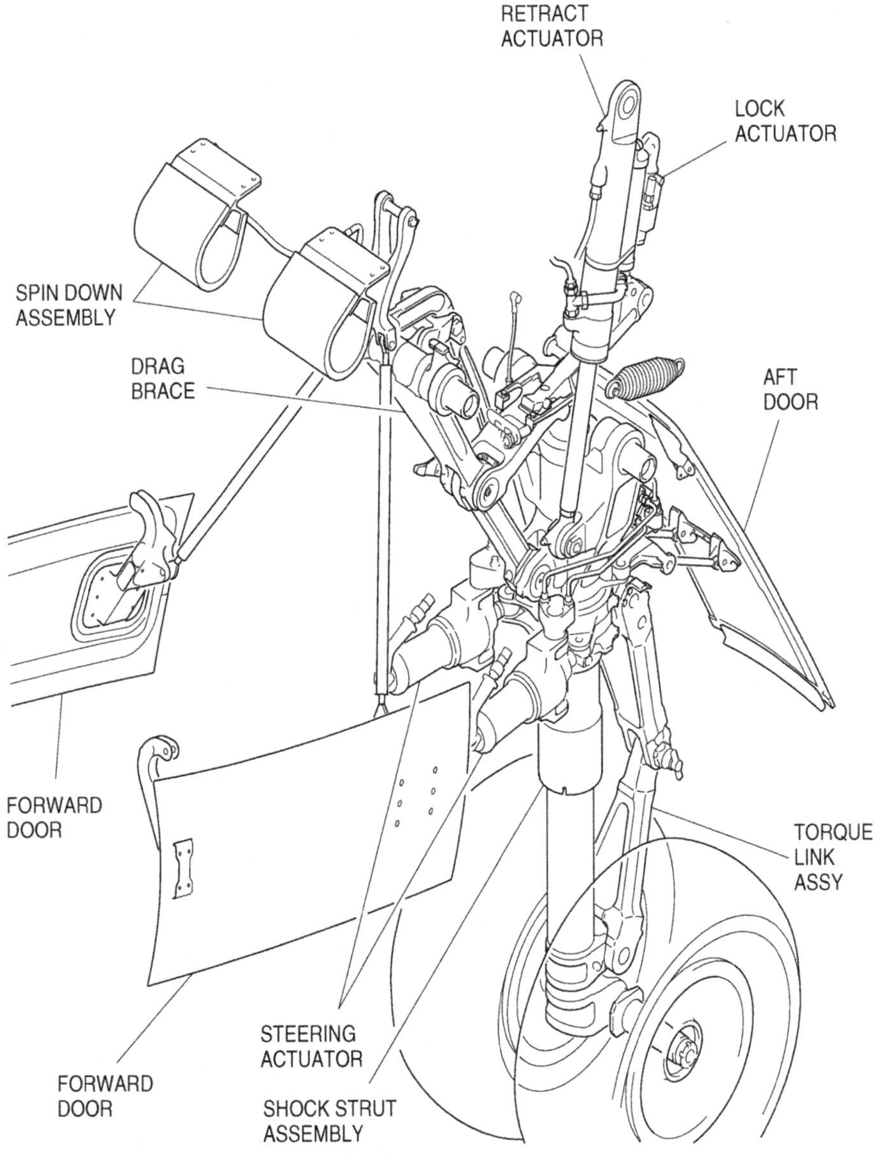

Figure 14 - 5: Nose Landing Gear

Landing Gear Extension and Retraction

Landing gear extension and retraction has two modes of operation:

- Normal extension and retraction
- Alternate extension

Components and Operation

Normal Extension and Retraction

Normal extension or retraction of the landing gear is initiated by LDG GEAR lever selection. The retraction or extension signal is sent to the PSEU which monitors various landing gear proximity sensing inputs and weight-on-wheels inputs. If the correct parameters are met, the PSEU energizes a selector valve to retract or extend the landing gear using hydraulic system 3 pressure.

The LDG GEAR lever is equipped with a solenoid lock which prevents an UP selection of the LDG GEAR lever with the airplane on the ground. In the event of a solenoid lock malfunction, a down-lock release (DN LCK REL) switch permits UP selection of the LDG GEAR lever by overriding the solenoid lock.

Retraction and extension of each landing gear is driven by a retract/ extend actuator. On early production 900 airplanes, an auxiliary actuator powered by hydraulic system 2, provides a backup means of extending the MLG.

Tension springs assisted by a down-lock actuator ensure that the main gear locks in t/he down position. The lock is released at the start of the retraction cycle. An uplock assembly locks the main gear in the retracted position. An uplock release actuator releases the uplock assembly at the start of the extension cycle.

The NLG locks in both the extended or retracted positions with a spring-loaded, over-center type locking mechanism. A lock actuator moves the locking mechanism out of the over-center condition at the beginning of each cycle.

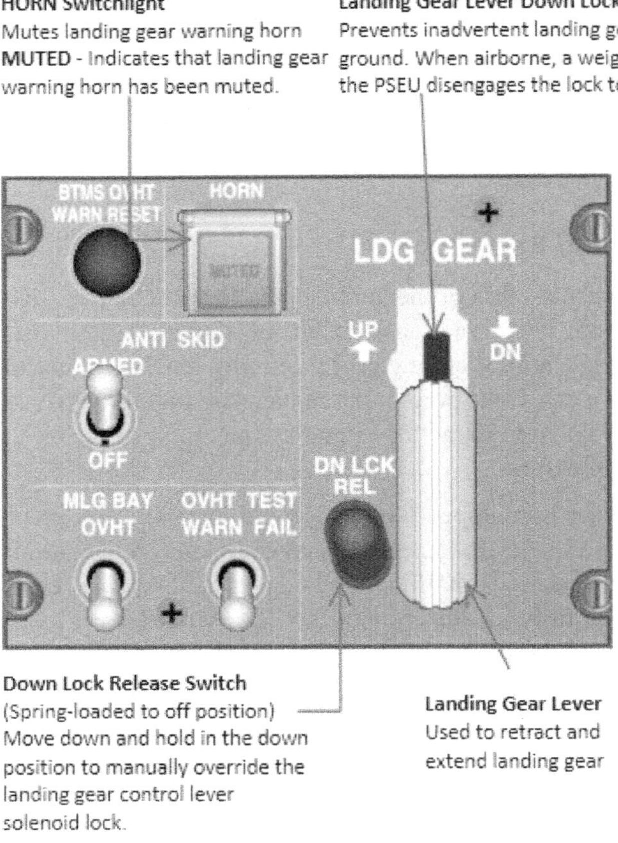

Figure 14 - 6: LDG GEAR Control Panel

(1) Retraction - Operation

Once the airplane is airborne with no weight-on-wheels signal, the PSEU commands and monitors the following events

- The LDG GEAR lever solenoid down-lock is released to permit UP selection of the LDG GEAR lever.
- The landing gear selector valve energizes the nose and MLG retract/ extend actuators, releases the down-locks and retracts the landing gear. Hydraulic pressure from the landing gear up line is routed to

activate the brake control valves to stop main wheel rotation. The tire spin-down assembly (removed) in the nose landing gear bay stops the nose-wheel rotation.

- The up-locks are engaged to secure the landing gear in the retracted position.

NOTE:
To prevent the landing gear from retracting when the airplane is on the ground, ground lock pins are inserted by the ground crew.

(2) Extension - Operation

The PSEU commands and monitors the following events:

- The LDG GEAR lever is manually selected to the ON position.
- The landing gear selector valve energizes the nose and MLG retract/ extend actuators, releases the up-locks and extends the landing gear.
- Down-locks are engaged to secure the landing gear in the extended position.

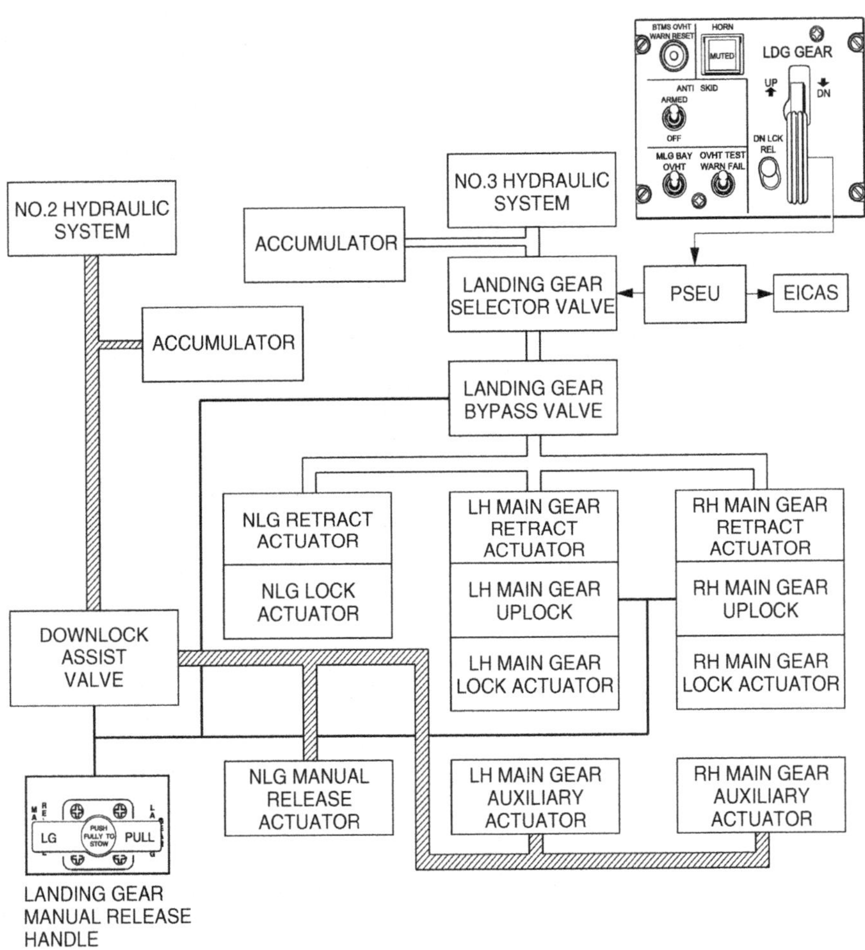

Figure 14 - 7: Landing Gear Retraction and Extension Aircraft 15001-15211

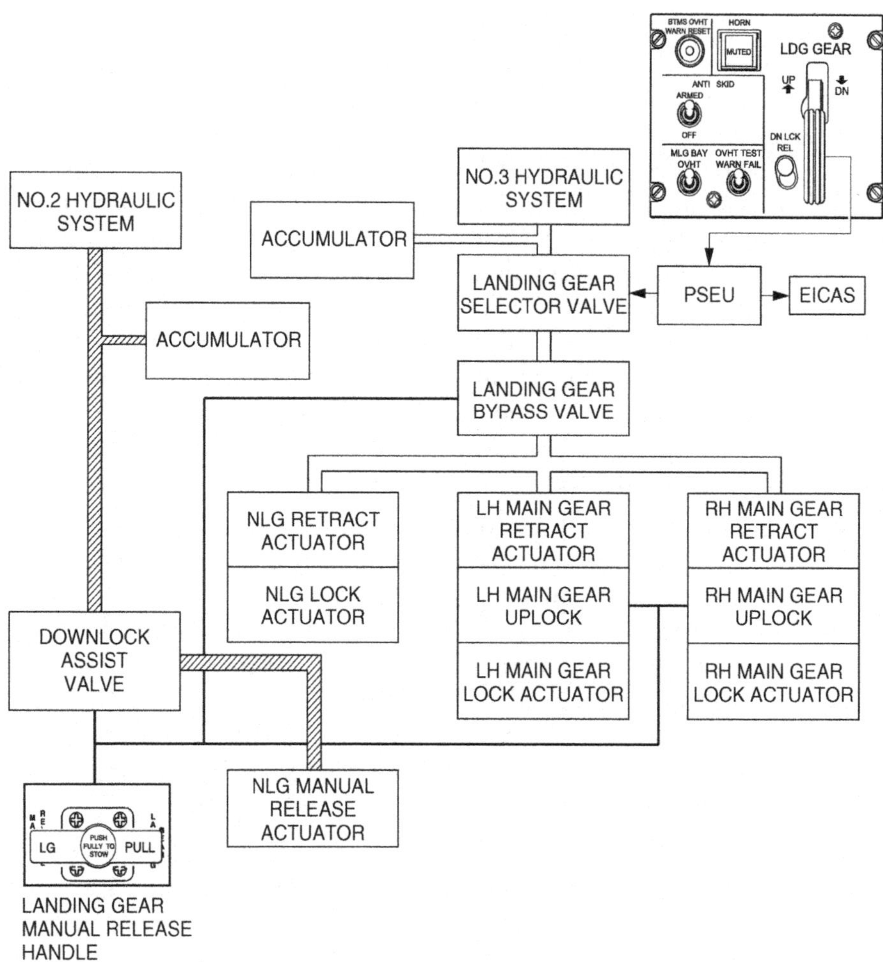

Figure 14 - 8: Landing Gear Retraction and Extension Aircraft 15212 and subsequent

Alternate Extension

Should a failure occur in the landing gear control system or in hydraulic system 3, the landing gear can still be extended by pulling the LANDING GEAR MANUAL RELEASE T-handle.

Aircraft With Auxiliary Actuators

When the LANDING GEAR MANUAL RELEASE T -handle is pulled to its full extension, the MLG up-locks are released by mechanical means. The LANDING GEAR MANUAL RELEASE T-handle is held in the operated position by a detent mechanism. At the same time a bypass valve dumps hydraulic system 3 pressure from the normal extension and retraction hydraulic circuits. This will permit the landing gear to partially extend under its own weight The manual release T-handle also positions a down-lock assist valve to direct hydraulic system 2 pressure to the MLG auxiliary actuators and to the nose gear uplock manual release actuator. The MLG is assisted to the down-and-locked position by the main gear auxiliary actuators and the NLG is assisted to the down-and-locked position by airflow and two tension springs.

Aircraft Without Auxiliary Actuators

When the LANDING GEAR MANUAL RELEASE T -handle is pulled to its full extension, the main and nose landing gear up-locks are released by mechanical means. The LANDING GEAR MANUAL RELEASE T-handle is held in the operated position by a detent mechanism. At the same time a bypass valve dumps hydraulic system 3 pressure from the normal extension and retraction hydraulic circuits. This will permit the MLG to extend to the down and locked position under its own weight. The NLG is assisted to the down-and-locked position by airflow and two tension springs.

To reset the LANDING GEAR MANUAL RELEASE T-handle, the operator must push the release button installed in the handle. This will release the detent mechanism and permit the handle to be moved to the stowed position.

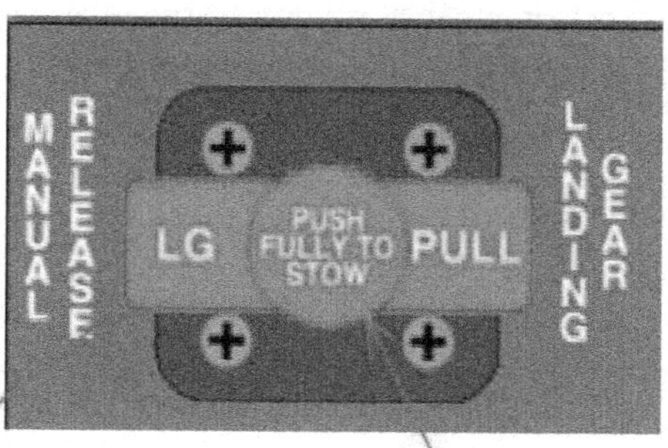

Landing Gear Manual Release

Used to manually lower the landing gear.

- Pull and hold handle in the fully extended position until EICAS indicates the nose and main landing gear are down and locked.

PUSH FULLY TO STOW button(red)

- Used to reset the manual release system (stow the LANDING GEAR MANUAL RELEASE handle) Pull the handle up, press button and then slowly return handle to the stowed position.

Figure 14 - 9: Manual Landing Gear Release Handle

Landing Gear Wheels and Brakes

The landing gear wheels and brakes include the following:

- Main and nose gear wheels and tires
- Main gear brake system
- Anti-skid system
- Parking brake
- Brake temperature monitoring
- Brake pressure monitoring

Components and Operation

Main and Nose Gear Wheels and Tires

There are two wheel/tire assemblies installed on each MLG axle and on the nose gear axle. Each assembly has an inflation valve and a pressure relief plug (overpressure valve). The main wheel assemblies have four heat sensitive fusible plugs that release excessive air pressure caused by heat build-up. The fusible plugs protect the main wheel against tire burst that could occur under heavy braking conditions.

Main Gear Brake System

Each wheel of the MLG is equipped with self-adjusting multi-disc brakes. The brakes of the inboard wheels are powered by hydraulic system 3 and the brakes of the outboard wheels are powered by hydraulic system 2.

Brake application is initiated by pressing the rudder pedals which are mechanically linked to the associated brake control valves. The brake control valves meter hydraulic pressure, proportional to the pedal pressure. to the four main wheel brake units, through four independent anti-skid control valves and four hydraulic fuses.

If a leak occurs in a brake line, the associated hydraulic fuse will close off the hydraulic line, preventing loss of the entire system fluid.\

With the loss of one hydraulic system, the airplane has 50% symmetric braking capability with full anti-skid control to the working brakes. In the event of a failure of both hydraulic systems 2 and 3, accumulators in each hydraulic system will provide reserve pressure for braking. During landing roll or rejected takeoff, reverse thrust and the ground spoilers will decelerate the airplane, if the brakes are degraded or fail completely.

Available inboard and outboard brake pressure is continuously monitored and displayed on the HYDRAULIC synoptic page, and any abnormal brake pressure detected is displayed as an EICAS message.

During landing gear retraction, hydraulic pressure is applied to the main wheel brake control valves to stop main wheel spin. A rubber spin-down pad assembly in the NLG wheel well provides resistance to stop the nose-wheel from spinning after gear retraction.

Two brake wear indicator pins, installed on each brake assembly, are used to provide a visual indication of brake wear.

NOTE:
The brake wear indicator pins must be checked with the brakes applied and hydraulic systems 2 and 3 pressurized.

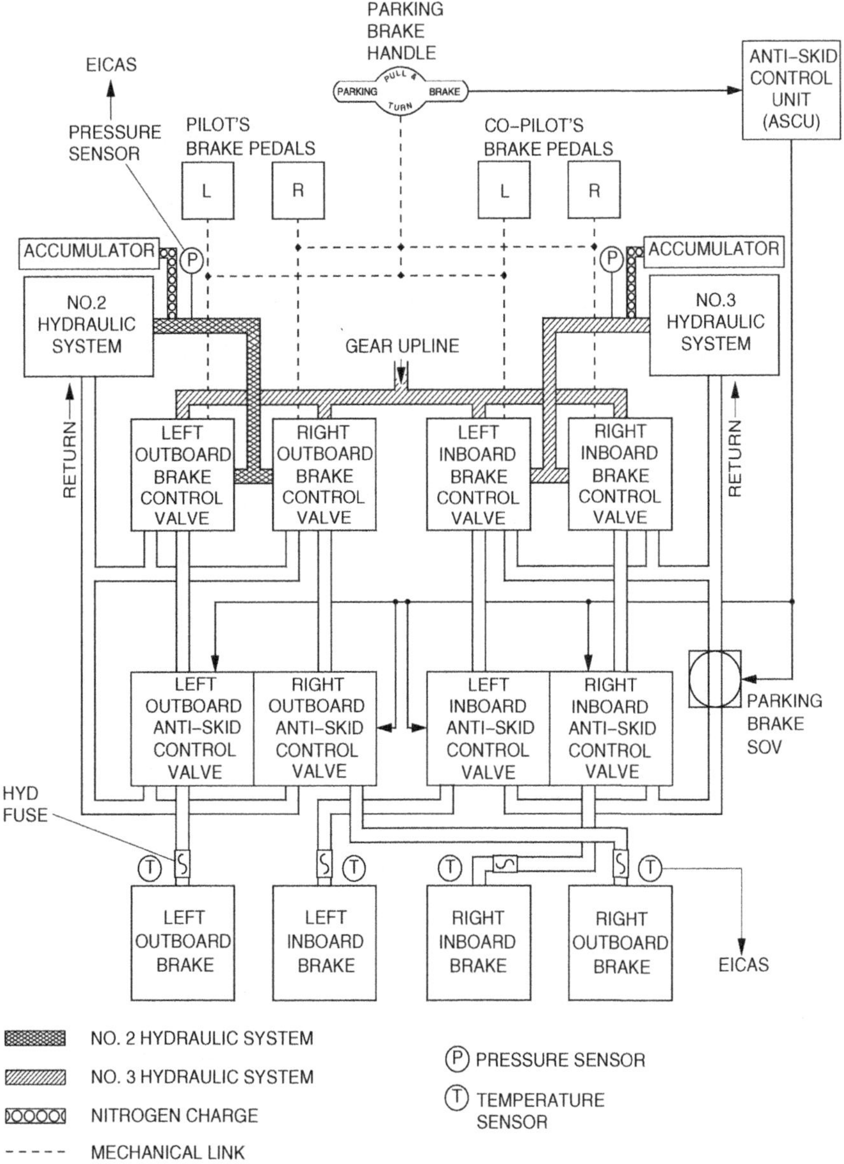

Figure 14 - 10: Brake System - Schematic

Anti-Skid System

The anti-skid system controls hydraulic pressure to the four main wheel brakes to provide anti-skid protection. The anti-skid system consists of a dual channel (inboard wheel control and outboard wheel control) anti-skid control unit (ASCU), four wheel speed transducers and two dual anti-skid control valves. The anti-skid system performs the following functions:

- Individual wheel anti-skid control: Prevents skids from developing.
- Touchdown protection: Prevents landing with locked wheels in the event that the pilot(s) are depressing the brake pedals during touchdown.
- Locked wheel protection: Allows a wheel to recover from a deep skid.

Selecting the ANTI-SKID switch, on the LDG GEAR control panel, to the ARMED position enables the ASCU (provided the parking brake is not engaged and both MLG are down and locked). In the event of a failure that causes loss of braking, manual braking is restored by selecting the anti-skid system OFF.

By monitoring each wheel speed individually, the ASCU can detect tire skidding. The ASCU independently reduces the braking pressure at the skidding wheel by modulating the pressure outputs of the appropriate anti-skid control valve. This modulation is controlled by the individual wheel speed and deceleration monitored through the wheel speed transducers.

In the air, with no weight-on-wheels signal, the anti-skid control valves dump pressure to prevent wheel lock-up on touchdown. The system becomes operational once a 35 knots wheel spin-up signal is present or a weight-on-wheels signal is present after a 5 second delay.

The ASCU continuously monitors the anti-skid system and any detected faults are displayed as an EICAS message.

Anti-Skid Inoperative - Landing Distance

Using the wheel brakes while employing anti-skid protection is a highly effective method of stopping the aircraft during a rejected takeoff or on landing. Aircraft braking performance is severely affected when operating the aircraft with limited or no anti-skid protection. Consult the QRH for the landing distance penalties incurred when operating with degraded anti-skid protection.

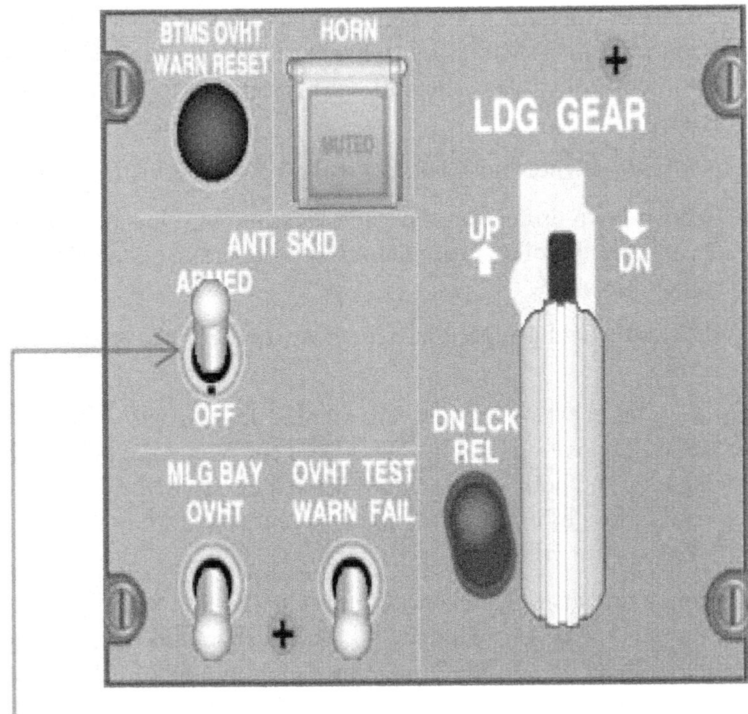

Figure 14 - 11: ANTI-SKID ARMED/OFF Switch

Parking Brake

Inboard brake control valves and the parking shutoff valve are used to provide braking when the airplane is parked. Pulling the parking brake handle while fully depressing both rudder pedals and turning the handle 90 degrees in either direction, locks both brake control valves in the applied position.

When the hydraulic systems are shut down, hydraulic pressure slowly leaks away via the anti-skid return lines. The parking brake shutoff valve closes when the parking brake is applied, ensuring that hydraulic system 3 accumulator pressure is maintained on the inboard brakes for a prolonged period of time.

Parking brake configuration and operational conditions are continuously monitored and any detected fault is displayed as an EICAS text and/or aural message.

Brake Temperature Monitoring System

The brake temperature monitoring system (BTMS) continuously monitors the temperature of each MLG wheel brake unit. The system consists of four temperature sensors, one on each brake unit. the anti-skid control unit (ASCU), and a BTMS OVHT WARN RESET switch.

The ASCU receives temperature signals from the sensors and transmits the temperature data to the data concentrator units (DCUs) for display on EICAS. The four color-coded brake temperature numerical readouts will be displayed on the EICAS status page whenever the landing gear is down, or when any brake overheat condition exists.

An aural warning sounds and a BRAKE OVHT warning message is displayed on the EICAS primary page if any of the brake units overheat. If a brake temperature sensor fails, the respective readout will turn to amber dashes. A BTMS OVHT WARN RESET switch, on the LDG GEAR control panel, is used to reset the system when the brake overheat condition no longer exists.

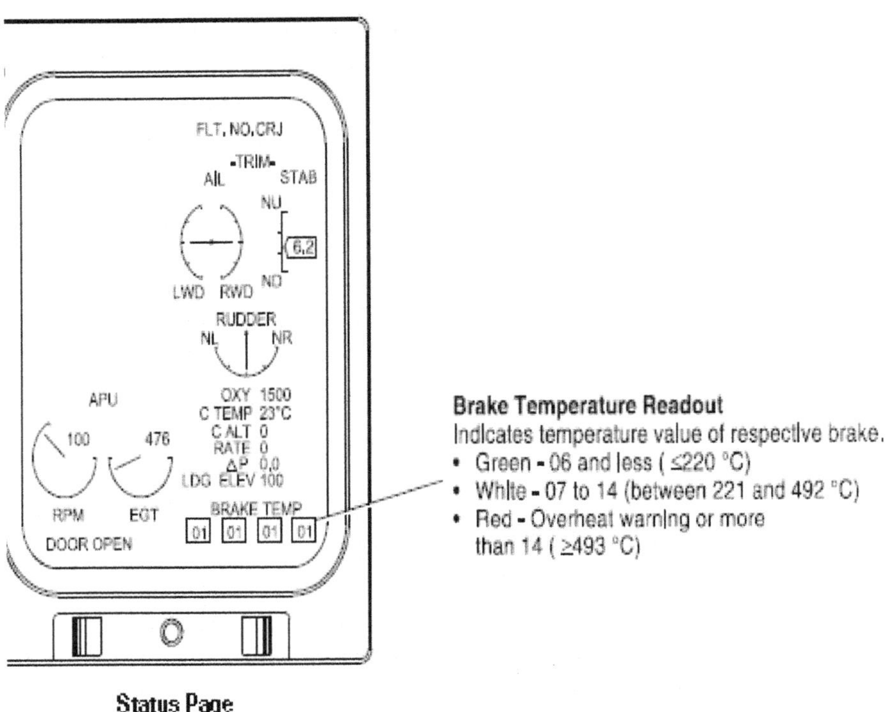

Figure 14 - 12: Brake Temperature Indications

Brake Pressure Monitoring System

The brake pressure monitoring system monitors the brake pressure in the inboard and outboard main wheel brake systems. Pressure for the inboard brake system is supplied by hydraulic system 3, and pressure for the outboard brake system is supplied by hydraulic system 2. The brake pressure readouts are displayed on the HYDRAULIC synoptic page. The system also provides IB or OB BRAKE PRESS caution messages on the EICAS primary page if the hydraulic pressure of the related brake system decreases below a predetermined value.

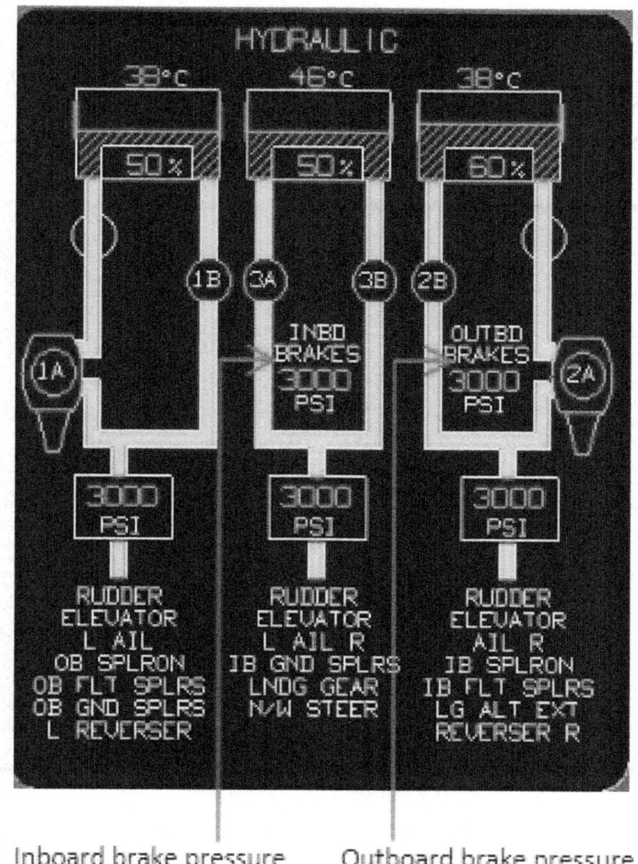

Figure 14 - 13: HYDRAULIC Page - INBD/OUTBD Brakes

Main Wheel Bins and Overheat Detection

Each MLG assembly retracts inward into wheel bins that are recessed into the MLG bay. The wheel bins isolate the wheels from the MLG bays and prevent tire-thrown debris from damaging the equipment located within the MLG bay. Each wheel bin contains a single-loop overheat detection wire to provide an EICAS warning should a wheel overheat.

Components and Operation

Wheel Bins

The wheel bins encase the MLG wheel when the gear is fully retracted. A brush seal on the circumference of the bin brushes against the tire to form an aerodynamic seal. The bins can be removed to allow maintenance personnel access to the MLG bay.

Figure 14 - 14: Brush Seal

Main Landing Gear Overheat Detection

The MLG overheat detection system consists of a single heat-sensing loop installed in each MLG wheel bin and a dual-channel overheat detection unit.

The detection unit and the heat-sensing loop continuously monitor the wheel bins for an overheat condition. Overheat warnings and system fault detection are displayed on the EICAS.

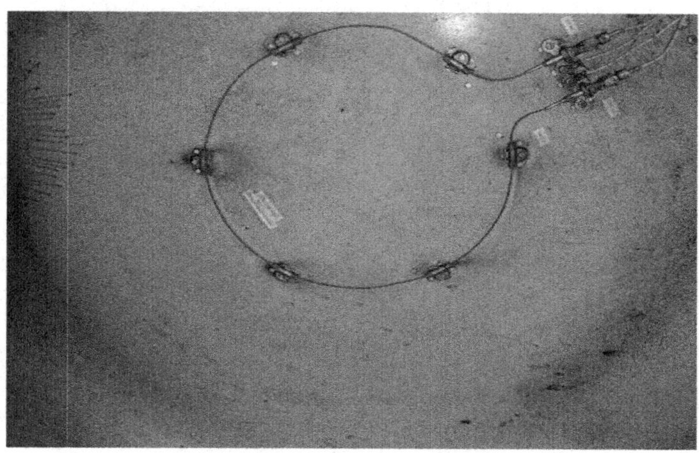

Figure 14 - 15: Main Landing Gear Overheat Detection Loop

MLG BAY OVHT Test

The MLG bay overheat detection system can be tested using the MLG BAY OVHT TEST switches on the LDG GEAR control panel. The WARN FAIL switch is used to do a continuity check of the loops. When the test is successful, the system generates a MLG BAY OVHT warning message and a "GEAR BAY OVERHEAT" aural warning.

The OVHT test switch simulates a failure of the loop, and a successful test generates a MLG OVHT FAIL caution message on EICAS.

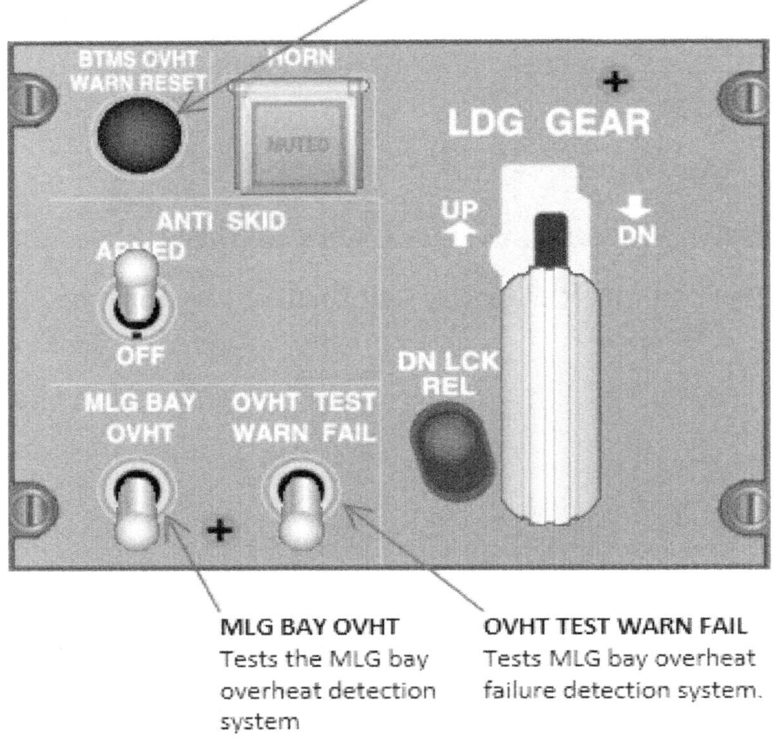

Figure 14 - 16: MLG Overheat Test Switches

Nose-wheel Steering System

The nose-wheel steering system provides directional control of the airplane on the ground for taxi, takeoff and landing operations

The nose-wheel steering system is controlled by a steering electronic control unit (ECU) and powered by hydraulic system 3. The nose-wheel steering (N/W STRG) arming switch is located on the pilot left side panel. Selecting the switch to the ARMED position arms the ECU.

Components and Operation

The ECU controls the nose-wheel position based on inputs from either the steering tiller on the pilot side console, or the rudder pedals. The steering tiller turns the nose-wheel up to 80 degrees either side of center, and is intended for low speed taxiing. Steering with the rudder pedals is limited to 8 degrees either side of center and is intended for high speed taxi and takeoff and landing rolls.

After takeoff, the steering ECU generates a straight ahead command, which centers the nose wheel prior to landing gear retraction. A centering cam on the nose-wheel strut maintains the nose-wheel center position when hydraulic power is shut down.

Powered steering using the steering tiller is available when the N/W STRG switch on the pilot side panel is ARMED and a nose weight-on-wheels signal is present.

If a failure is detected by the steering ECU or hydraulic system 3 is lost, the system reverts to shimmy-damping mode which allows free castering of the nose-wheel. The pilot then maintains ground directional control through differential braking and differential thrust.

The steering ECU continuously monitors the nose-wheel steering system, and any detected faults are annunciated on the EICAS in a caution or status message. Fault detection will result in steering system shutdown which will revert the system to shimmy-damping mode.

NOTE:
Prior to landing, if the nose-wheel steering tiller is moved more than 2 degrees, the STEERING INOP caution message may come on.

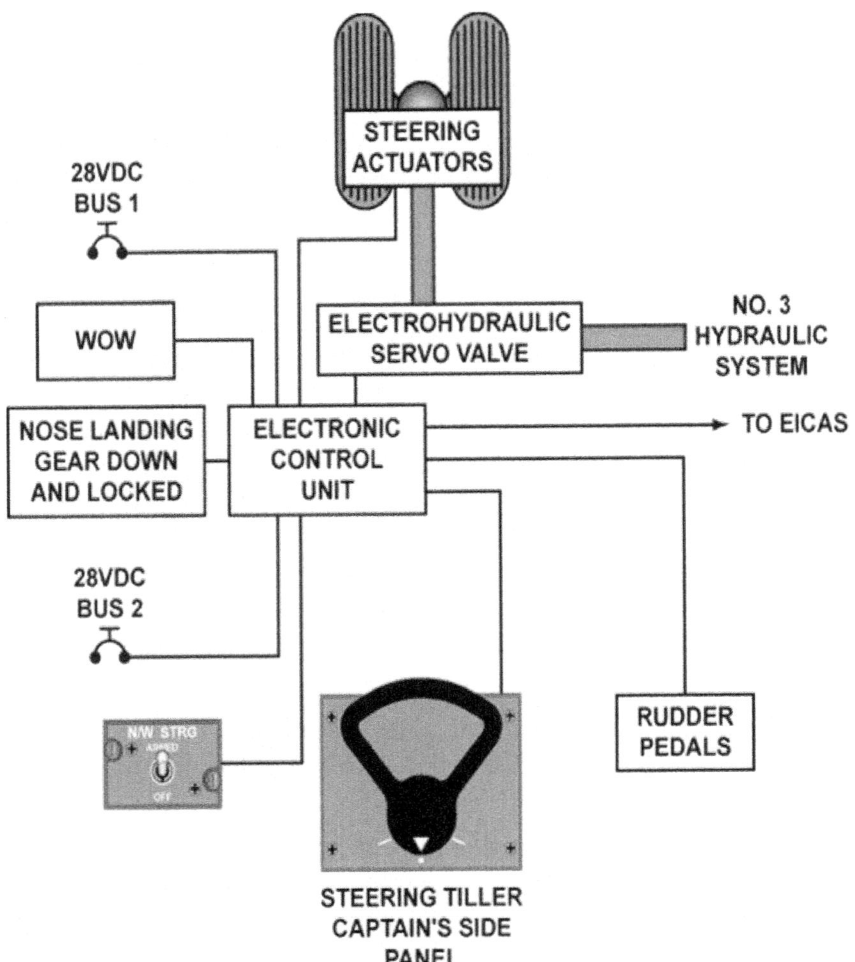

Figure 14 - 17: Nose-wheel Steering System

Landing Configuration Warning System

The warning system advises the pilot when the landing gear is incorrectly configured (not down and locked) during approach and landing. The warning system monitors: airspeed. position of the thrust levers, flaps and landing gear positions. vertical speed (VS) and radio altimeter (RA) inputs.

When an incorrect gear configuration (not down and locked) is presented, the landing gear horn sounds and the EICAS landing gear indications flash amber. Under some conditions, the gear warning indications can be inhibited. Pressing the guarded HORN MUTED switch-light on the LDG GEAR control panel resets the amber landing gear indications, mutes the warning horn, illuminates the HORN MUTED switch-light and posts a HORN MUTED status message.

Components and Operation

The landing gear warning horn will sound 2 minutes after ground to air transition with any landing gear not down and locked, if one of the following conditions exist:

1. Radio altitude is less than 500 feet AGL with both thrust levers at less than maximum landing setting or with flaps greater than 30

OR

2. Both thrust levers are at less than maximum landing setting or any-one thrust lever is at IDLE with the landing gear warning horn MUTED

AND

- Airspeed is less than 170 KIAS with flaps greater than 30 or
- airspeed is less than 190 KIAS with flaps/slats at 0

AND

- Radio altitude or thrust lever is not valid

OR

3. Radio altitude is less than 1000 feet AGL with vertical speed less than -400 ft/min

AND

- No windshear warning or a windshear warning with a windshear monitor failure

OR

- Radio altitude is less than 1000 feet AGL with vertical speed or EGPWS not valid

NOTE:
The landing gear warning horn may be MUTED with one thrust lever at IDLE and the landing gear not in the down and locked position.

The "TOO LOW GEAR" aural warning sounds if any landing gear is not down and locked with the radio altitude less than 500 feet AGL and the indicated airspeed at less than 190 KIAS.

Proximity Sensing System

The proximity sensor system (PSS) includes the proximity sensor electronics unit (PSEU) and associated proximity sensors and proximity switches installed throughout the airplane. The PSS provides five basic functions:

1. Normal landing gear positioning control:
   - The PSS provides the control signals for the landing gear extension and retraction
2. Landing gear position indication:
   - The PSS monitors the landing gear position and provides the gear position status to the EICAS
3. Weight-on-wheels indication:
   - The PSS monitors landing gear strut compression and provides air or ground status information to various airplane systems
4. Fuselage door indication:
   - The PSS monitors the fuselage door positions, door latches and lock status and provides indication of the status of the doors to the EICAS
5. Thrust reverser indication:
   - The PSS monitors and reports the stowed/unstowed position of the left and right thrust reversers and supplies the position status to the EICAS

Components and Operation

Landing Gear Proximity Sensor System

The PSEU monitors the status of the proximity sensors to determine the position and operation of the MLG, NLG and the NLG doors. The PSEU processes the sensor inputs, generates outputs that are used to control landing gear position, report status and provides control data for other airplane systems.

Continuous and periodic tests are performed by the PSEU to monitor the airplane systems health and status. Landing gear position and status are displayed on the EICAS primary page. The landing gear position indication is removed 30 seconds after the landing gear is in the up and locked position with the flaps at 0.

The landing gear warning horn will sound if the airplane is not properly configured and if any of the landing gear is not down and locked. A horn MUTE switch on the landing gear control panel is used to cancel the warning horn under certain airplane configurations.

Figure 14 - 18: PSEU Switch

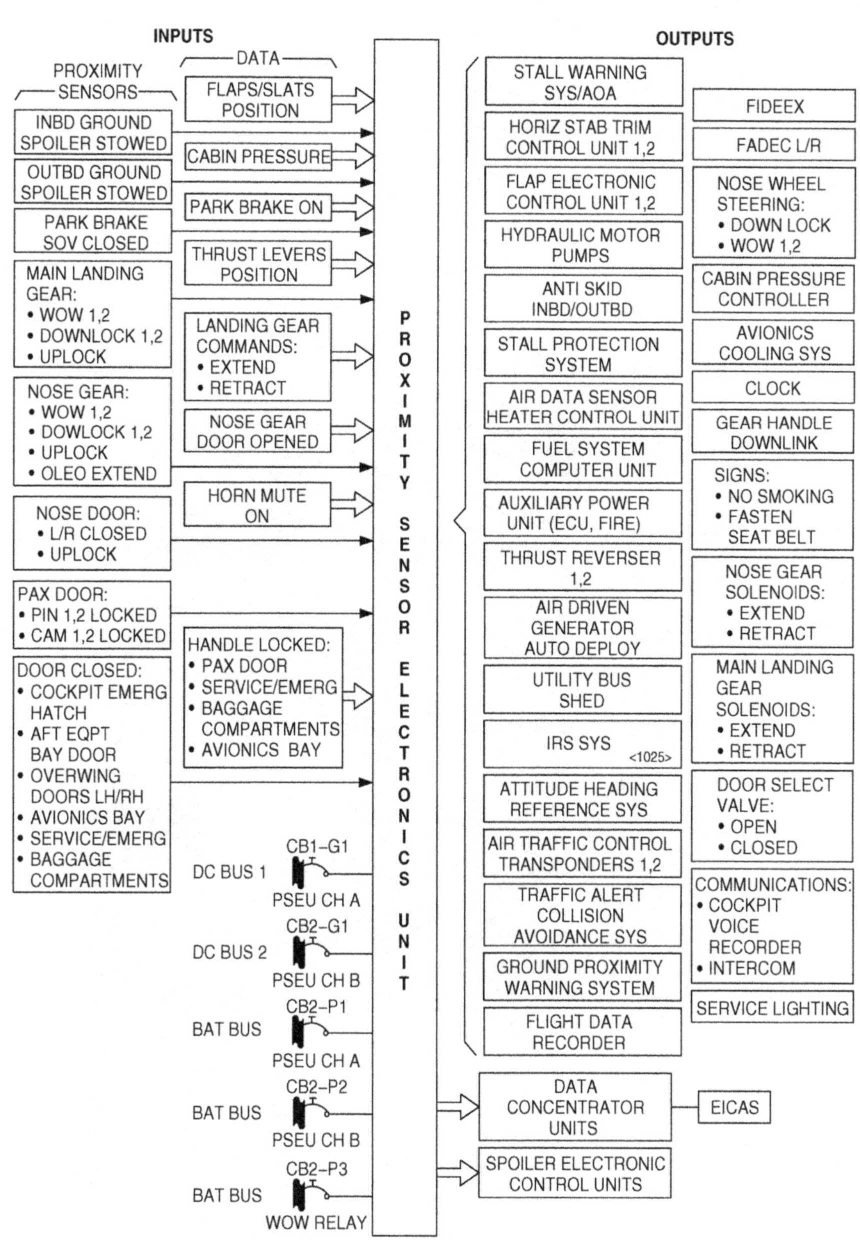

Figure 14 - 19: Proximity Sensing System - Schematic

Tail Bumper Assembly

Located on the bottom of the aft fuselage just forward of the tail cone, the tail bumper prevents the tail of the airplane from making ground contact in the case of a high rotation pitch angle during takeoff.

The tail bumper assembly consists of a:

- Skid plate
- Shock absorber
- Strike indicator

The strike indicator located on the forward end of the tail bumper assembly gives a visual indication that a tail strike has occurred.

Figure 14 - 20: Tail Bumper Assembly

Controls and EICAS Messages

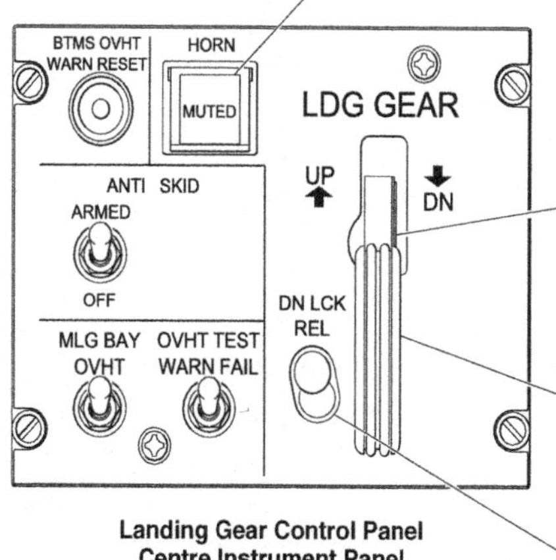

HORN
Used to mute landing gear warning horn.
- MUTE (white) light indicates that landing gear warning horn has been muted.

Landing Gear Lever Down Lock
Prevents inadvertent landing gear up selection when on ground.
When airborne, a weight–off–wheels signal from the PSEU disengages the lock to permit a gear up selection.

Landing Gear Lever
Used to retract and extend landing gear.

Landing Gear Control Panel
Centre Instrument Panel

DN LCK REL
Used to manually release the landing gear lever down lock.

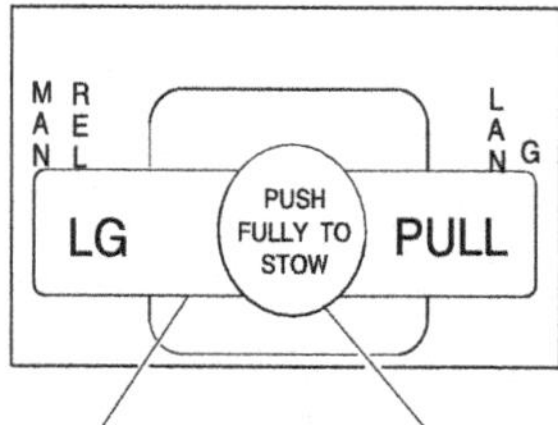

Landing Gear Manual Release
Used to manually lower the landing gear.
- Pull and hold handle in the fully extended position until EICAS indicates the nose and main landing gear are down and locked.

PUSH FULLY TO STOW button (red)
- Used to reset the manual release system (stow the LANDING GEAR MANUAL RELEASE handle) Pull the handle up, press button and then slowly return handle to the stowed position.

NOTE
Considerable force is required to operate the landing gear manual release system.

Figure 14 - 21: Landing Gear - Controls

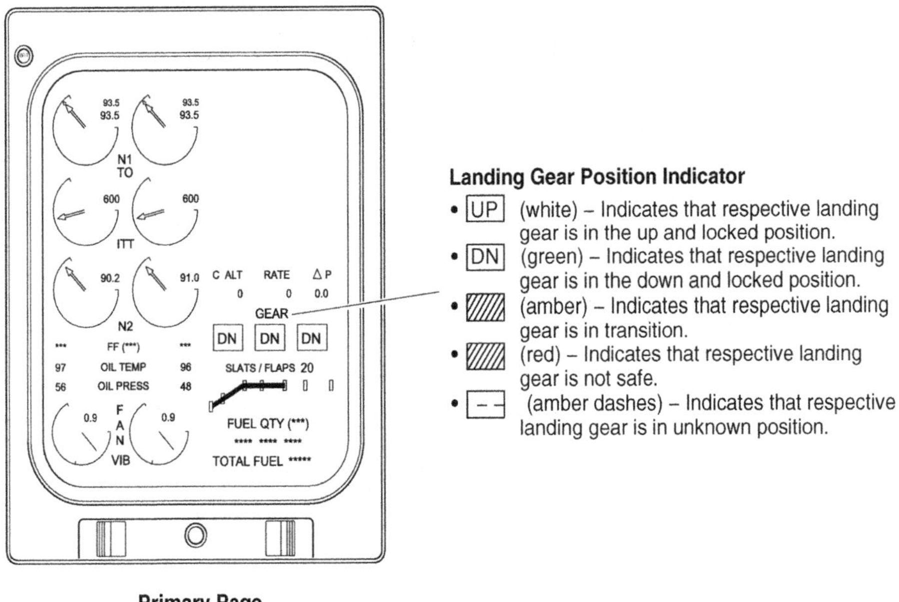

Primary Page

Figure 14 - 22: Landing Gear - EICAS Indications

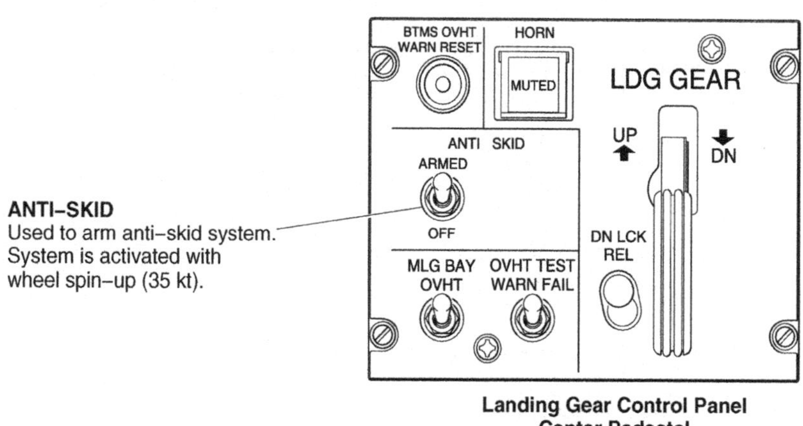

Landing Gear Control Panel
Center Pedestal

Figure 14 - 23: Anti-Skid System Controls

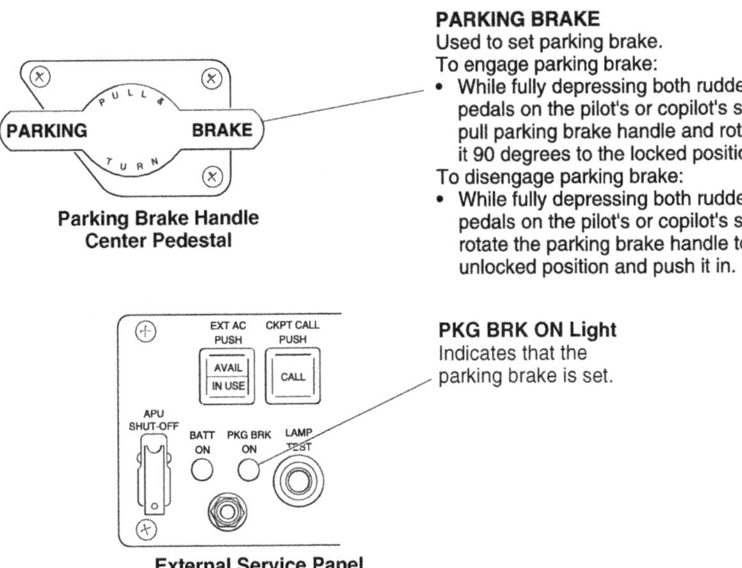

Figure 14 - 24: Parking Brake Controls and Indications

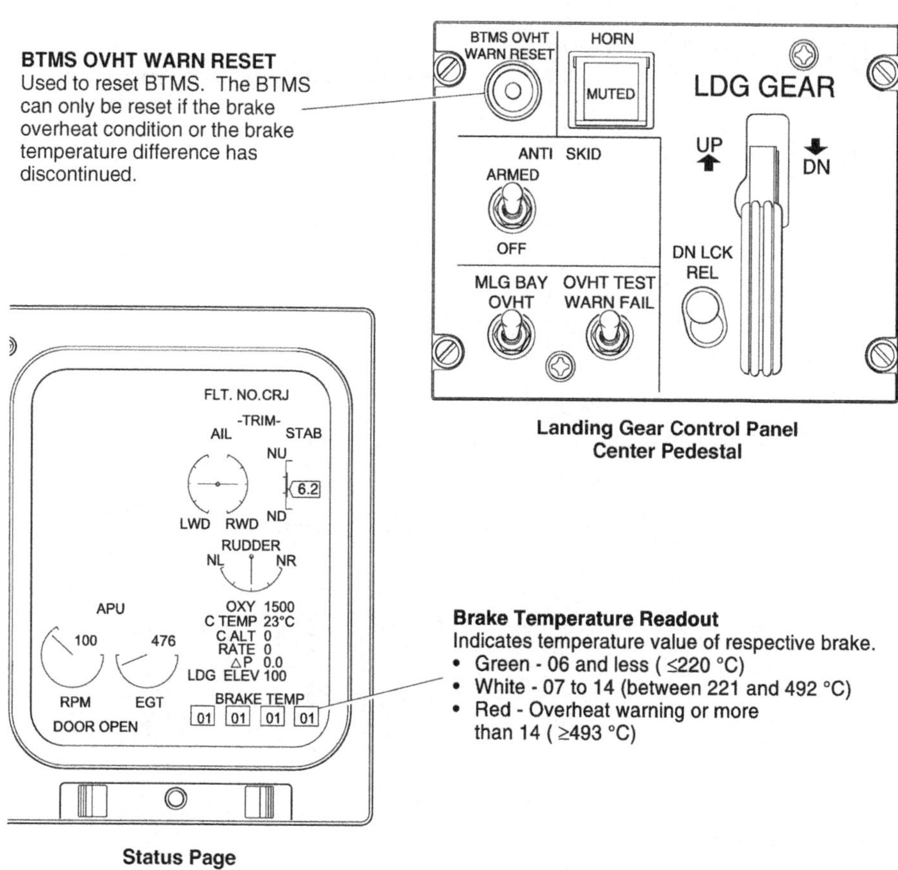

Figure 14 - 25: Brake Temperature Monitoring System

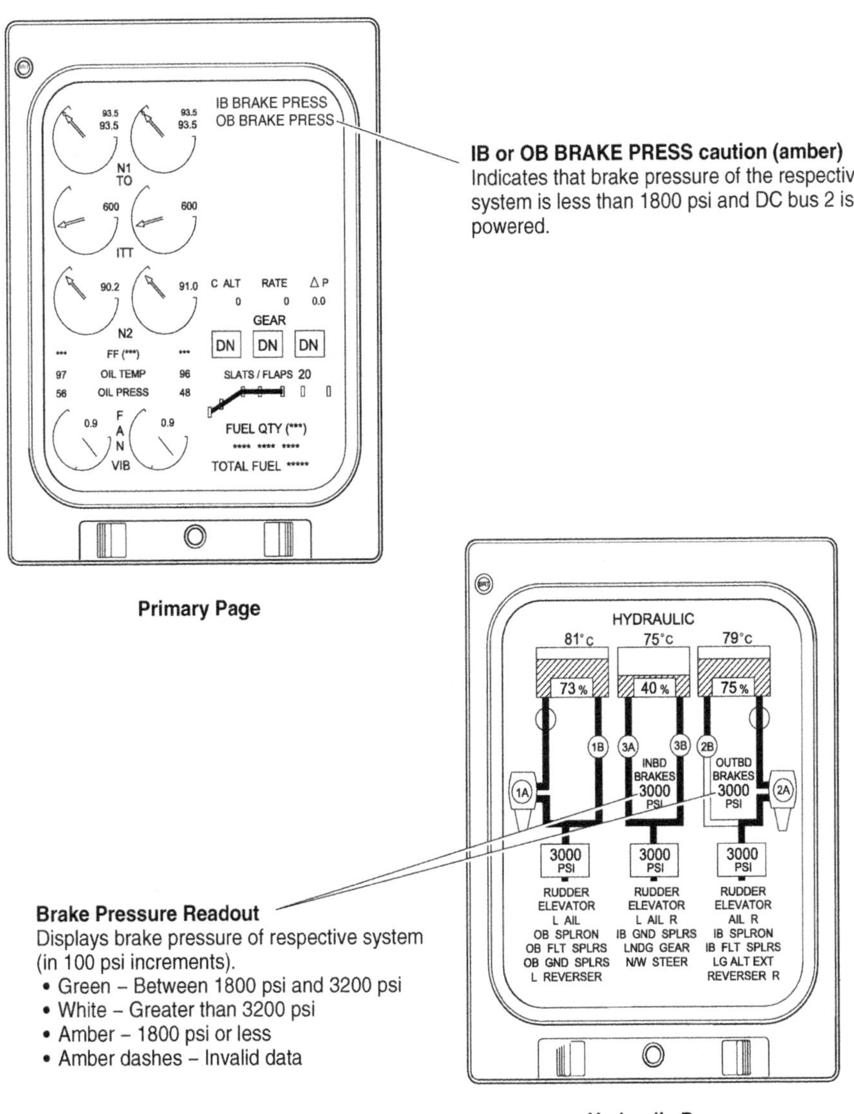

Figure 14 - 26: Brake Pressure Indications

Master Warning	Aural	Cause
BRAKE OVHT	"Brakes"	Over-temperature condition exists on any brake.
GEAR DISAGREE	"Gear Disagree"	Landing gear position does not match landing gear lever selector position.
MLG BAY OVHT	"Gear Bay Overheat"	Overheat condition exists in one or both main landing gear bays.
NOSE DOOR OPEN	"Nose Door"	Nose landing gear bay door are open for more than 10 seconds after gear has locked or with airspeed greater than 250 kts.
PARKING BRAKE	"Config Brakes"	Parking brake is set with the airplane in air (no WOW) or configured for takeoff with N_1 greater than 70%.

Master Caution	Cause
A/SKID INBD	Indicates that the inboard channel of the anti-skid system has failed, parking brake shut-off valve failed closed or loss of ASCU output.
A/SKID OUTBD	Indicates that the outboard channel of the anti-skid system has failed or loss of ASCU output.
IB (OB) BRAKE PRESS	Brake pressure of the respective system is less than 1800 psi and DC Bus 2 is powered.
PARK BRAKE SOV	Indicates that the parking brake shutoff valve has failed open with inboard brake pressure greater than 800 psi and the parking brake set.
PROX SYSTEM	Indicates loss of both PSEU channels.
PROX SYS CHAN	Indicates loss of one PSEU channel or input/ output of a critical system.
STEERING INOP	Indicates that the steering control unit has detected a fault.
WOW INPUT	WOW sensing circuit has failed or is in disagreement with the other sensing circuits.

WOW OUTPUT	WOW outputs channel 1 and 2 do not agree.
Status Message	**Cause**
A/SKID FAULT	Indicates loss of redundancy of ASCU, loss of weight-on-wheels input, spin down fail or loss of internal communication.
L or R MLG FAULT	Indicates that left or right actuator shuttle valve or the pressure switch have failed in the closed position (for MLG down-locked) OR left or right actuator pressure switch has failed in the open position (for MLG up-locked)
HORN MUTED	Landing Gear warning horn has been manually muted.
MLG FAULT	Indicates PSEU has detected a fault in the main landing gear shuttle valve.
PROX SYS FAULT 1	Indicates a failure of any one PSEU input or output related to a critical aircraft system.
PROX SYS FAULT 2	Indicates that any one non-critical sensor or an input or output is failed or unreasonable.
STEERING DEGRADED	Indicates possible intermittent loss of steering due to nose wheel bouncing. ***NOTE:*** Aft CG and/or lightweight are possible conditions for this message to come on.
Advisory Message	**Cause**
PARKING BRAKE ON	Parking brake is set on ground with: both N_1 less than 70%, inboard brake pressure greater than 800 psi, parking brake SOV closed.

System Schematic

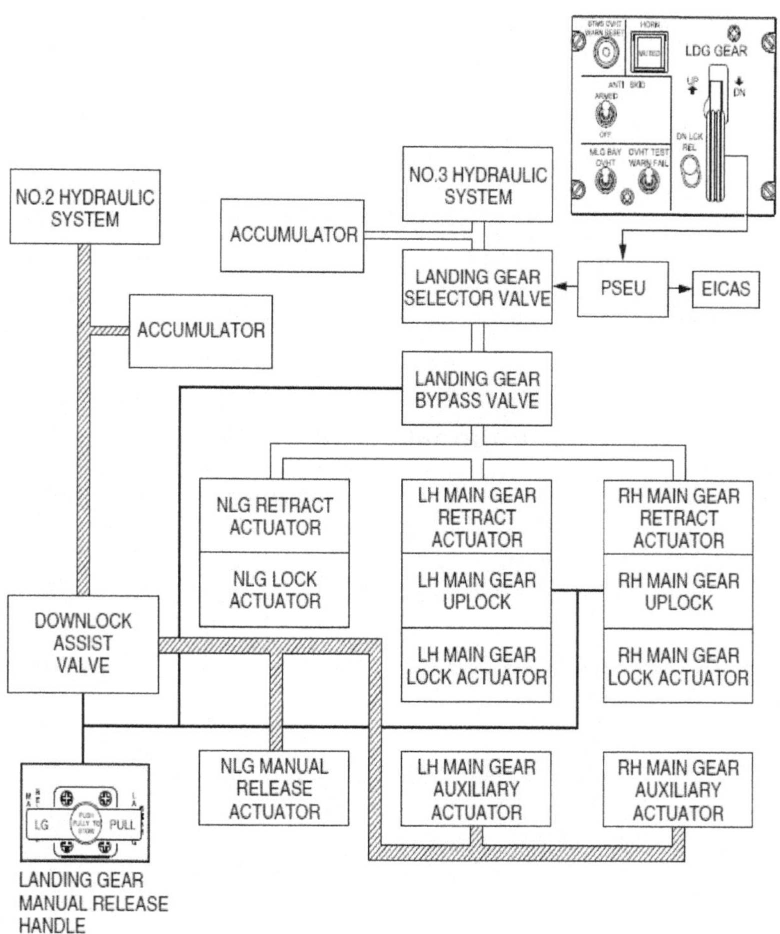

Figure 14 - 27: Landing Gear Retraction & Extension Schematic

Intentionally Left Blank

Chapter 15: Aircraft Lighting

Intentionally Left Blank

Chapter 15: Aircraft Lighting

General

The aircraft lighting system provides interior and exterior illumination of the aircraft. In addition, lights provide information and guidance to passengers in normal and emergency situations.

The lighting system includes:

- External lighting
- Flight deck lighting
- Passenger compartment lighting
- Service compartment lighting
- Emergency lighting

The flight deck and external lighting is selected at lighting control panels located on the flight deck.

Emergency lights are controlled from the flight deck and may also be controlled from the forward flight attendant's panel. When armed, they illuminate automatically when normal electrical power is lost. EICAS messages display the status of the emergency lights and passenger signs.

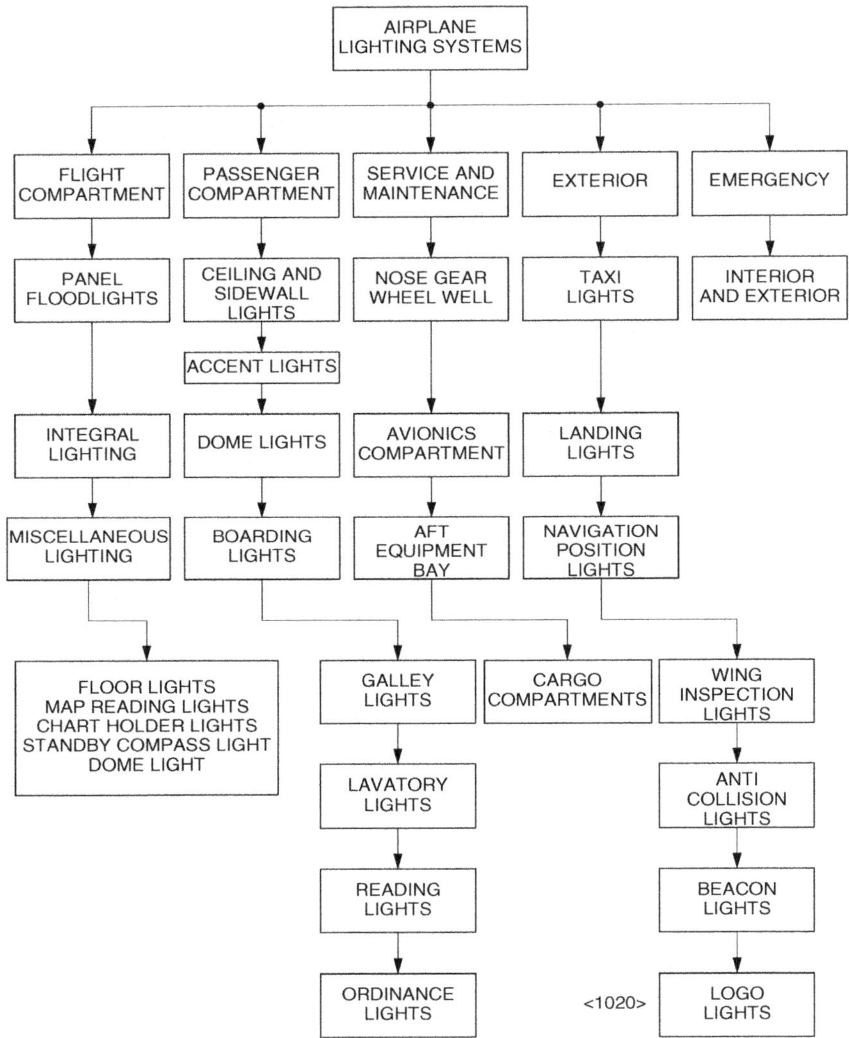

Figure 15 - 1: Lighting Systems - Block Diagram

Lighting for Aircraft Visibility

The following external lights enhance aircraft visibility to crew in other aircraft and ATC controllers:

- Navigation lights
- Anti-collision strobe lights
- Rotating beacon
- Logo lights (optional)
- Wing inspection lights

Figure 15 - 2: EXTENAL LTS Panel

Components and Operation

External lights are controlled through switches on the overhead panel.

The NAV switch controls the navigation lights that are located in each wing tip and the tail cone. In each wing tip there are two bulbs (primary and secondary). Normally the primary bulb is illuminated. Should the primary bulb fail, power is applied to the lower wattage and smaller secondary or backup bulb. There are two white rear navigation lights, one in the vertical stabilizer fairing and the other in the tail cone.

Figure 15 - 3: NAV Lights (Left, Right, Tail (top)), Strobe Light

The BEACON switch controls the red optional upper and lower beacon lights.

Figure 15 - 4: Beacon

The STROBE switch controls anti-collision strobe lights in the wing tips (under the same lens as the navigation lights) and in the tail area of the aircraft. The flight data recorder (FDR) operates when either the STROBE, BEACON switch is selected ON, or weight off wheels is detected.

The LOGO switch controls the lights that illuminate the company logo on the vertical fin of the aircraft. The LOGO Lights are installed on the upper surface of each engine pylon.

WING INSP switch controls the inspection lights that illuminate the wing area.

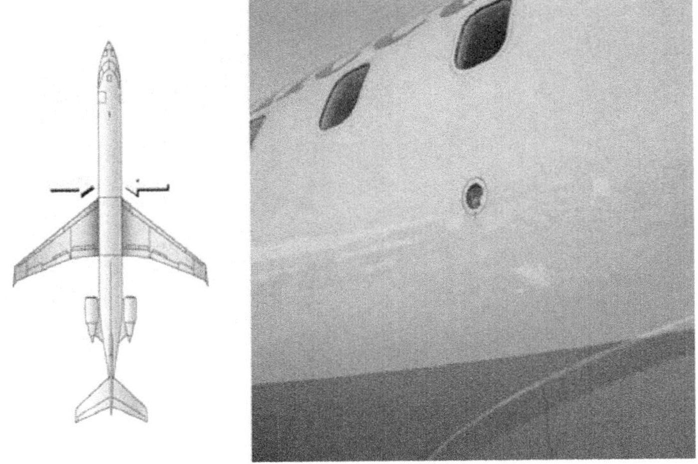

Figure 15 - 5: Wing Inspection Light

External Illumination

To illuminate the exterior of the aircraft and the surrounding area, there are recognition/taxi lights, a nose landing light, wing landing lights and wing inspection lights. All the lights are controlled from the LANDING LTS control panel, except the wing inspection lights, which are on the EXTERNAL LTS panel.

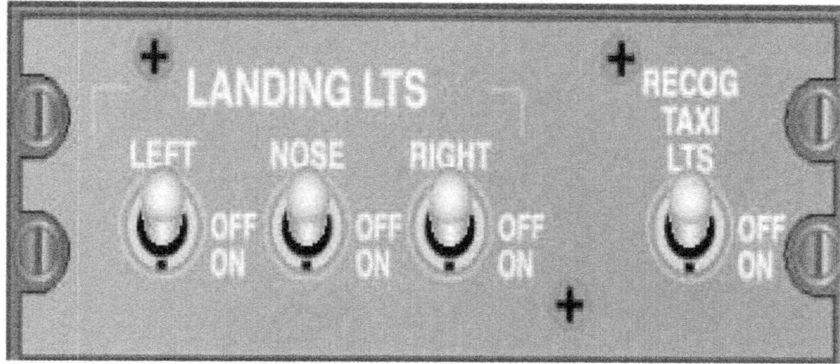

Figure 15 - 6: LANDING LTS Control Panel

Components and Operation

Each wing root contains a dual purpose recognition/taxi light, on the inboard portion of the light housing. It illuminates the ground close to the aircraft and makes the aircraft more visible in the air. The RECOG/TAXI LTS switch controls the recognition/taxi lights.

Figure 15 - 7: RECOG/TAXI LTS

Each wing root also contains a landing light, located on the outboard portion of the light housing, to illuminate the runway environment. The LEFT LANDING LTS switch turns on the left landing light and the left recognition/taxi light. Similarly, the RIGHT LANDING LTS switch turns on the right landing light and the right recognition/taxi light. There is a landing light on the nose gear that is controlled by the NOSE LANDING LTS switch. Wing inspection lights, controlled by the WING switch on the EXTERNAL LTS panel, are used to check the wing leading edges for possible ice accumulation.

Figure 15 - 8: Landing Lights

Flight Deck Lighting Panels

Flight deck lighting provides illumination of the instrument panels and flight deck area. There are four flight deck lighting panels:

- Miscellaneous lights (MISC LTS) panel
- Center pedestal lighting panel
- Pilot and copilot lighting panels (identical panels)

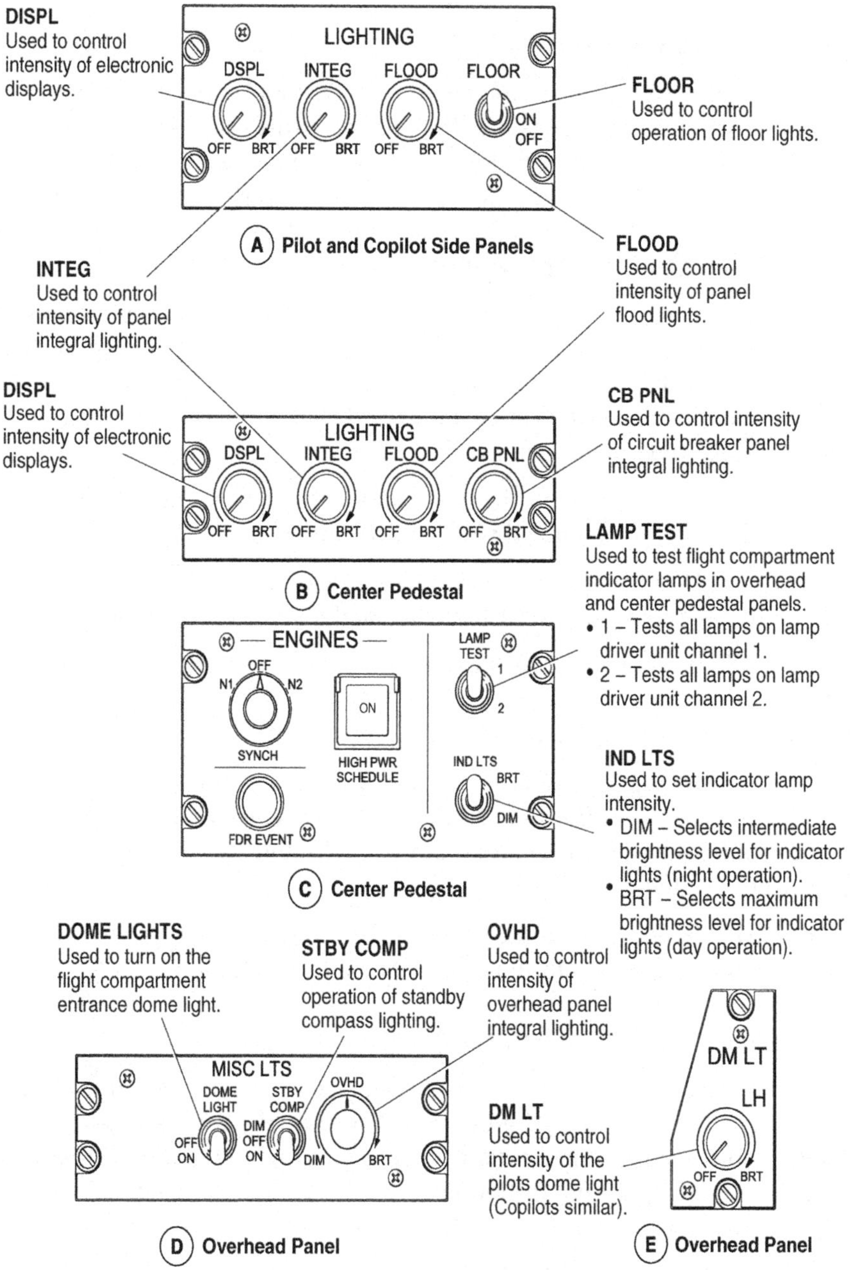

Figure 15 - 9: Flight Deck Lighting - Controls

Components and Operation

Miscellaneous Lights Panel

The DOME LIGHT switch on the MISC LTS panel controls the overhead floodlight at the entrance to the flight deck. On the same panel, the STBY COMP light switch controls the compass light, and an OVHD dimmer knob controls the integral lighting for various overhead control panels.

Center Pedestal Lighting Panel

The center pedestal LIGHTING panel consists of four dimmer knobs. The DSPL knob controls the CRT intensity of ED 1 and ED 2 as well as the FMS and the radio tuning unit (RTU) displays. The brightness of the individual displays can also be adjusted through the use of a recessed knob on the top of each screen.

The INTEG knob controls the panel and instrument integral lighting. Integral lights are contained within each control panel faceplate.

The FLOOD knob controls the center instrument panel floodlights. Floodlights are incandescent lights located above the instrument panels.

The CB PNL knob controls the circuit breaker panel's integral lighting.

Pilot and Copilot Lighting Panels

The pilot and copilot LIGHTING controls on each side panel are identical. Their functions are similar to the switches on the center pedestal lighting panel, except that the knobs and switches apply to the respective side lighting.

The DSPL switch controls the PFD and MFD light intensity. There is also a FLOOR light switch to illuminate the area at each pilot's feet.

Chart Holder and Map Reading Lights

Chart holder lights and map reading lights are located at the pilot and copilot stations. The map lights are mounted on the window post and are activated when the unit is extended from the unit holder. The light can be adjusted to provide spot or flood illumination. Light intensity is adjusted by a dimmer knob that is integral to each unit. A map reading light is provided for the jumpseat occupant.

Individually-controlled reading lights are installed on the outer edge of each overhead panel. Light intensity is adjusted by a dimmer knob.

A chart holder is mounted on each control column. The chart light is controlled by a dimmer (OFF to BRIGHT) knob.

Lamp Test Switch

A two-position (LAMP TEST 1/2) switch on the center pedestal is used to test the flight deck indicator lamps in the overhead and center pedestal panels. This switch has no effect on EFIS, EICAS and RTU displays.

Indicator Lights Bright/Dim Switch

A two-position (BRT/DIM) switch on the center pedestal controls the intensity of the flight deck indicator lamps on the overhead and center pedestal panels: DIM for night operations; BRT for day operations. This switch has no effect on EFIS, EICAS and RTU displays.

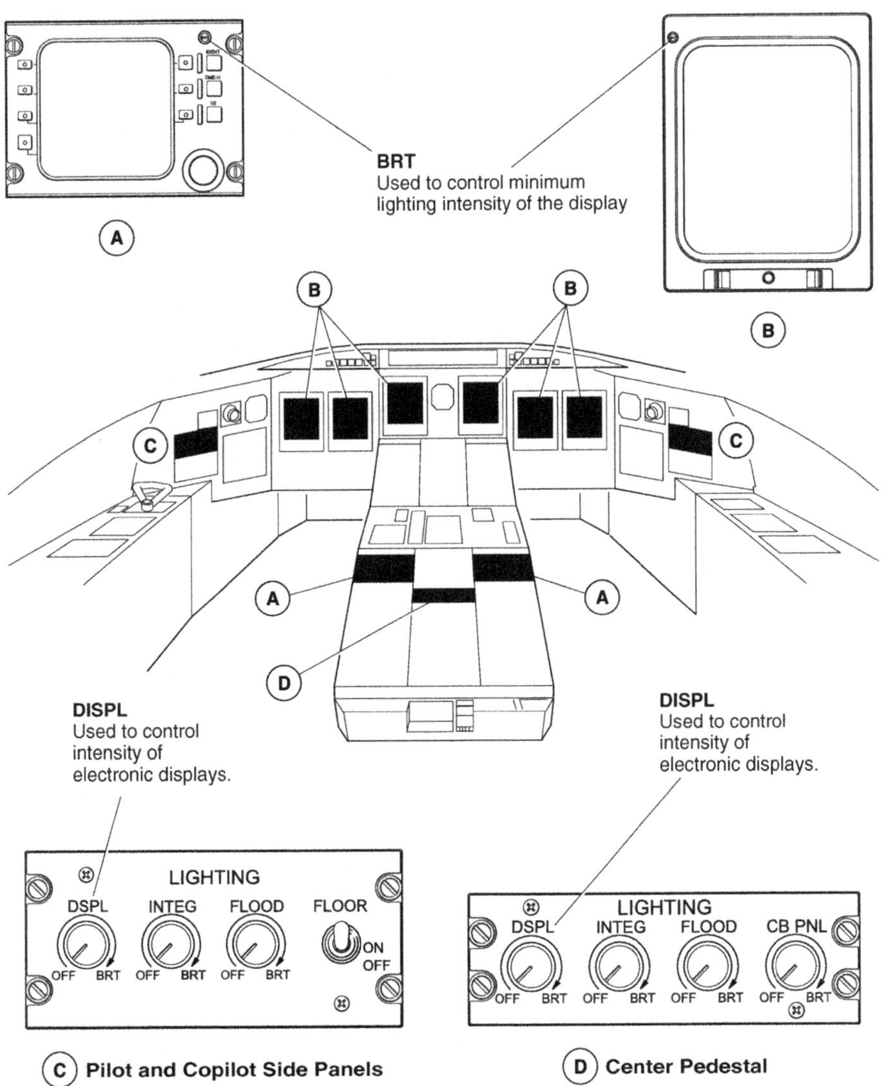

Figure 15 - 10: CRT Lighting - Intensity Adjustment

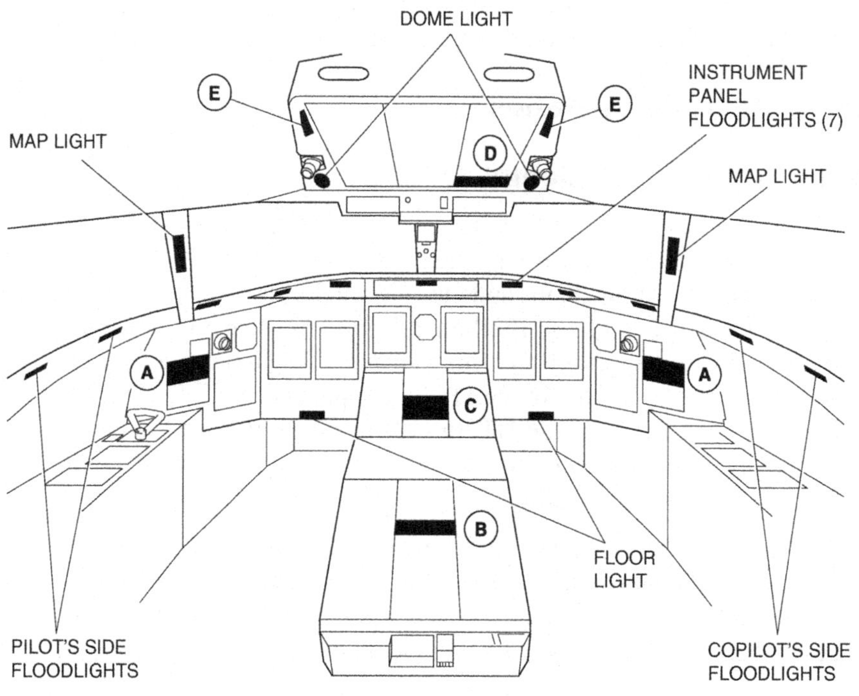

Figure 15 - 11: Flight Deck Lighting - General

Passenger and Service Compartment

The passenger cabin contains numerous lights for passenger comfort and safety such as floodlights, reading lights and illuminated passenger signs.

Components and Operation

Ceiling and Sidewall Lights

The ceiling and sidewall lighting consists of strips of white, LED wash lights. The ceiling lights are located along inboard top of the passenger compartment overhead bins and the sidewall lights on the lower outboard side of the overhead bins. The lights supply up-wash and down-wash illumination to the ceiling and sidewall areas. The lights are controlled by two switch-lights on the forward flight attendants panel. The switches are identified as CEILING LIGHT, ON/OFF and SIDEWALL LIGHT, ON/OFF. A DIM/BRT selector next to each switch-light is used to control the respective lighting intensity.

To ensure that limited passenger compartment lighting is available in the event of a total electrical power failure, every fourth ceiling LED strip of the left cabin ceiling lights will remain powered from the DC essential bus (which remains powered when the ADG is deployed).

The entrance lights illuminate the main entrance door area. They are controlled by the ENTRANCE LIGHT switch-lights on the forward attendant's panel.

When the ON switch-light is pressed in, the entrance lights are illuminated at low intensity and the DIM portion of the BRT/DIM switch-light is illuminated. When the BRT/DIM switch-light is pressed, the BRT portion is illuminated, the stair lights are on and the entrance lights are at high intensity.

Accent Lights

Accent lights are mounted in light assemblies, installed in each ceiling panel, that run the length of the passenger compartment. The accent lights consist of two LED spotlights that provide illumination of the overhead bin handles and area. The accent lights are controlled by a switch identified as ACCENT LIGHT, ON/OFF located on the forward flight attendants panel. Also mounted in each light assembly is a green zone indicator light or an emergency floodlight. The zone lights and emergency floodlights alternate through each ceiling panel light assembly. The emergency floodlights will be powered, in an emergency, from an emergency lights battery pack.

Reading Lights

Two LED down-wash reading lights are installed in each passenger service unit (PSU). They supply personal lighting for each passenger and are controlled by independent switches on each PSUs. All the passenger reading lights can be turned on and tested using the PSU READING LIGHT, TEST, OFF switch-lights on the forward flight attendants panel.

Each flight attendant station is also equipped with an LED reading light that is controlled by a switch-light on the respective flight attendant's panel.

Passenger Signs (Ordinance Lights)

Ordinance lights provide information to the passengers during flight. They are controlled through the PASS SIGNS panel located on the overhead panel in the flight compartment. Ordinance lights consists of the following:

Fasten Seat Belts and Return to Seat Symbol

return to seat symbol is installed in the lavatories. They are controlled by the SEAT BLTS switch located on the PASS SIGNS panel. The SEAT BLTS switch has three positions, ON/OFF/AUTO. When the switch is set to ON, a chime transmits on the passenger compartment speakers and the fasten seat belt symbols are turned on.

When the SEAT BLTS switch is set to AUTO, the seat belt and return to seat symbols illuminates and the chime comes on automatically when:

- Landing gear is selected down
- Flaps not in the zero degree position
- Cabin altitude reaches 8500 feet

No Smoking Symbol

A lighted no smoking symbol is installed in each passenger service unit (PSU), and are controlled by the NO SMKG switch located on the PASS SIGNS panel. The NO SMKG switch has three positions, ON/OFF/AUTO. When the switch is set to ON, a chime transmits on the passenger compartment speakers and the no smoking symbols are turned on.

When the NO SMKG switch is set to AUTO, the no smoking symbols and chime comes on automatically when:

- Landing gear is selected down
- Cabin altitude reaches 8500 feet

WI-FI Symbol

A lighted Wi-Fi sign is installed in each passenger service unit (PSU and in the main entrance. They are controlled by a Wi-Fi SIGN, ON/OFF switch-light located on the forward flight attendant panel.

NOTE:

The XJ 900s "NO SMKG" switch and PSUs have been upgraded to a "WI-FI/No Smoking" switch and the PSU indicates a "WI-FI/No Smoking" light.

Lavatory Lighting

The lavatory is illuminated by three fluorescent lights (two in the vanity and one above the counter). The lights come on dim when airplane power is applied. With the lavatory door locked, the vanity light assembly will come on bright.

Galley Lighting

Galley lighting is provided by six fluorescent lights in the galley ceiling panel. The lights are controlled by two switch-lights labeled GALLEY LIGHT, ON/OFF and BRIGHT/DIM, on the galley control panel. Lights in the stowage compartments and wardrobe are controlled by micro-switches in the doors, so that the lights come on when the door is opened.

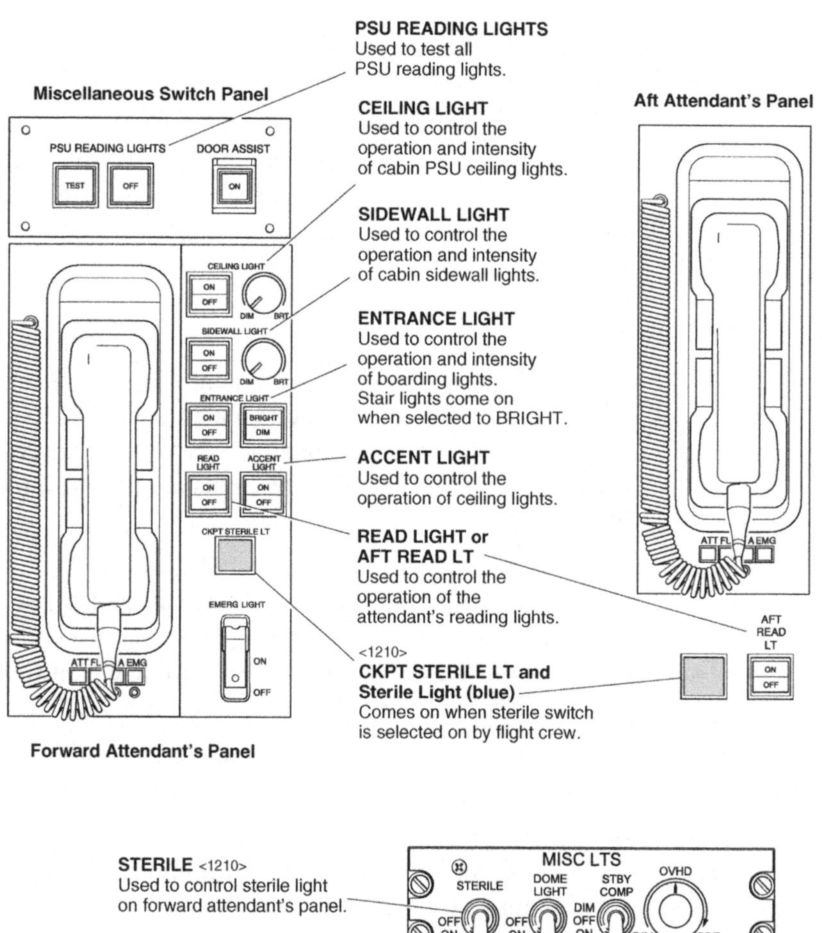

Figure 15 - 12: Flight Attendant - Control Panels

Service and Maintenance Lighting

Service lighting is provided for the cargo compartments. Maintenance lighting is provided for the landing gear bays, APU compartment, aft equipment compartment and the underfloor avionics compartment.

Service lighting is provided for the cargo compartments and external loading area. Maintenance lighting is provided for the landing gear bays, APU compartment, aft equipment compartment and the underfloor avionics compartment.

Service Lighting

Two lights illuminate the forward cargo compartment. The forward cargo compartment lights are controlled by a switch located at the inside forward edge of the forward cargo door opening. Activation requires a weight-on-wheels signal. This ensures that the lights are off when the airplane is in flight.

Two lights illuminate the aft cargo compartment. The aft cargo compartment lights are controlled by a switch located at the inside forward edge of the cargo door. Activation requires a weight-on-wheels signal. This ensures that the lights are off when the airplane is in flight.

External loading area lighting consists of a forward cargo compartment loading area light and an aft cargo compartment loading area light. The lights are designed to illuminate the cargo compartment loading areas.

The forward cargo compartment loading area light and switch is installed within the forward cargo compartment. The light illuminates the loading area and the ground immediately below the loading area, when the forward cargo door is open.

The aft cargo compartment loading area light is installed under the left engine pylon and angled to illuminate the loading area and the ground immediately below the aft cargo door. The light switch is located inside the aft cargo compartment.

Maintenance Lighting

Six flood lights are installed down the length of the underfloor avionics compartment. The lights are controlled by a switch located in the compartment.

Two lights and a control switch are installed in the aft equipment compartment.

Two lights and a control switch are installed on the APU rear bulkhead to illuminate the APU compartment area.

Two high intensity halogen lights are installed in each main landing gear bay. Each light has a control switch located next to it. A single high intensity halogen light and switch is installed in the nose landing gear bay.

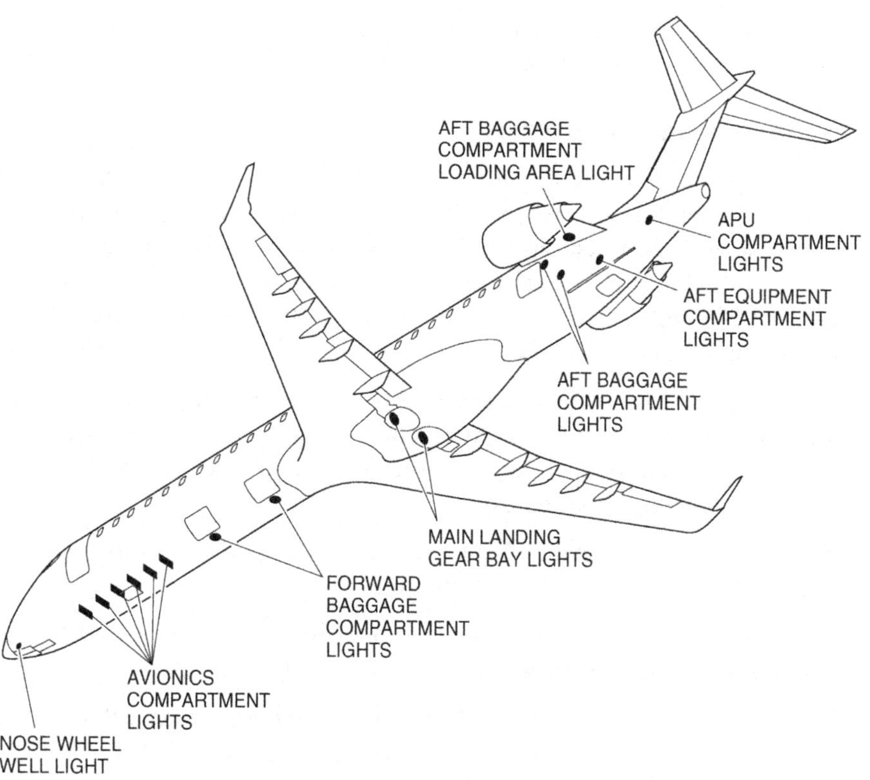

Figure 15 - 13: Maintenance and Service Lights

Emergency Lighting

Emergency lighting in the passenger cabin contributes to the safe egress of passengers and crew during an emergency ground evacuation or ditching situation. Cabin emergency lighting consists of:

- Wall and floor level exit signs
- Ceiling and floor level emergency flood lights
- Floor track photo luminescent tape

Exterior fuselage-mounted emergency exit lights provide illumination of the aircraft's immediate surroundings.

Components and Operation

Photo-luminescent tapes are affixed to either side of the passenger cabin's center aisle and illuminate the paths to all emergency exits. The photo luminescent floor track tapes must be exposed to a light source, either natural or cabin lights in the bright position for at least 15 minutes to become luminous. Exposure of 30 minutes to light will give the maximum duration for a dark cabin period.

Switches on the flight deck or at the forward flight attendant panel control all other interior and exterior emergency lights. Five self-contained rechargeable battery packs are capable of powering the emergency lights for approximately 10 minutes.

The battery packs contain 6 Volt NiCad batteries that are supplied with a trickle charge from the DC essential bus.

The EMER LTS switch is mounted on the flight deck's overhead center panel. It is a three-position switch labeled ON/OFF/ARM.

- When the switch is set to ON, the associated passenger cabin and exterior emergency exit lights illuminate and the EMER LTS ON status message is presented on EICAS.
- When the switch is positioned to OFF. the emergency lights are extinguished and the EMER LTS OFF caution message is presented on EICAS.

- When the switch is selected to ARM and power is lost to either the AC or DC ESS BUS, the battery packs supply power to illuminate the associated passenger cabin and exterior emergency exit lights. The EMER LTS ON status message is presented on EICAS.

A red guarded two-position (ON/OFF) EMER LTS switch is located at the forward flight attendant's panel. The switch allows the flight attendant to test and control the emergency lighting.

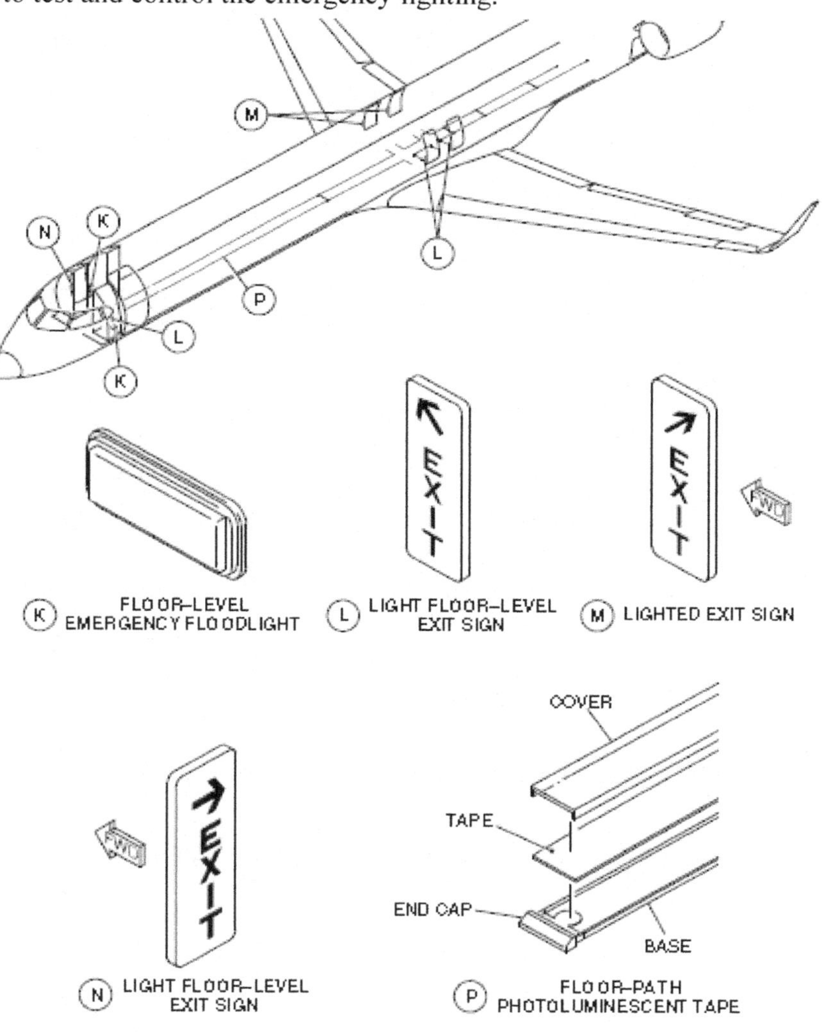

Figure 15 - 14: Internal Emergency Lights

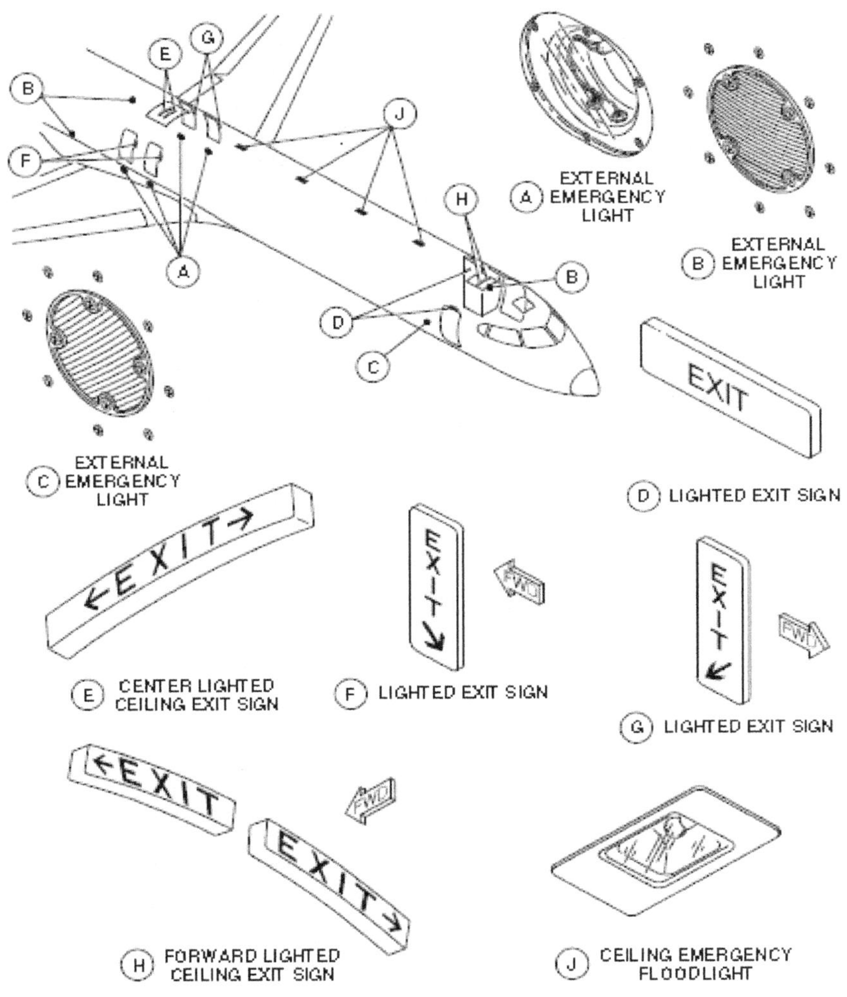

Figure 15 - 15: External Emergency Lights

EICAS Indications

Master Caution	Cause
EMER LTS OFF	Emergency lighting system has been selected off (battery pack voltage is less than 4.5 volts or lights are off).
Status Message	**Cause**
EMER LTS ON	Emergency lighting system is operational and battery pack voltage is greater than 4.5 volts.
NO SMOKING	NO SMOKING has been selected ON either manually or automatically.
SEAT BELTS	SEAT BELTS has been selected ON either manually or automatically.

System Schematic

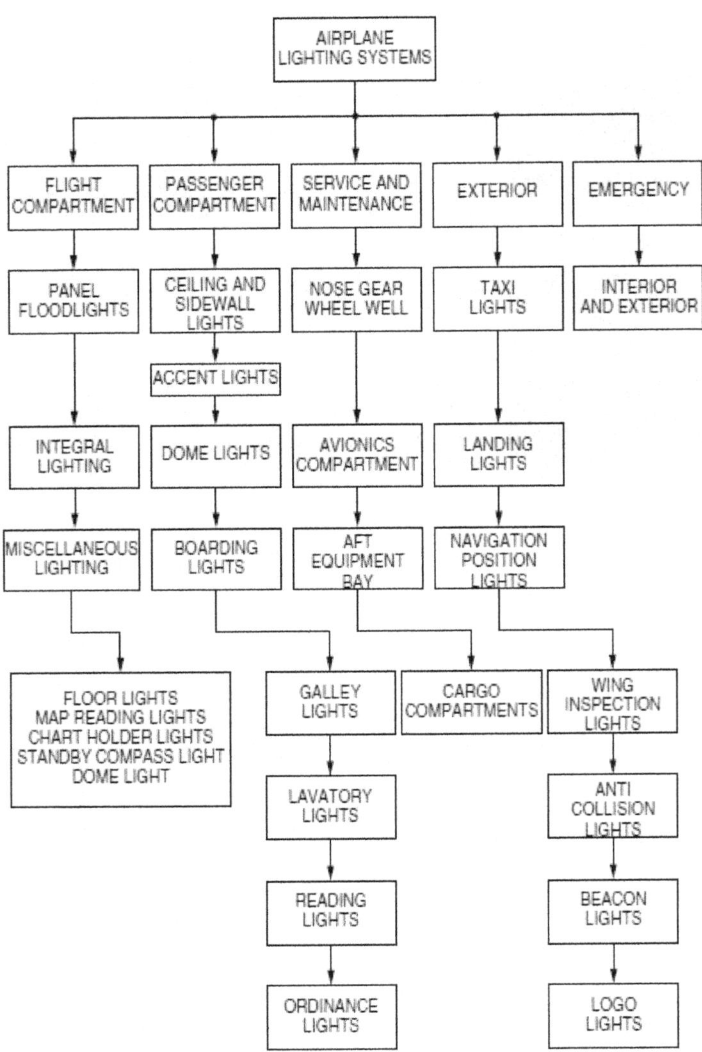

Figure 15 - 16: Aircraft Lighting Block Diagram

Chapter 16: Navigation

Intentionally Left Blank

Chapter 16: Navigation

General

The Bombardier CRJ 900 is equipped with conventional navigation systems. They include the following:

- Dual VHF VOR/LOC/GS navigation systems
- Dual DME navigation systems
- Dual ADF navigation systems
- Dual marker beacon receivers
- Dual ATC transponder systems
- Enhanced ground proximity warning system (EGPWS)
- Weather radar system
- Traffic alert and collision avoidance system (TCAS)
- Global positioning system (GPS)
- Flight management system (FMS)
- Inertial Reference System (IRS)

NOTE:
The ADF radios, ATC transponders, TCAS and VHF navigation are tuned at RTU 1 and RTU 2. The FMS control display unit provides remote tuning and control of communication, navigation and A TC transponder systems. The weather radar has its own control panel.

VHF Navigation Radios

Radio tuning units control the NAV 1 and NAV 2 VHF navigation radios. The navigation data is displayed on the PFDs and the MFDs in HSI, FMS MAP or NAV SECTOR formats. The VHF NAV system receives and monitors the VOR, LOC, GS and marker beacon signals.

Effectivity:

Airplanes 15001 thru 15316:

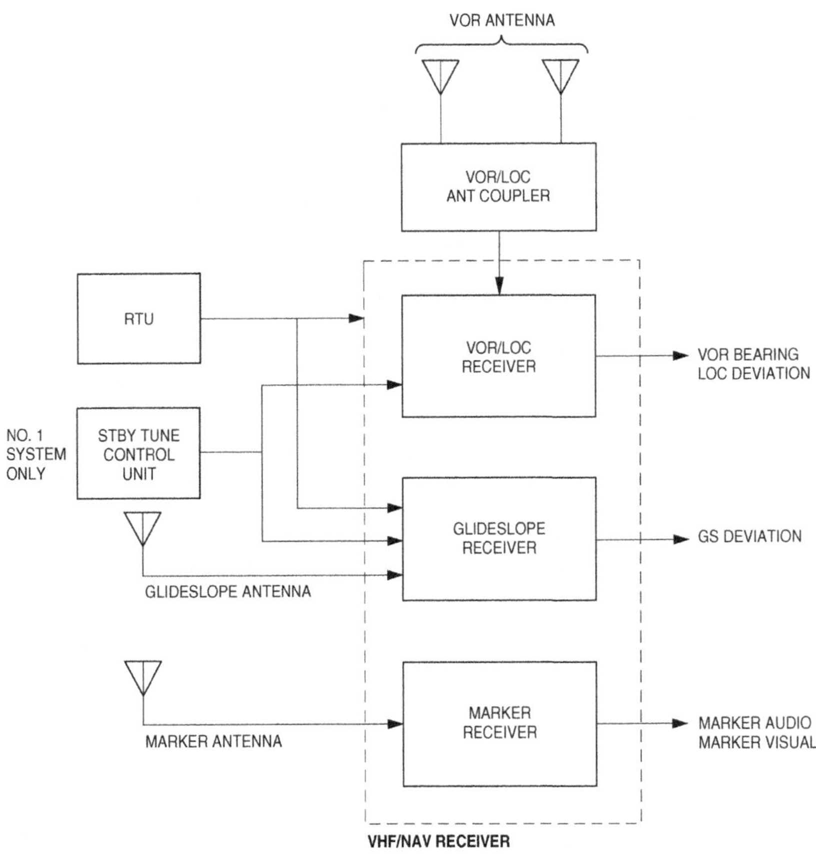

Figure 16 - 1: VHF/NAV Block Diagram - AC 15001-15316

Effectivity:

Airplanes 15317 and subsequent:

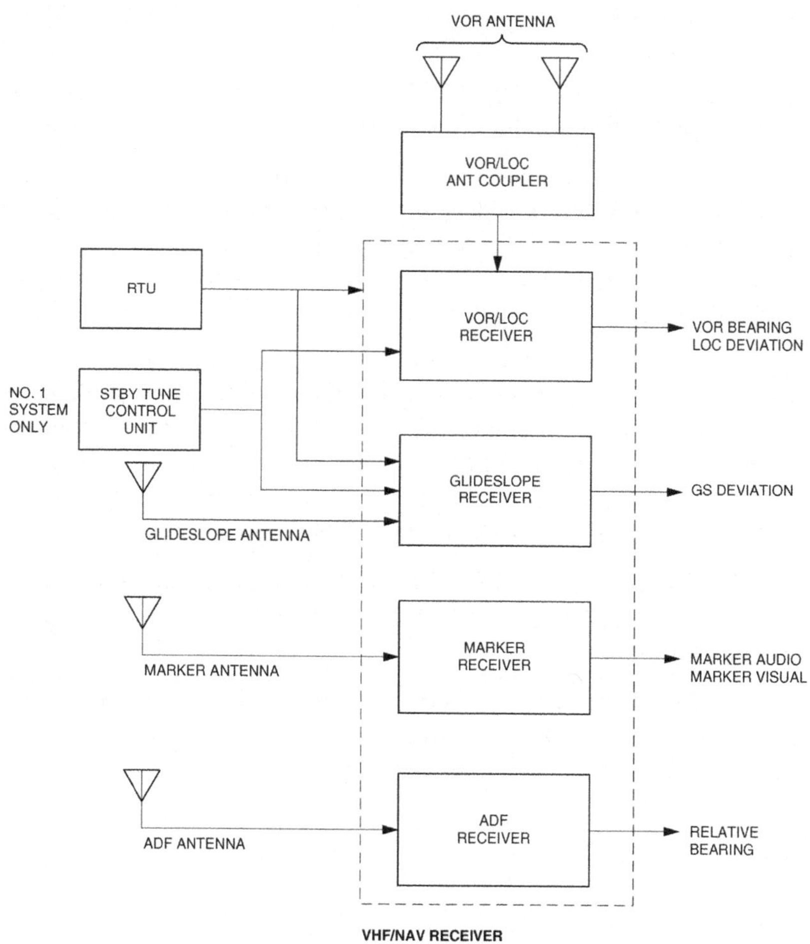

Figure 16 - 2: VHF/NAV Block Diagram - AC 15317 and subsequent

Components and Operation

VHF Navigation Radios

Two VHF navigational radio receivers are installed and are controlled via the RTUs. Each radio is capable of receiving VOR, ILS, GS and marker beacon signals.

Reversionary control of the VHF navigation radios on these aircraft is accomplished through the FMS CDU.

In VOR operation, the NAV receivers supply en-route and terminal area guidance by monitoring selected VOR stations. In LOC and ILS modes, the NAV receivers supply final approach guidance data. When the navigation receiver is tuned to a localizer frequency, the paired glide-slope frequency is automatically tuned.

The FMS computer automatically tunes DME channels 2 and 3 of each DME receiver. Channel 1 is manually tuned at the RTU or via the FMS radio page. Channel 1 can be used for DME hold. If AUTO tuning is selected at the CDU, the FMS will automatically tune all channels for the receiver. AUTO tuning of the receiver is not available if the NAV SOURCE is not FMS or DME HOLD has been selected.

The FMS text page or VOR/DME STATUS, can be displayed on either MFD. The VOR/DME STATUS page lists the VOR/DME stations that are being monitored.

Two DME antennas are located on the bottom of the fuselage just forward of the wing leading edge.

Marker Beacons

Marker beacon indications are presented on the PFDs during station passage. Two marker beacon antennas are installed on the bottom of the fuselage just forward of the aft equipment bay hatch.

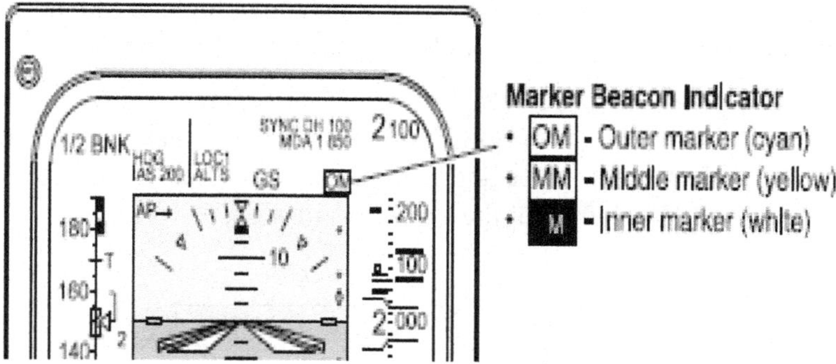

Figure 16 - 3: Outer, Middle, and Inner Marker Beacon Indications

Audio Control Panel

The audio control panel (ACP) allows audio monitoring of the DME, ADF, marker beacon or navigation radios. Audio monitoring is selected at designated receive switches on the ACP. When selected, the receive switch illuminates and can be used to adjust the volume of the individual radio. Any number of radios can be monitored at the same time.

The VOICE/BOTH switch allows the pilot to separate the simulcast voice and CW transmissions on VOR, ILS or ADF transmissions. When BOTH is selected, the voice and CW station identifier are heard. Selecting VOICE filters out the CW transmission.

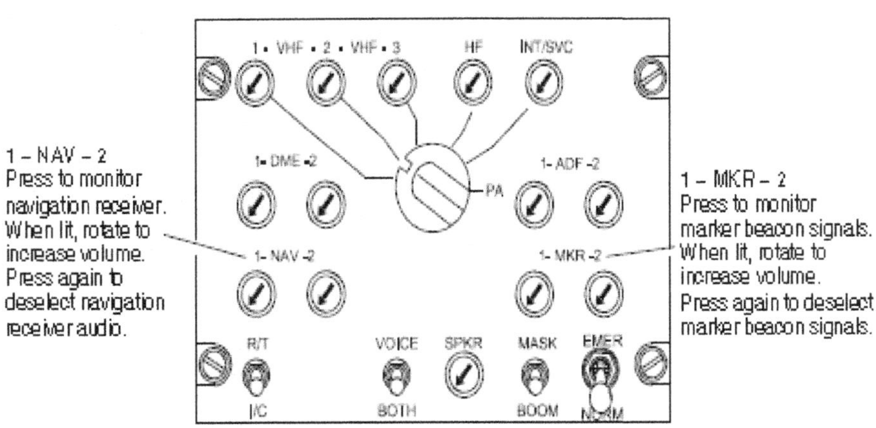

Figure 16 - 4: Audio Control Panel

FMS RADIO TUNING Page

The FMS control display unit (CDU) is capable of tuning ADF radios, ATC transponder and channel 1 of the NAV radios. The active, preset and recall COM frequencies are present. The active AUTO or MAN tuning of the NAV frequencies is presented in cyan.

FMS FREQUENCY DATA Page

The FREQUENCY DATA page provides communication frequencies associated with airports to quickly find and tune a desired frequency. The data page lists the departure (origin), destination, and alternate airports contained in the active flight plan, along with space for a pilot-entered airport The pilot can select a frequency and tune the radio to that frequency by using the CDU scratch pad.

Radio Tuning Units (RTU's)

The RTU top page displays VHF navigation and communications, active (ACT), preset (PRE) and RECALL frequencies. It also displays the active ADF frequency, transponder code, TCAS status and DME HOLD frequency (when selected).

Main Page

The main page is accessed by pressing twice on the associated left line key. From the main page, radio frequencies can be tuned and displayed. Also, marker sensitivity and DME frequency can be monitored and changed.

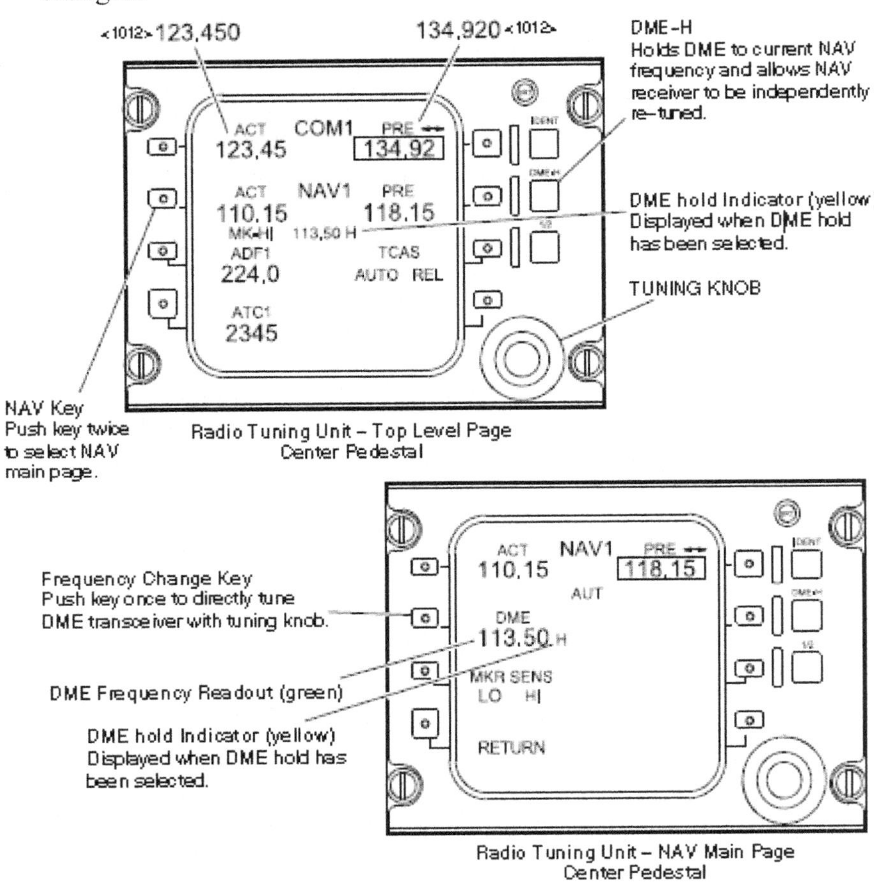

Figure 16 - 5: RTU AC 15001-15276

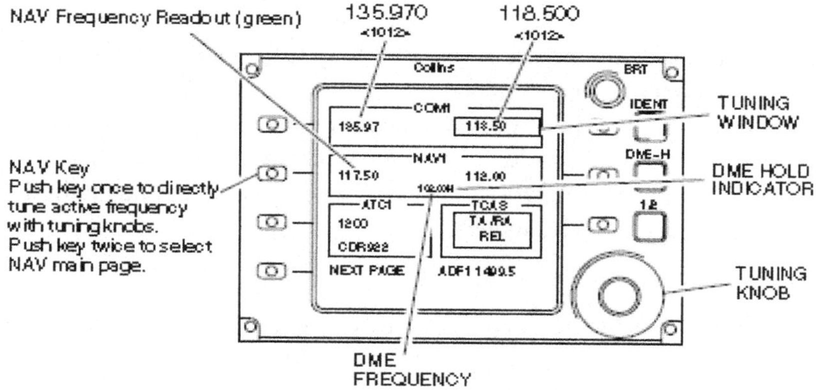

Radio Tuning Unit - Top Level Page
Centre Pedestal

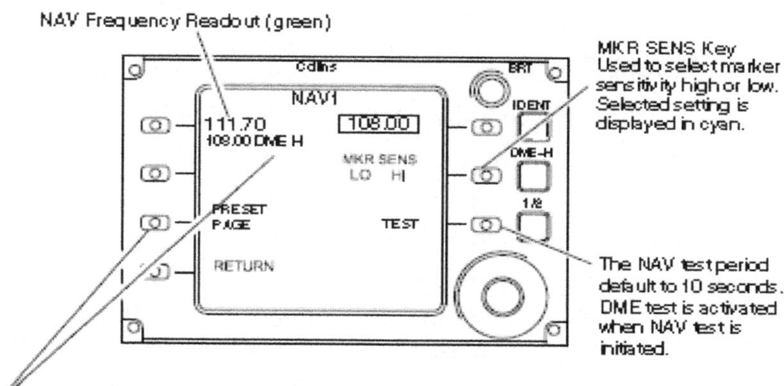

Figure 16 - 6: RTU AC 15277 and subsequent

DME Hold

DME hold allows the pilot to use DME channel 2 for distance measuring. The DME hold feature provides the pilot with range data and allows channel 1 to be tuned to a VOR for bearing information. When the DME-H pushbutton on the RTU is pressed, the DME frequency paired with the active VOR frequency is selected to hold. When a new active VOR frequency is selected, the DME function stays on the frequency paired with the previous VOR frequency.

The hold frequency is identified with:

- An amber H below the active frequency on the RTU top page
- An amber H next to the DME frequency on the RTU main page
- On the PFD and MFD displays, an amber H that replaces the NM indication and the removal of the station identifier

Tuning Select Panel

Backup tuning is accomplished through the FMS CDU.

The FMS TUNE INHIBIT switch-light on the tuning select panel permits the pilot to inhibit the AUTO tune function of the FMS and the manual (MAN) tune function of the FMS radio page.

During normal operations the FMS TUNE INHIBIT switch-light is deselected (OFF), permitting AUTO tuning of the VHF NAV. To inhibit the AUTO tune function use the FMS TUNE INHBIT switch.

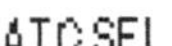

ATC SEL
Used to select ATC transponders.
- 1 – ATC 1 transponder is activated and ATC 2 transponder is on standby.
- STBY – Both transponders are on standby.
- 2 – ATC 2 transponder is activated and ATC 1 transponder is on standby.

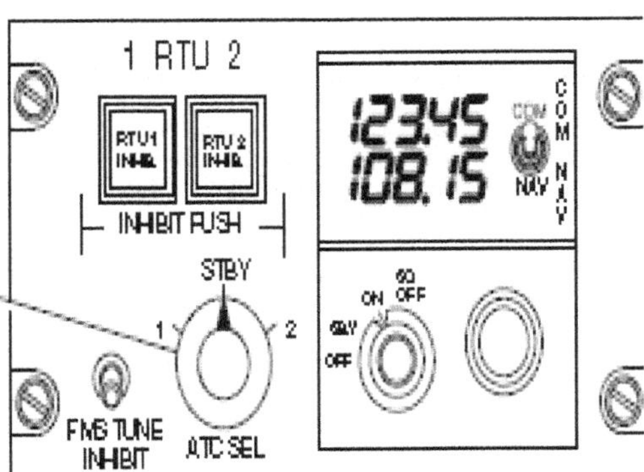

Display Control Panel - NAV SOURCE Knob

The NAV SOURCE knob on the display control panel (DCP) selects the navigation source. Navigation source data is indicated on the PFD. Navigation sources include VOR, LOC or FMS.

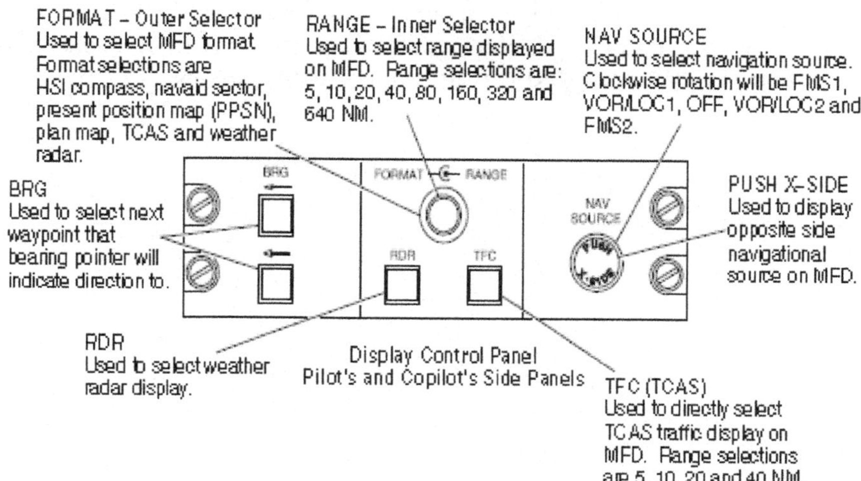

Display Control Panel
Pilot's and Copilot's Side Panels

Nav X-Side Function

When the center of the NAV SOURCE knob is pressed, the cross-side NAV data is posted in amber on the on-side PFD.

Primary Flight Displays (PFDs)

When the RTU is tuned and receiving a VHF NAV frequency and the NAV SOURCE is set to the VOR/LOC position, the primary flight display shows:

- VOR station identifier in the course display window
- Green course information on the HSI (course selection is made at the CRS knob of the flight control panel)
- Active or armed lateral VOR or LOC mode in the FMA NAV source, mode and course data are displayed on the HSI, NAV SECTOR and FMS

MAP formats of the MFDs. VOR or ADF information is displayed by selecting the BRG switches on the pilot and copilot DCPs. VOR 1 or ADF 1 is displayed as a single magenta pointer. VOR 2 or ADF 2 is displayed as a dual cyan pointer.

The bearing pointers will not appear on the PFD or MFD unless a valid VOR or ADF frequency is received. Bearing pointers can be displayed on the MFD in HSI, NAV SECTOR or FMS MAP formats.

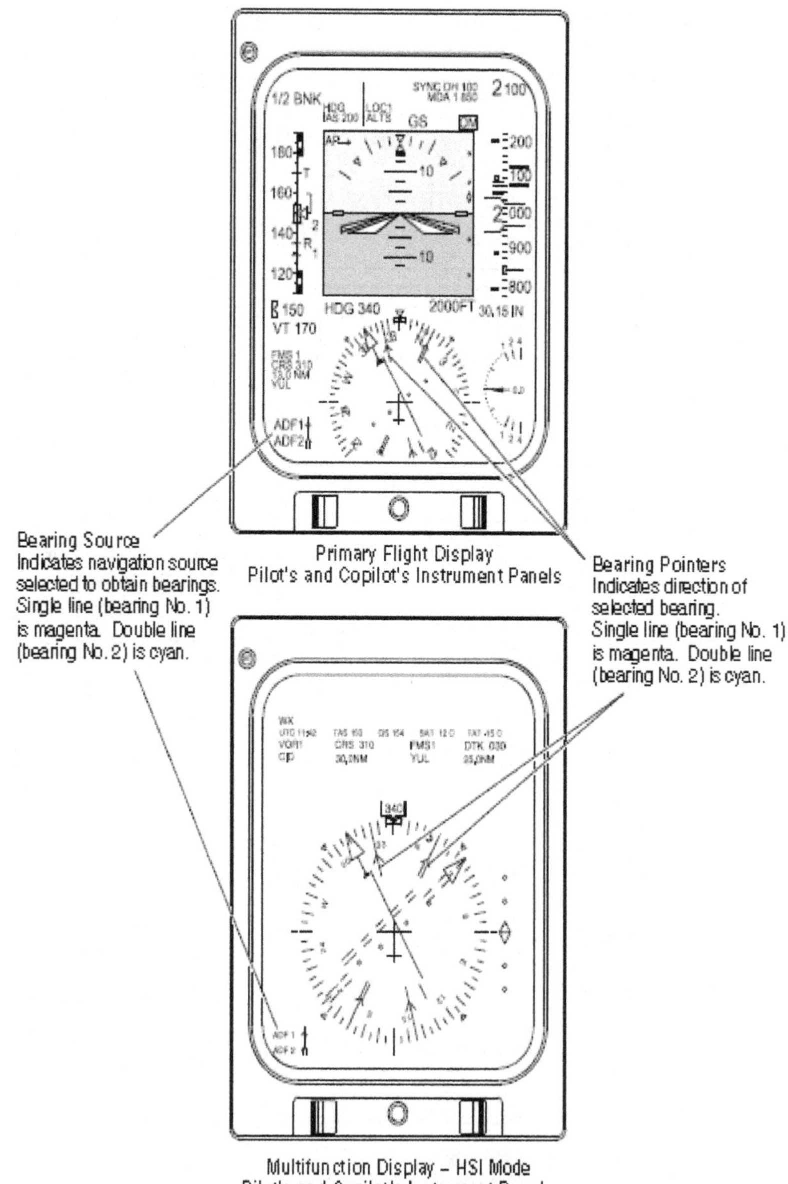

Figure 16 - 7: VHF Navigation - Bearing Source

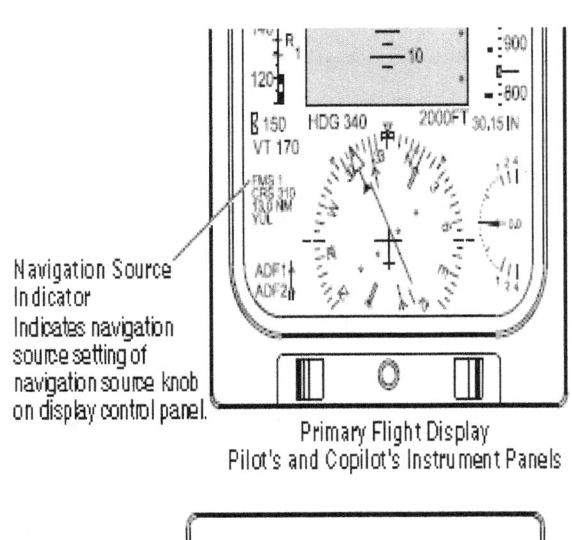

Primary Flight Display
Pilot's and Copilot's Instrument Panels

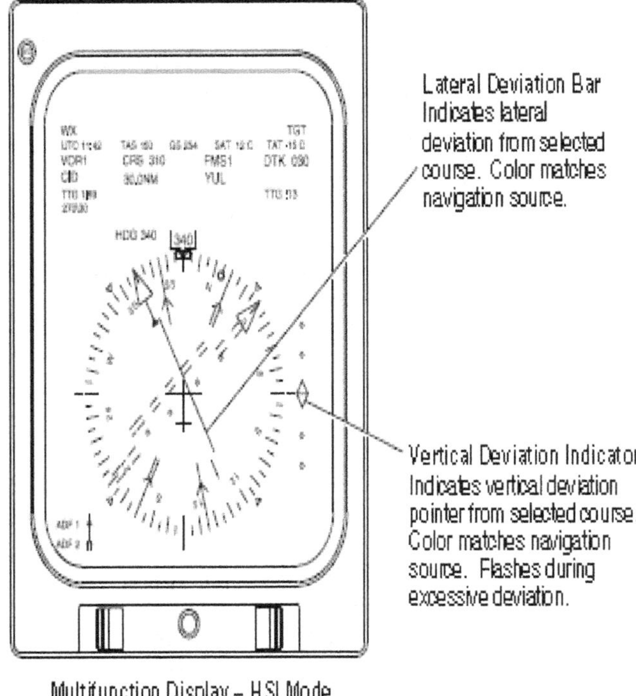

Multifunction Display – HSI Mode
Pilot's and Copilot's Instrument Panels

Figure 16 - 8: VHF Navigation - Deviation/Source Indicator

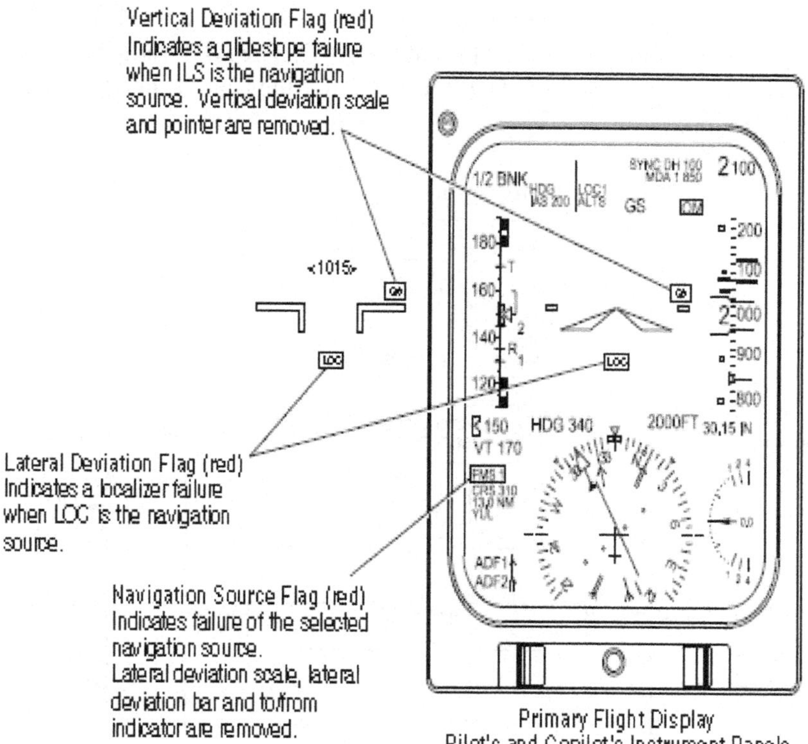
Vertical Deviation Flag (red)
Indicates a glideslope failure when ILS is the navigation source. Vertical deviation scale and pointer are removed.
Lateral Deviation Flag (red)
Indicates a localizer failure when LOC is the navigation source.
Navigation Source Flag (red)
Indicates failure of the selected navigation source.
Lateral deviation scale, lateral deviation bar and to/from indicator are removed.
1/2 BNK
GS
180
160
140
120
150
VT 170
HDG 340
2000FT
30,15 IN
ADF1
ADF2
200
100
2000
900
800

Primary Flight Display
Pilot's and Copilot's Instrument Panels

ADF Navigation

Two ADF radios provide relative bearing and station identification data to the pilots. The information is displayed on the PFD and MFD and can be monitored aurally by selection at the ACP. The ADF radios are tuned at the RTU or the FMS CDU. Other ADF settings can be changed on the ADF main page.

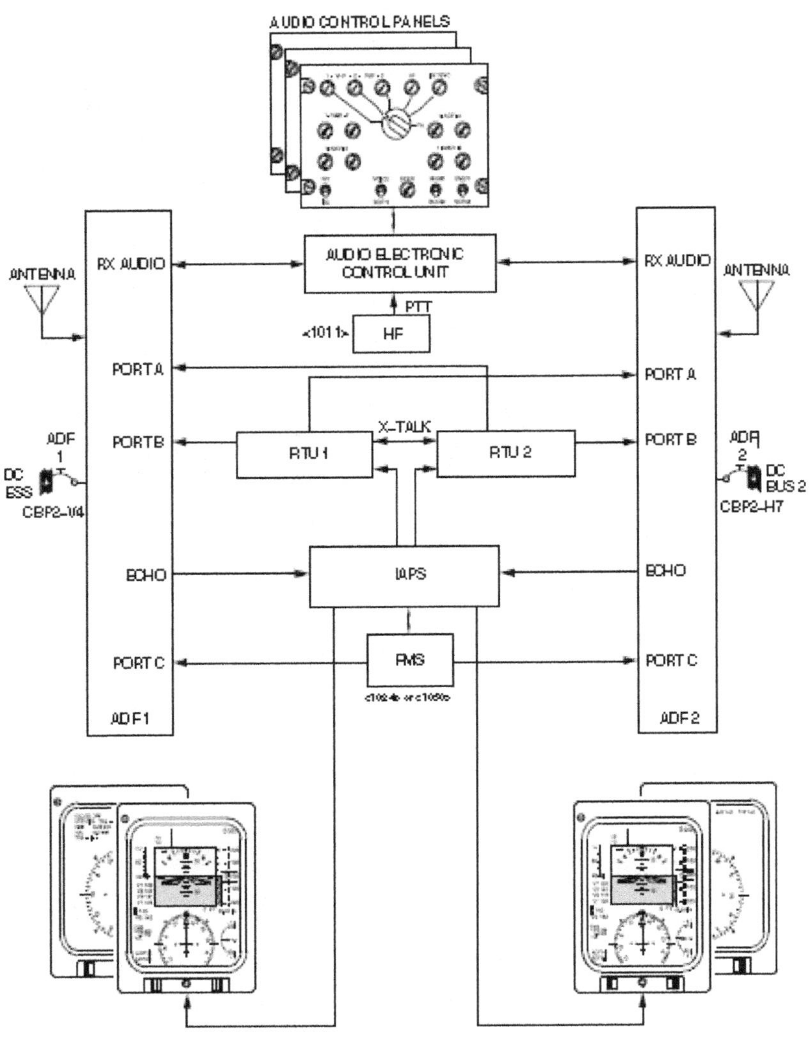

Figure 16 - 9: ADF System Interface 15001 - 15316

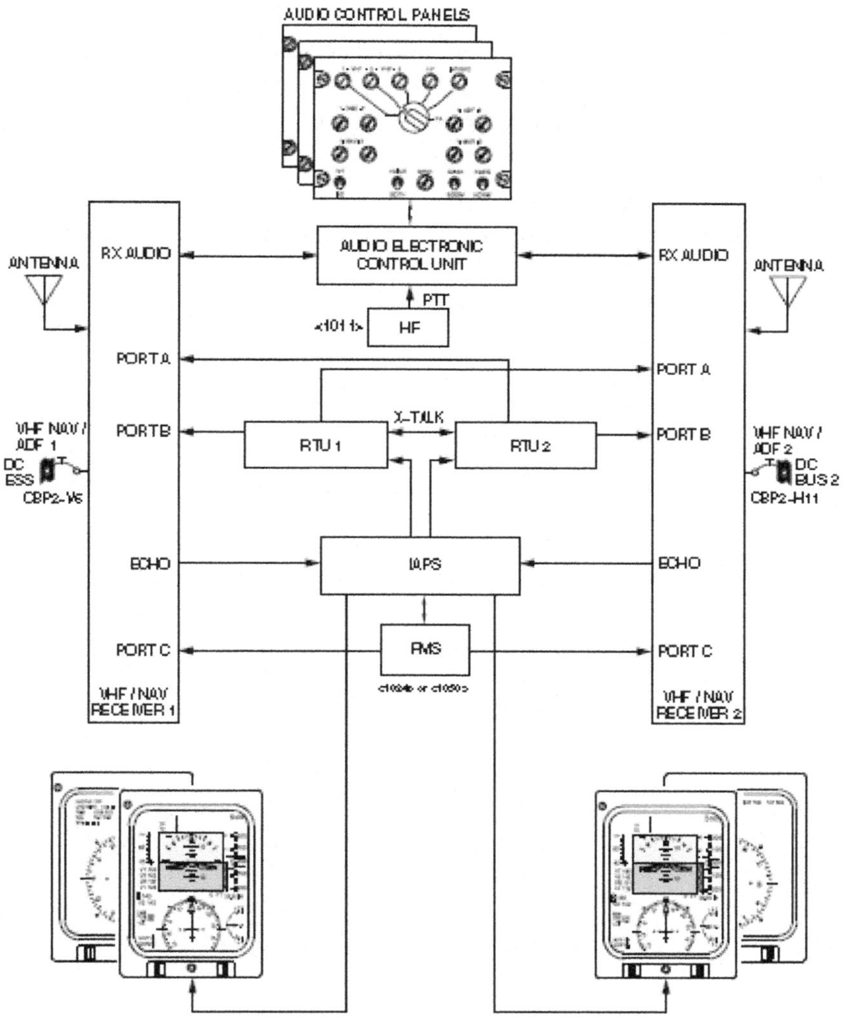

Figure 16 - 10: ADF System Interface - AC 15317 and Subsequent

RTU Tuning- ADF

Top Page

Frequency selection is accomplished at the RTU by selecting the line key opposite the ADF frequency. A second press of the line key brings the ADF main page into view.

Main Page

The ADF main page allows the following functions to be accessed:

- Frequency selection
- ADF or antenna mode selection
- BFO ON or OFF

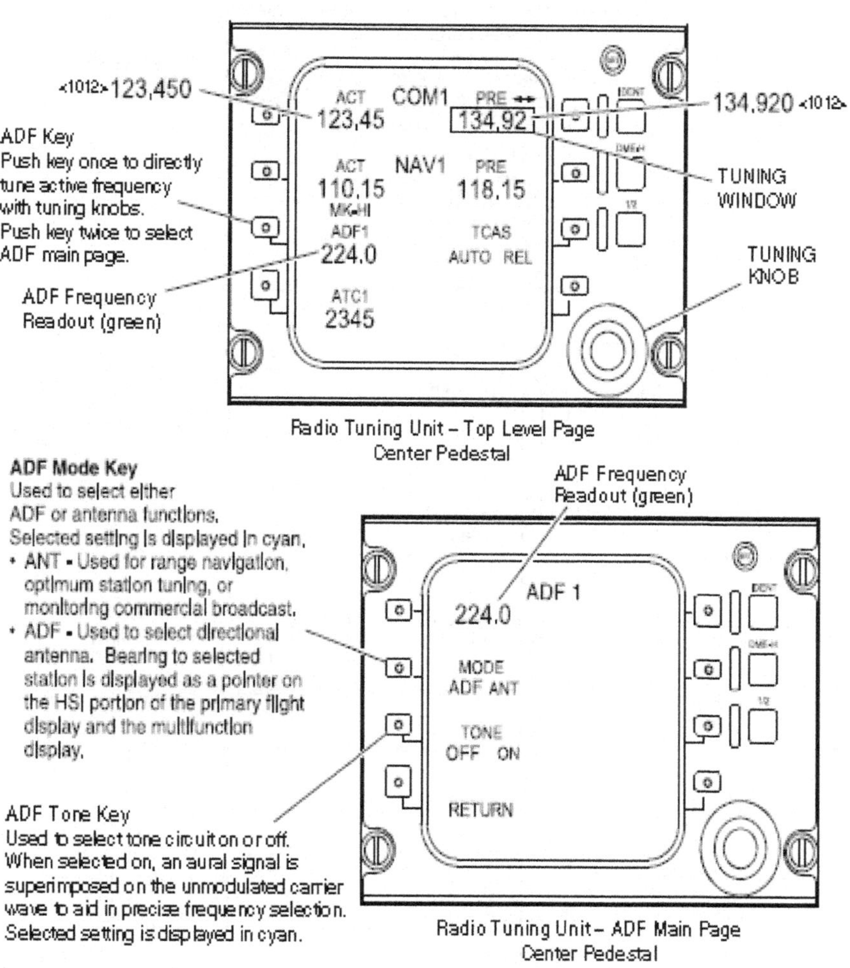

Figure 16 - 11: ADF Tuning - AC 15001 - 15276

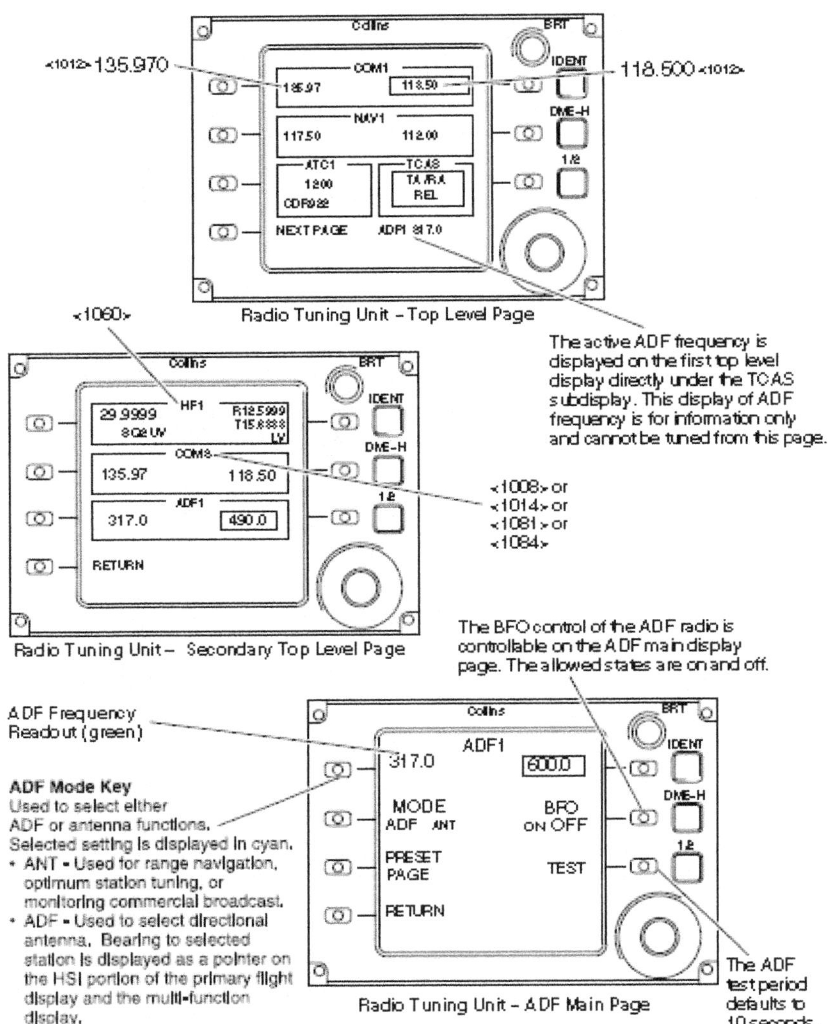

Figure 16 - 12: ADF Tuning - AC 15277 and Subsequent

PFD and MFD Presentation

ADF information is displayed by selecting the BRG switches on the pilot and copilot DCPs. ADF 1 is displayed as a single magenta pointer. ADF 2 is displayed as a dual cyan pointer.

The ADF pointers will not appear on the PFD or MFD unless a valid ADF frequency is received. ADF pointers can be displayed on the MFD in HSI, NAV SECTOR, FMS MAP formats

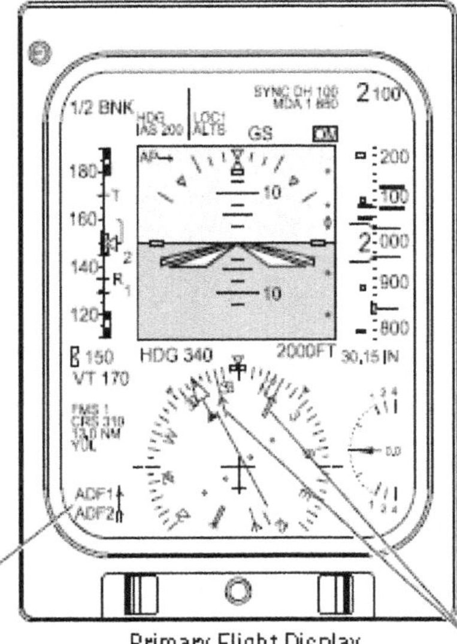

Bearing Source
Indicates navigation source selected to obtain bearings. Single lined (bearing No. 1) is magenta. Double lined (bearing No. 2) is cyan.

Primary Flight Display
Pilot's and Copilot's Instrument Panels

Bearing Pointers
Indicates direction of selected bearing.
Single lined (bearing No. 1) is magenta. Double lined (bearing No. 2) is cyan.

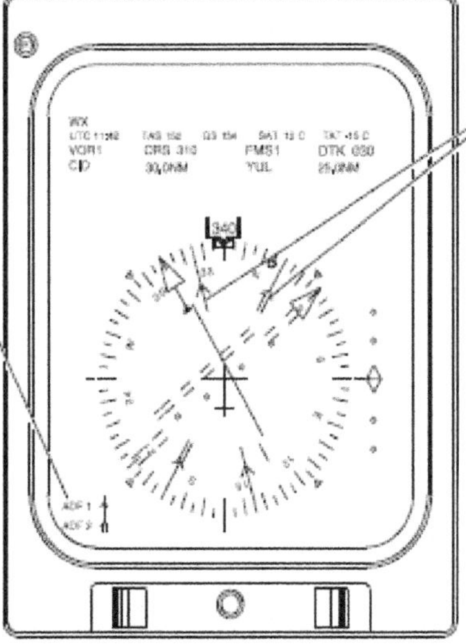

Multifunction Display – HSI Mode
Pilot's and Copilot's Instrument Panels

ATC Transponder

The two air traffic control transponders (ATC 1 and ATC 2) provide ground radar beacon systems with coded identification responses in the following modes:

- Mode A - Airplane identity reporting
- Mode C - Altitude reporting
- Mode select (S) - Data link with other Modes transponders for the traffic collision avoidance system (TCAS)
- ADS-B Out

Mode S data link includes air-to-air, ground- t o -air (data up-link or COM A), air-to-ground (data down-link or COM B), and multi-site (ground station to ground station messages.

Active transponder selection is made using the ATC SEL knob on the tuning select panel (new production airplanes) or the backup tuning unit (legacy airplanes).

Transponder codes are set on the top level page of the radio tuning units and can also be set using the FMS control display unit. ATC identification is selected using the IDENT button on the radio tuning unit (RTU).

Mode C altitude reporting selection is made on the RTU - ATC main page.

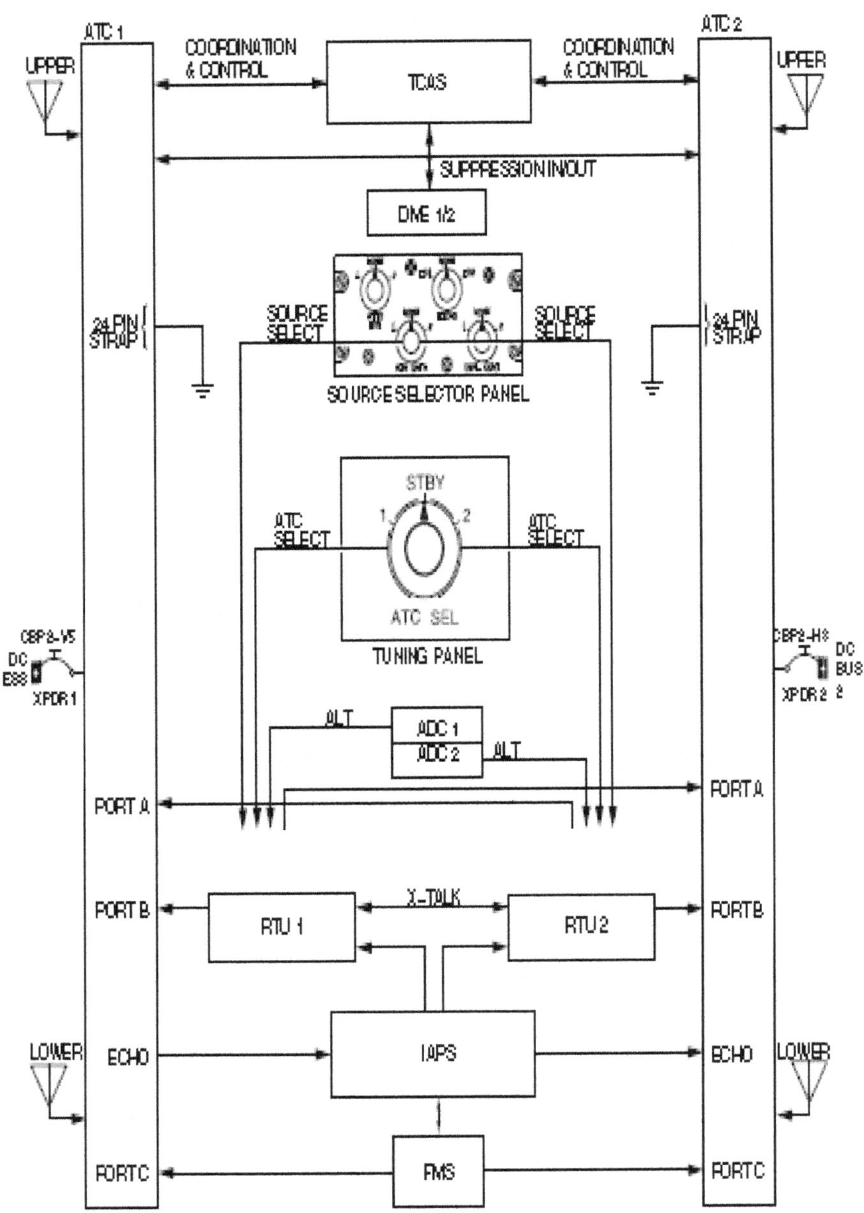

Figure 16 - 13: ATC Transponder System Interface - AC 15001-15316

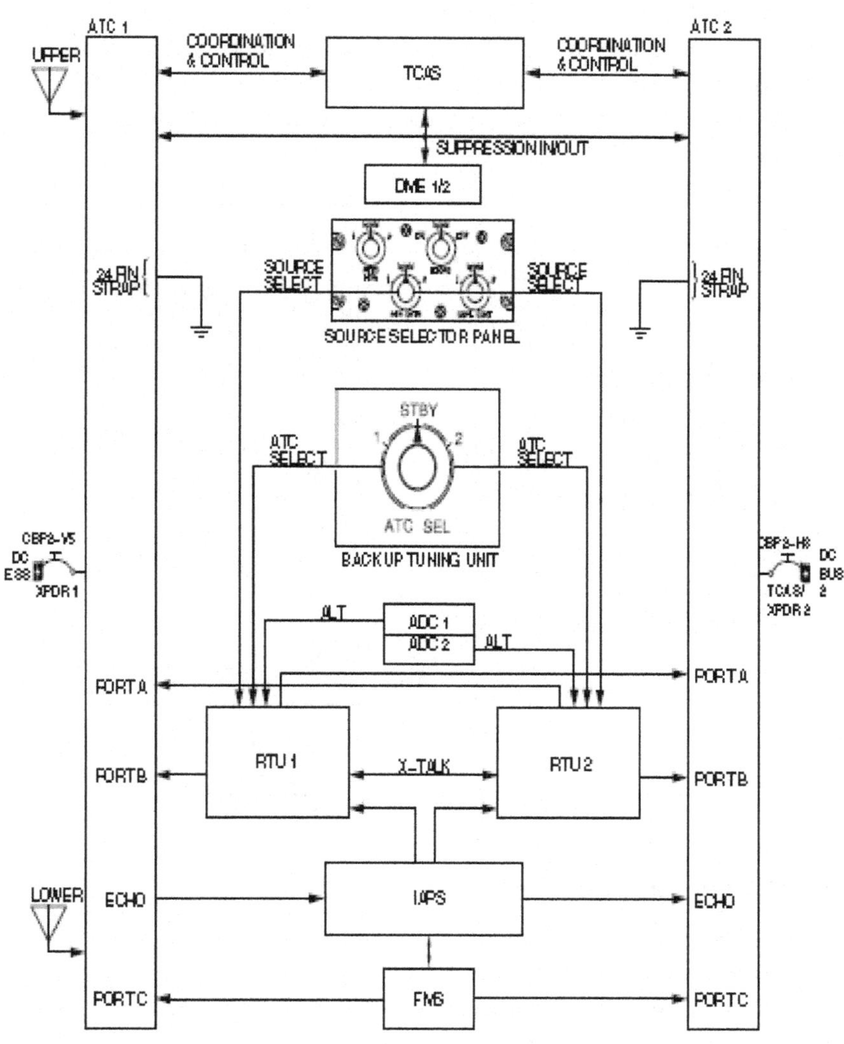

Figure 16 - 14: ATC Transponder System Interface - AC 15317 and subsequent

Components and Operation

ATC Selector Knob

The ATC selector knob has three positions: 1, STBY and 2. When ATC 1 transponder is selected it uses altitude information from ADC 1 to respond to Mode C and S interrogations. ATC 2 receives altitude information from ADC 2. When either transponder is selected, the RTU four-digit identification code turns green. When in STBY the RTU code is white. The system can receive but cannot reply to interrogations when on the ground.

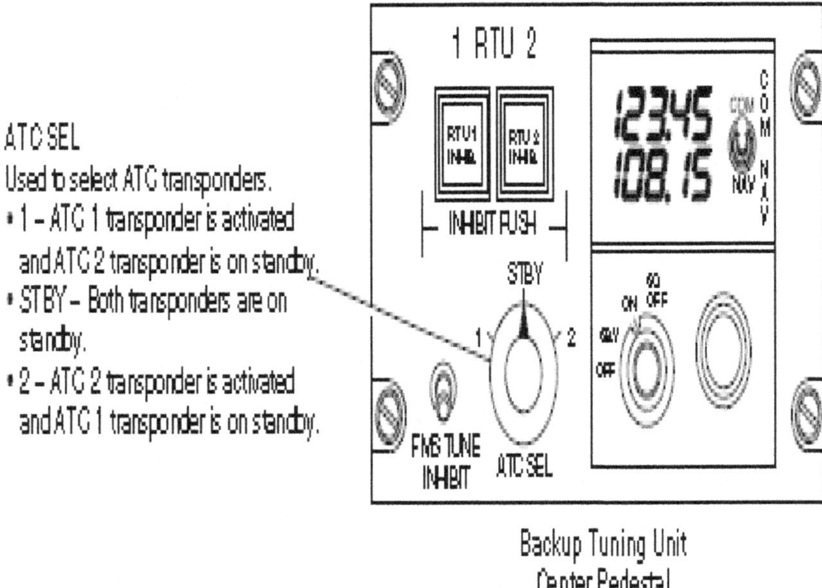

Backup Tuning Unit
Center Pedestal

Mode S Transponder

Mode S has the capability to display either a four squawk code or the flight identification (FLT ID) on line 4 of the RTU top level page. Selection of either the squawk code or the FLT ID for display on the top level page is made on the ATC main page.

To access the ATC main page from the top level page, the ATC line select key is pressed twice. Once the main page is displayed, the DISPLAY line select key is pressed to select either the SQUAWK or FLT ID (the selected function will be displayed larger). The selected function is then displayed on line 3 of the main page, line 4 of the top level page and on the FLT ID page.

To modify the squawk code or the FLT ID on the top level page, the ATC line select key is pressed, which will cause a tune window to surround the left character. The small tuning knob is then used to change the character displayed in the tune window. The RTU then waits 2 seconds after knob rotation stops before locking in the new character. Rotating the large tune knob cycles the tune window from character to character.

Automatic Dependent Surveillance-Broadcast (ADS-B Out)

Automatic Dependent Surveillance-Broadcast (ADS-B) Out is a system in which the airplane automatically transmits its own GPS position to the Air Traffic Control (ATC), along, but not limited to, heading, ground track, ground speed, altitude and identification.

The system doesn't require any pilot or external input and is dependent of the airplane's navigation system. The airplane's flight parameters are periodically broadcast, enabling it to be tracked.

The ADS-B Out system uses the transponder to broadcast airplane flight parameters to the ground using the Mode S 1090 MHz extended squitter capability with a refresh rate of 0.5 seconds. The onboard precision GPS receivers are the primary sources of position and velocity related parameters.

In a dual GPS installation, the transponders use data from the GPS that indicates the best radius of containment in a transmitted position squitter. The ADS-B information can be used by ATC to augment the existing primary and secondary transponder-based radar or be used where ATC radar coverage is limited. ADS-B information used by ATC allows improved aircraft separation, better situational awareness and traffic avoidance.

The information broadcast by an ADS-B Out equipped airplane can also be received by other nearby aircraft. Aircraft that can receive ADS-B information have ADS-B In.

Our company is incorporating two different modifications of ADS-B Out. One will interface fully with EICAS and failure modes of the ADS-B Out system are indicated on the EICAS primary and status pages. The second type of modification will have an ADS-B FAIL annunciator light installed below the standby instrument panel. Failure will be indicated by the annunciator being illuminated while the failed side is selected on the transponder as well as the transponder code displaying yellow.

RTU - ATC Top Page

The ATC line key on the RTU top page is selected to move the white tune window to the ATC line. Using the frequency selector knob on the RTU changes the code within the tune window.

The top page displays the following cyan messages:

- STBY - Both ATC transponders are in the standby mode, the transponder code is white
- ALT OFF - Mode C is selected OFF
- ID - The ID key has been selected
- R - The transponder is responding to an interrogation
- STBY, ID and RPLY appear in cyan beside ATC code
- ALT OFF appears in cyan below ATC code RTU - ATC Main Page

The ATC main page is accessed at the ATC line key.

The cyan messages STBY, ID and R are reproduced and displayed when appropriate. The cyan messages STBY, ID and RPLY are reproduced and displayed beside the ATC code when appropriate.

The ALT line key selects the Mode C altitude reporting Mode ON or OFF.

The reporting altitude is removed when Mode C is selected to ALT OFF.

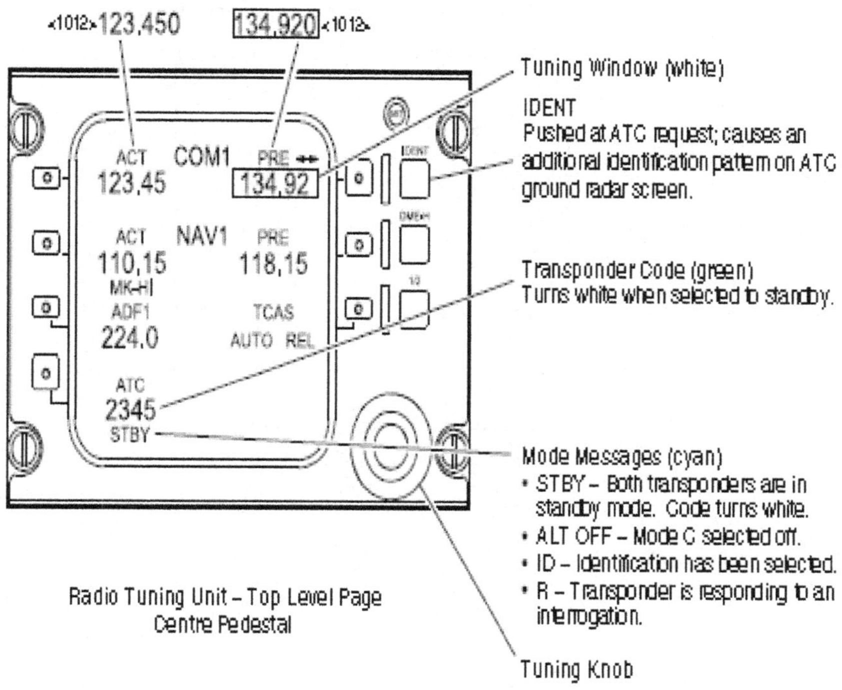

Figure 16 - 15: ATC RTU Main Page - AC 15001 - 15276

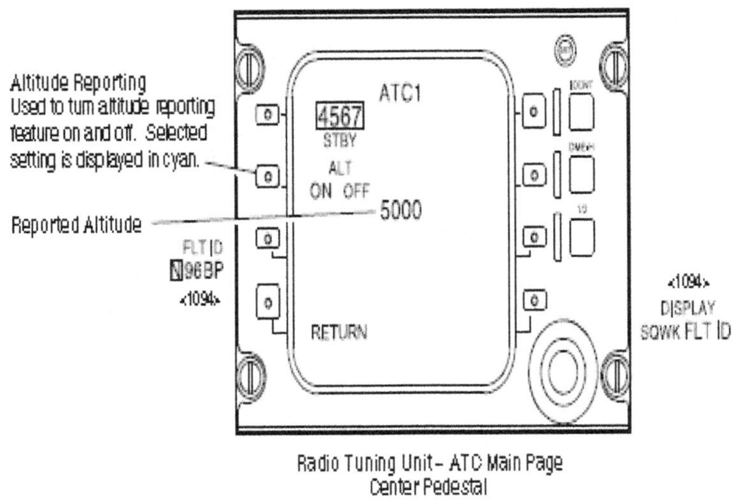

Radio Tuning Unit – ATC Main Page
Center Pedestal

Figure 16 - 16: ATC RTU Main Page - AC 15001 - 15276

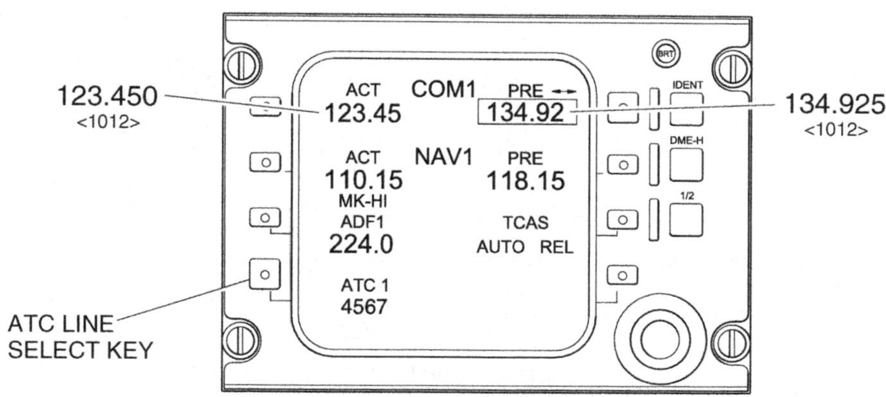

Radio Tuning Unit – Top Level Page

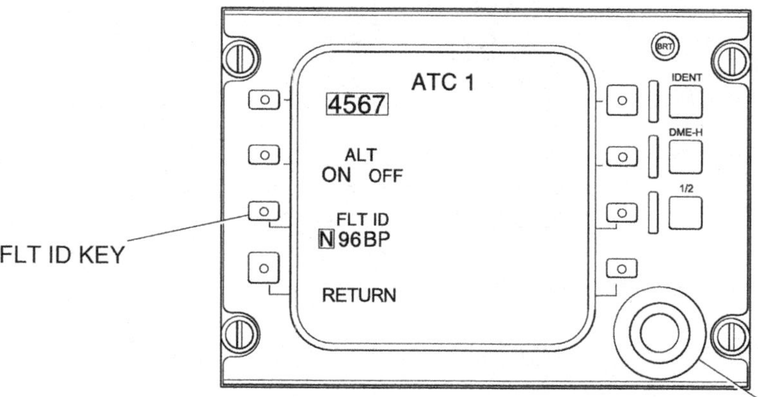

Figure 16 - 17: RTU with ADS-B Installed

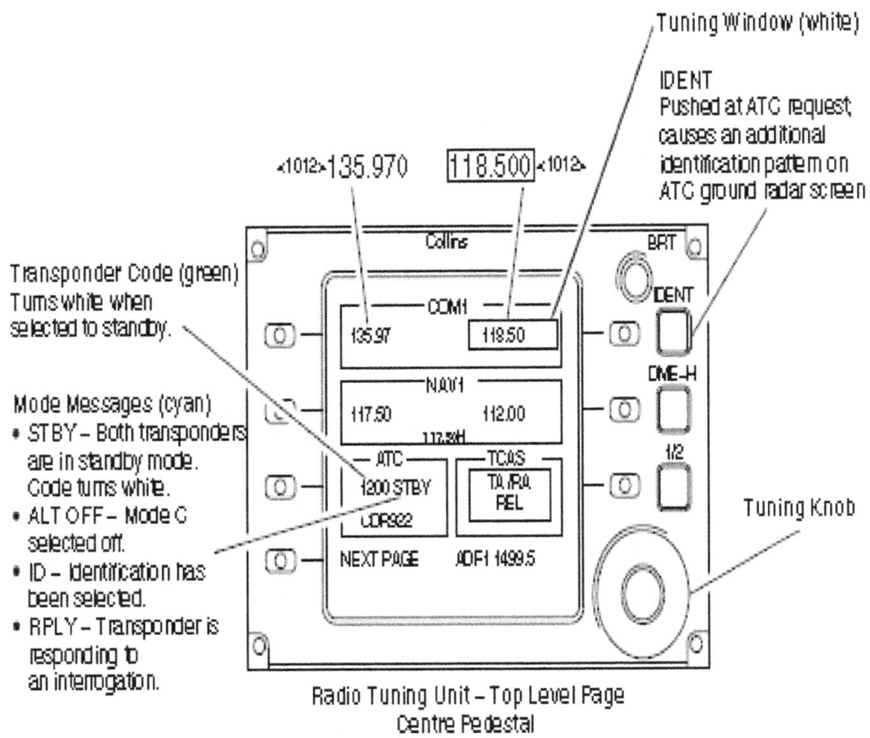

Figure 16 - 18: ATC RTU Top Page- AC 15277 and Subsequent

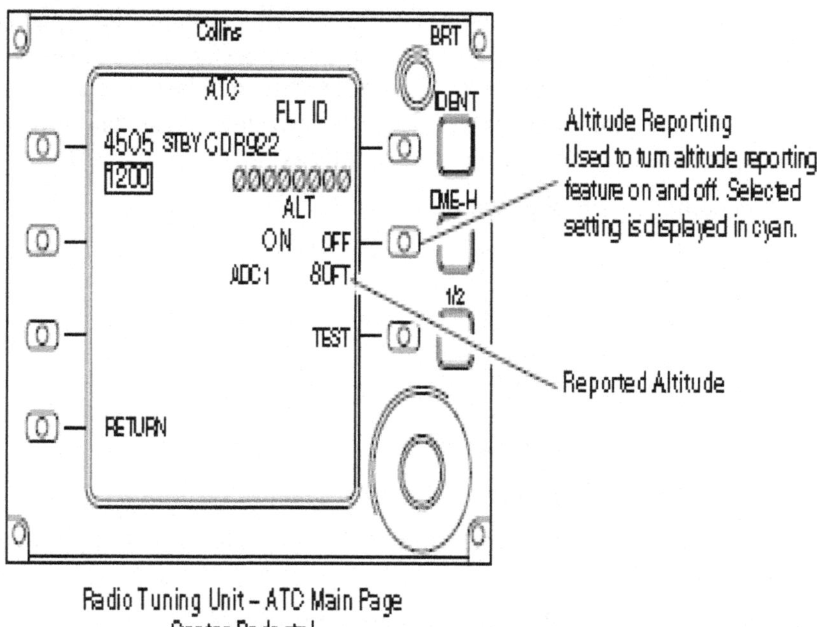

Figure 16 - 19: ATC RTU Main Page - AC 15277 and Subsequent

Enhanced Ground Proximity Warning System

The purpose of the enhanced ground proximity warning system (EGPWS) is to reduce the incidences of controlled flight into terrain (CFIT) and provide advance warning of severe windshear. Unwanted alerts will be very rare if the flight crew maintains situational awareness with respect to terrain and follow correct avoidance procedures for windshear activity.

The EGPWS computer supplies the following features:

- EGPWS Modes
    - Mode 1 Excessive descent rate
    - Mode 2 Excessive terrain closure rate
    - Mode 3 Altitude loss after take-off
    - Mode 4 Unsafe terrain clearance
    - Mode 5 Below glide slope alert
    - Mode 6 Call-Outs (descent below minimums, altitude call outs, and bank angle alert)
    - Mode 7 Windshear detection and alerting
- Terrain clearance floor and obstacle databases with MFD terrain presentation

NOTE:
In the event of a momentary loss of AC electrical power, the TERRAIN FAIL status message may be displayed while the GPS satellites are reacquired (approximately 75 seconds) and the FMS airplane position is re-entered.

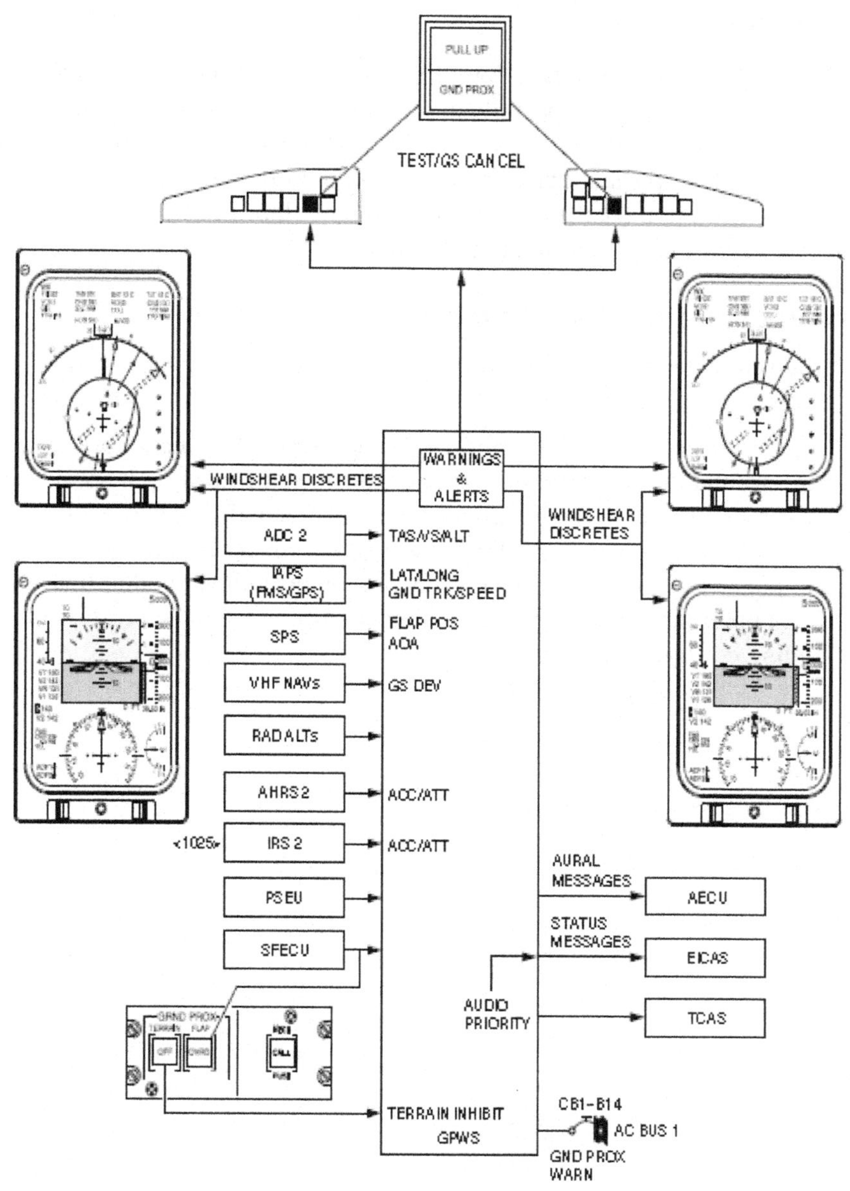

Figure 16 - 20: EGPWS Interface

Components and Operation

EGPWS Databases

The EGPWS provides the pilot with a greater awareness of the terrain ahead of the aircraft, especially around airports. The EGPWS computer generates alerts and warnings by comparing the actual aircraft's position to terrain features and obstacles found in the EGPWS databases. The EGPWS has two databases: terrain clearance floor (TCF) and terrain/obstacle awareness alerting and display (TAAD).

Terrain Clearance Floor Database

The TCF database includes worldwide coverage of all airports with runways longer than 3500 feet.

The TCF database provides an increasing terrain clearance protection zone around the center point of the nearest airport runway. The profile of the circular protection envelope is unique to each airport and is directly related to the distance from the runway. This enhanced feature complements existing Mode 4 protection by providing an alert that is based upon insufficient terrain clearance, even when the aircraft is in landing configuration.

Terrain/Obstacle Awareness Alerting/Display Database

The terrain awareness alert function of the EGPWS computes and predicts potential conflicts between the aircraft's flight path and terrain ahead of the aircraft during all phases of flight.

Awareness alerts are based upon the terrain awareness alerting display database (TAAD) and actual aircraft altitude and position. The TAAD database is limited to North America at this time. The database covers all of the USA but only portions of Canada, Mexico and the Bahamas.

Included in the database is a library of all obstacles that are higher than 100 feet AGL.

The EGPWS computes the terrain closure rate by monitoring radio altitude, vertical speed and by comparing the aircraft's position to the TAAD database. The obstacle avoidance logic uses flight path angle and ground speed to generate awareness alerts of obstacle ahead of the aircraft.

GND PROX Switch-lights and Audio Alerts

Audio alerts and red or amber flashing glareshield switch-lights are used to present EGPWS information to the pilot.

The GND PROX switch-lights are used to test the system and visually present ground proximity warnings. The red PULL UP section of the switch-light flashes when an EGPWS PULL UP warning is generated. The amber GND PROX section of the switch-light flashes for all other ground proximity warnings. There is no switch light indication for windshear.

The audio warnings are prioritized in accordance to their impact upon flight safety. An audio message will be interrupted to allow a higher priority message to sound. Windshear warnings take priority over all other alerts and warnings except a stall warning. windshear warnings are presented on the PFD.

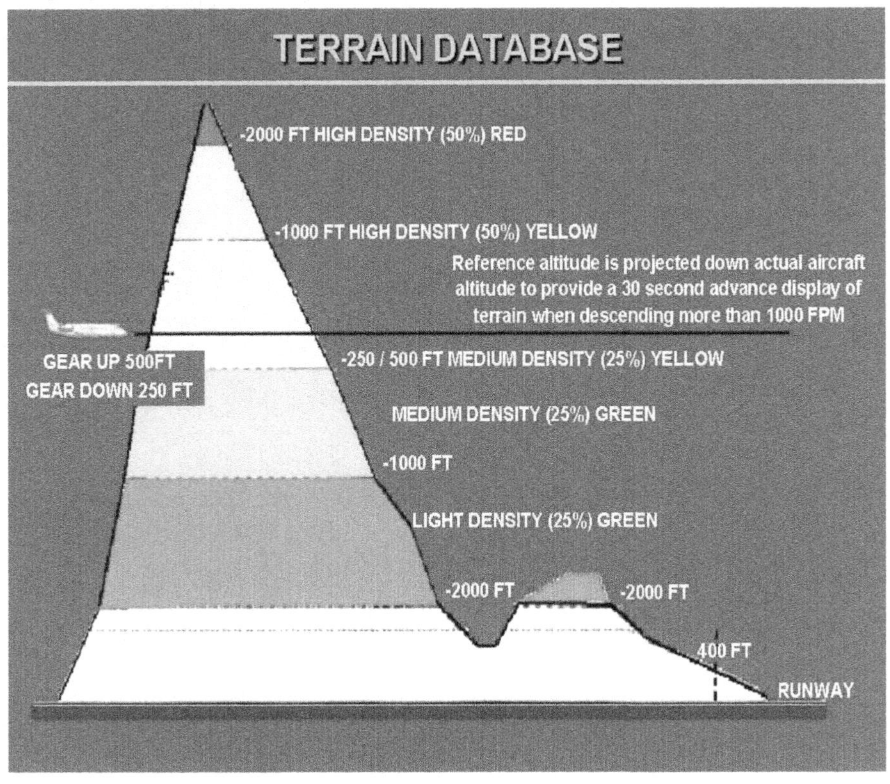

Figure 16 - 21: Terrain Database

Multi function Display Terrain Presentation

Terrain information is presented on the multifunction displays and can be overlaid on the following MFD formats:

- NAV sector
- FMS map

The terrain function is active at all times but is not necessarily displayed.

Weather radar and terrain share the same MFD formats and cannot be displayed at the same time. Radar or terrain presentation on the MFD is selected by pressing the RDR/TERR pushbutton on the DCP. When the radar is deselected, the cyan TERRAIN advisory message appears on the radar line at the top of the MFD.

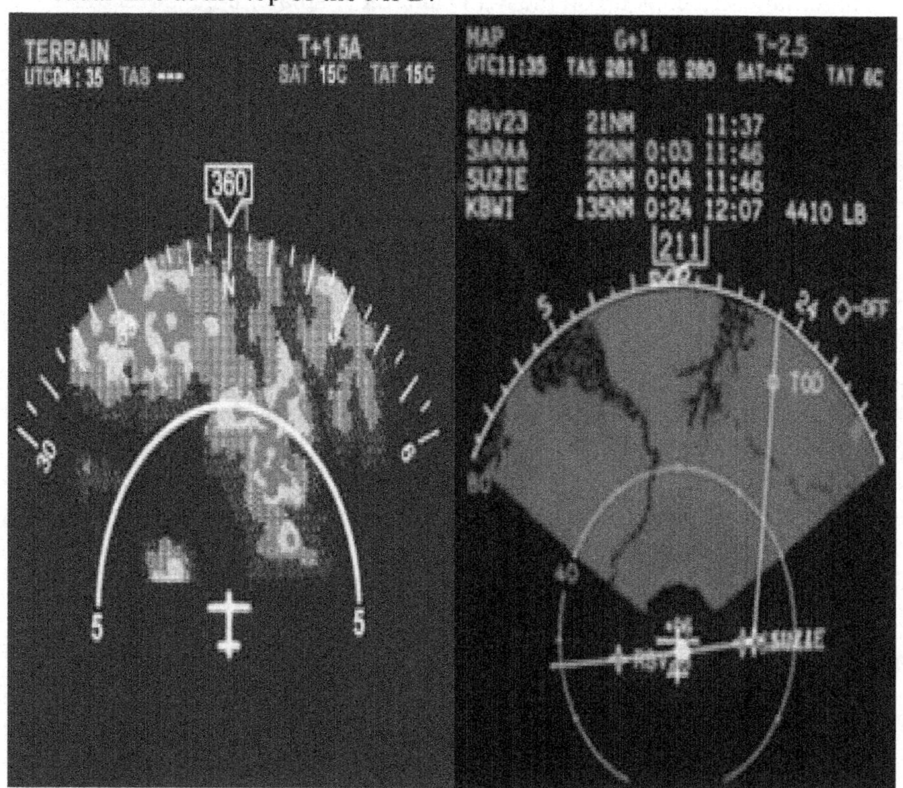

Figure 16 - 22: MFD Terrain Mode

MFD Terrain Color Logic

On the MFD, color and color densities are used to inform the pilot of the vertical separation of aircraft and terrain/obstacles. Shading or low-density colors are used to identify terrain and obstacles that present no immediate hazard to the flight.

Solid or high-density red and yellow colors are used in conjunction with audio alerts and flashing GND PROX switch-lights to advise of a potential flight hazard.

Auto Alert Function

When conflicting terrain or an obstacle is detected, the terrain overlay automatically pops up on the MFD and the display range is reset to 10 miles. The "CAUTION TERRAIN" or "CAUTION OBSTACLE" audio message sounds, the amber GND PROX switch-light flashes on the glareshield and the conflicting terrain or obstacle is shown on the MFD as a solid amber color. Warning time is approximately 60 seconds to impact.

If radar is in use and an auto terrain/obstacle alert occurs. the radar presentation is replaced by the terrain overlay and the cyan TERRAIN message appears on the MFD radar line in the top left corner of the display.

If the pilot fails to respond to the amber caution alert. the red terrain or obstacle warning is activated. The audio message "TERRAIN, TERRAIN, PULL UP" or "OBSTACLE, OBSTACLE, PULL UP'" sounds; the red PULL UP portion of the GND PROX switch-light flashes on the glareshield, and the conflicting terrain or obstacle is shown on the MFD as a solid red color.

Warning time is approximately 30 seconds to impact.

Terrain System Inhibits

The EGPWS TERRAIN function can be manually inhibited on the EGPWS/MECH CALL control panel.

Selecting the TERRAIN switch-light inhibits the terrain function of the EGPWS and presents the TERRAIN OFF status message on EICAS. If landing flaps are not available, selecting the FLAP OVRD switch-light inhibits the EGPWS for landing.

Mode 1: Excessive Descent Rate

Mode 1 is active for all approach phases of flight and is independent of the aircraft configuration. Mode 1 has two boundaries. Penetration of the outer boundary activates the amber GND PROX switch-lights and generates the "SINK RATE" audio alert.

The outer boundary is desensitized when the aircraft is above the glide-slope beam to prevent unwanted alerts when the aircraft is capturing the glide-slope or repositioning to the center line from above.

Penetration of the inner boundary activates the flashing red PULL UP switch-lights. The audio warning "WHOOP, WHOOP, PULL UP" sounds continuously and is not silenced until the rate of descent has been corrected.

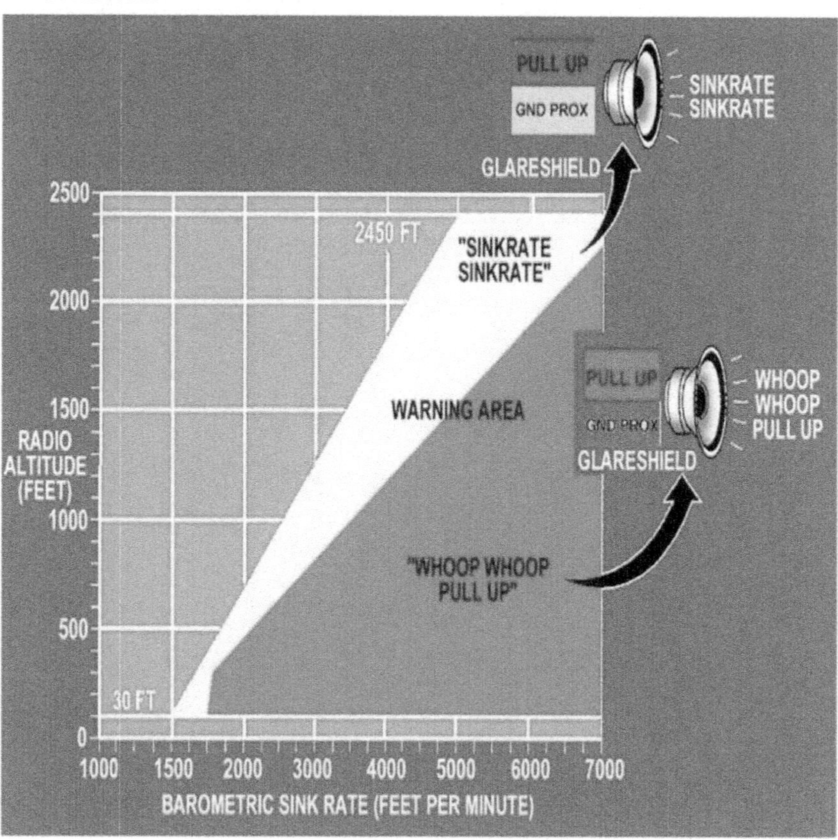

Figure 16 - 23: Mode 1

Mode 2: Excessive Terrain Closure Rate

Mode 2 alerts are generated when the aircraft is closing with terrain at an excessive rate. It is not necessary for the aircraft to be descending to produce a Mode 2 alert.

Level flight or even a climb toward obstructing terrain can result in a hazardous terrain closure rate. The EGPWS computes the terrain closure rate by monitoring radio altitude, vertical speed, barometric altitude, aircraft configuration, and by comparing the aircraft's position to the EGPWS databases.

Mode 2 has two submodes referred to as Mode 2A and Mode 2B. The active submode is determined by aircraft configuration.

Mode 2A: Excessive Closure Rate - Flaps Not LDG Config

Mode 2A is enabled when flaps are not in the landing configuration. I f the aircraft penetrates the Mode 2A alerting envelope, the audio message 'TERRAIN, TERRAIN" sounds and the GND PROX switch-lights flash amber. If the aircraft continues to penetrate the envelope, the audio message "WHOOP. WHOOP, PULL UP" sounds continuously and the PULL UP switch-lights flash red until the warning envelope is exited. Upon exiting warning envelope, the red PULL UP switch-lights are replaced by the amber GND PROX switch-lights and the audio message reverts to a

continuous "TERRAIN, TERRAIN". The amber GND PROX switch lights will remain illuminated while the aircraft is within the envelope.

The upper boundary of the Mode 2A alert envelope varies as a function of aircraft speed. As airspeed increases from 220 KIAS to 310 KIAS, the boundary expands to provide increased alert times at higher speeds.

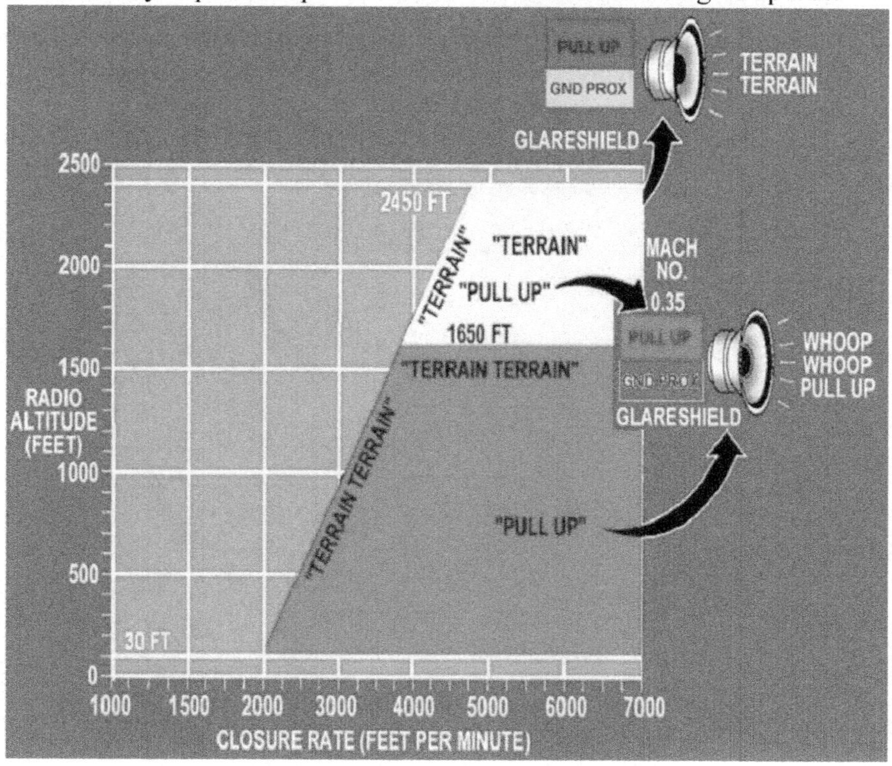

Figure 16 - 24: Mode 2A

Mode 2B: Excessive Closure Rate

Mode 2B provides a desensitized alert envelope to permit normal landing or approach maneuvering without producing unwanted alerts when close to the terrain. Mode 2B is enabled when:

- Flaps are selected for landing
- Aircraft is performing an ILS approach and is within 2 dots of both the localizer and glide-slope center lines
- During the first 60 seconds after takeoff

If the aircraft penetrates the Mode 2B envelope with either the gear or flaps not in the landing configuration, the audio message "TERRAIN. TERRAIN" sounds and the amber GND PROX switch-lights flash. If the aircraft continues to penetrate the envelope, the red PULL UP switch-lights flash and the audio message "WHOOP, WHOOP, PULL UP" is repeated continuously until the warning envelope is exited.

If the aircraft penetrates the Mode 2B envelope with both gear and flaps in the landing configuration the audio message "TERRAIN, TERRAIN" is repeated until the envelope is exited.

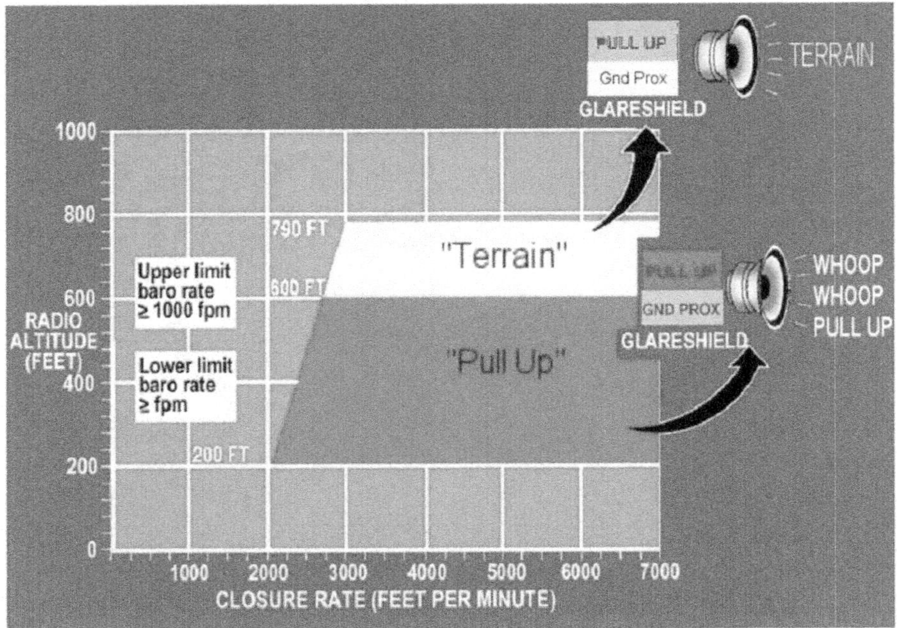

Figure 16 - 25: Mode 2B

Mode 3: Altitude Loss After Takeoff

Mode 3 provides alerts when the aircraft altitude decreases immediately after takeoff or during a missed approach procedure. The amount of altitude loss permitted before an alert is generated is a function of the height of the aircraft above the terrain.

Mode 3 is enabled after takeoff or during a go-around when the landing gear is up and the flaps are in other than the landing position. Mode 3 is disabled when the EGPWS computer determines that the aircraft has gained sufficient altitude and is no longer in the takeoff or go-around mode.

When the aircraft penetrates the Mode 3 boundary the audio message "DON'T SINK" sounds and the amber GND PROX switch-lights flash.

The flashing GND PROX switch-lights remain active until a positive rate of climb is reestablished.

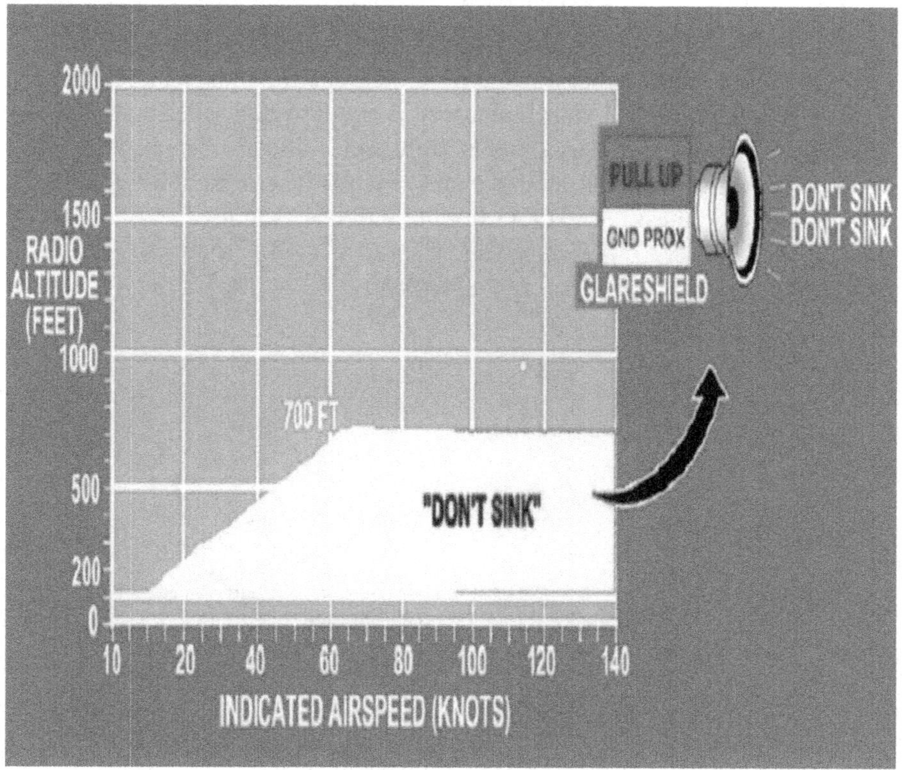

Figure 16 - 26: Mode 3

Mode 4: Unsafe Terrain Clearance

Mode 4 provides alerts and warnings for insufficient terrain clearance climb out, cruise, initial descent or approach. Alerts and warnings are based upon radio altimeter data and by comparing the aircraft's position with EGPWS computer databases. Increased aircraft speed will automatically expand the warning envelope to provide improved warning times. Mode 4 exists in three forms: 4A, 4B, and 4C.

Mode 4A: Unsafe Terrain Clearance - Gear UP

During cruise, Mode 4A prevents inadvertent flight into terrain where the terrain is not rising significantly or the aircraft is not descending at an excessive rate. During approach, Mode 4A provides alerting for protection against an unintentional gear-up landing.

The standard upper boundary for Mode 4A is 500 feet RA.

If the aircraft penetrates the boundary with the landing gear retracted, the audio message "TOO LOW GEAR" sounds and the amber GND PROX switch-lights flash. Above 190 KIAS, the upper boundary increases linearly with airspeed. When airspeed is equal to or greater than 250 KIAS the upper boundary is set at 1000 feet (radio altimeter). When the upper boundary is penetrated, the amber GND PROX switch-lights flash and the audio message "TOO LOW TERRAIN" sounds.

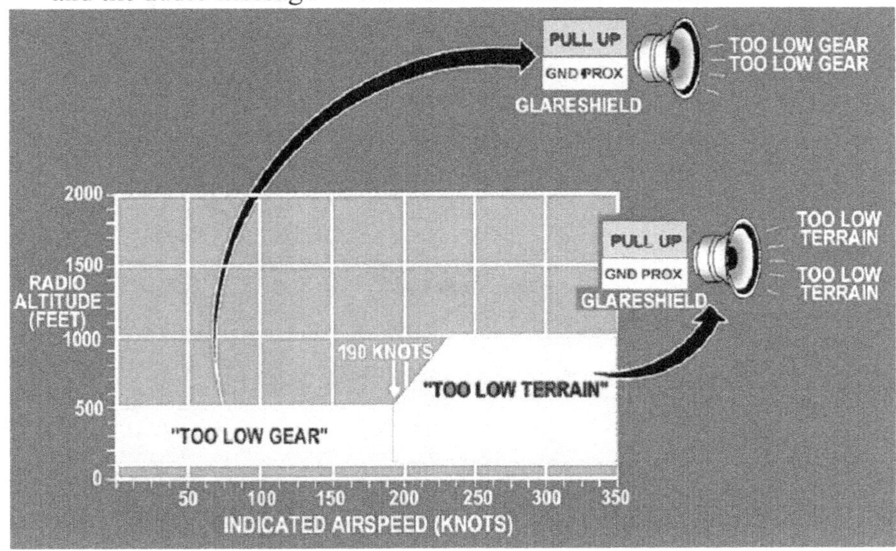

Figure 16 - 27: Mode 4A

Mode 4B: Unsafe Terrain Clearance - Gear ON/No Flap

Mode 4B provides audio alerts and visual warnings when the aircraft is on approach with the landing gear down but the flaps not positioned for landing.

When the upper boundary of 1000 feet (radio altimeter) is penetrated with the landing gear down but flaps not positioned for landing, the amber GND PROX switch-lights flash and the audio message 'TOO LOW TERRAIN" sounds.

When the aircraft speed decreases to less than 159 KIAS and the lower boundary of 245 feet (radar altimeter) is penetrated, the amber GND PROX switch-lights flash and the "TOO LOW FLAP" audio alert sounds.

The amber GND PROX switch-lights continue to flash and the audio alerts sound continuously until the flight condition has been corrected. The "TOO LOW GEAR" alert takes priority over "TOO LOW FLAPS".

Selection of the FLAP OVRD switch-light will silence the "TOO LOW FLAPS" audio alert.

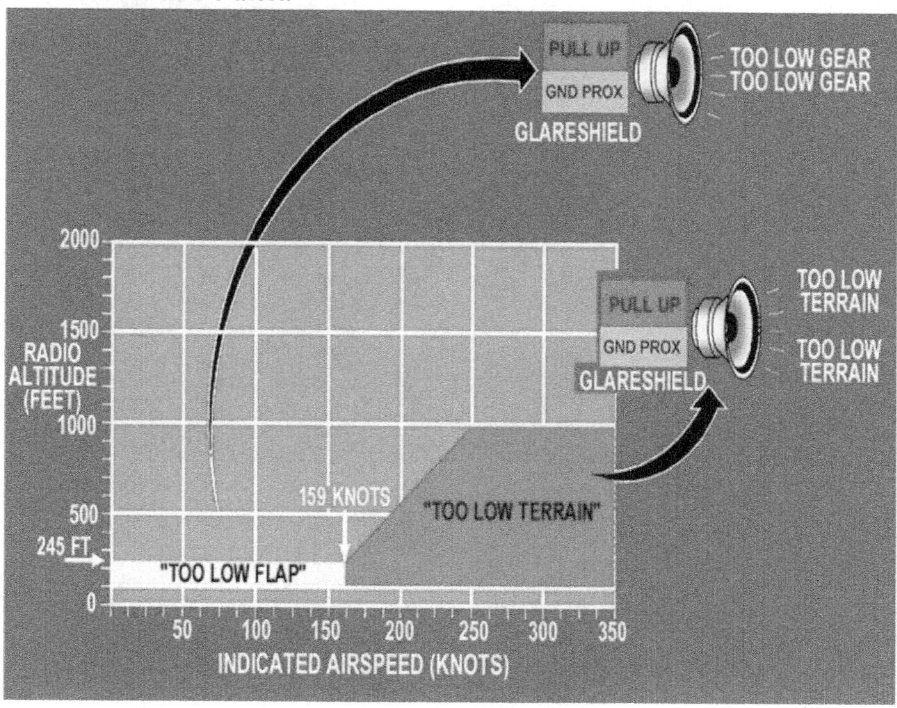

Figure 16 - 28: Mode 4B

Mode 4C: Unsafe Terrain Clearance at Takeoff

Mode 4C is based on a minimum terrain clearance or floor that increases with radio altitude during takeoff It is active after takeoff or go-around when the landing gear and flaps are not in the landing configuration.

Mode 4C alerts the pilot when the terrain is rising more steeply that the aircraft is climbing by flashing the amber GND PROX switch-lights and sounding the "TOO LOW TERRAIN" audio warning.

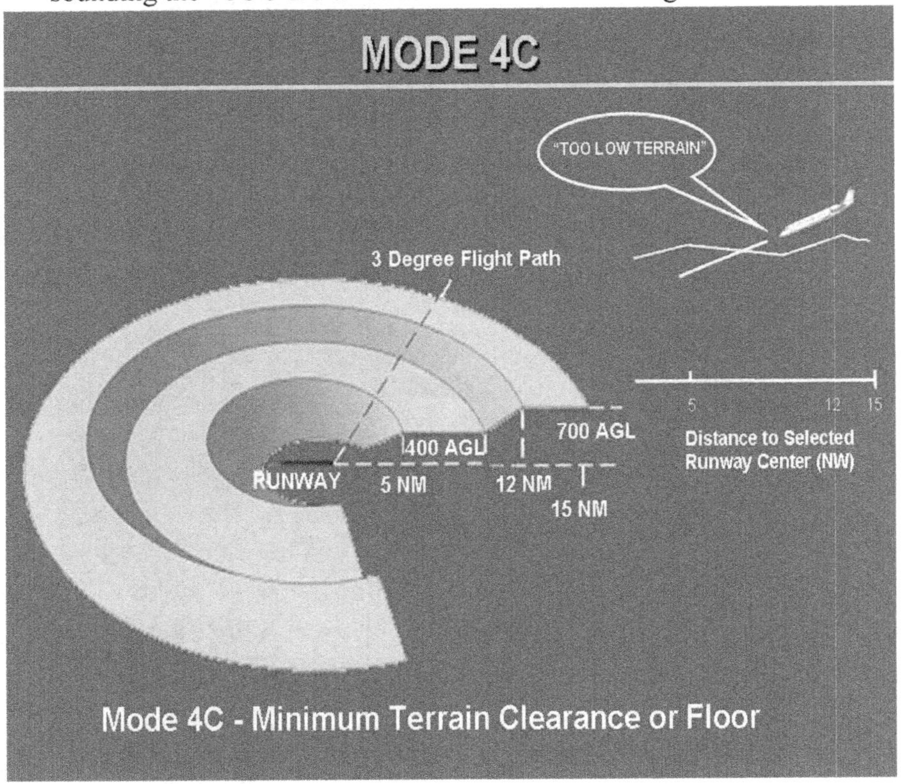

Figure 16 - 29: Mode 4C

Mode 5: Below Glide-slope Deviation Alert

Mode 5 alerts the flight crew when the aircraft's flight path descends below the ILS glide-slope. There are two Mode 5 alerts.

The first alert occurs between 1000 feet and 150 feet (radio altitude) when the aircraft is more than 1.3 dots below the ILS glide-slope. It is called a 'soft' alert because the audio volume of the "GLIDESLOPE" alert is approximately half that of the second alert. The amber GND PROX switch-lights flash continuously.

The second alert boundary occurs between 300 feet and 150 feet (radio altitude) with greater than 2 dots deviation below the glide-slope. It is called a loud or hard alert because the "GLIDESLOPE" audio alert is approximately twice as loud as the first alert. The amber GND PROX switch-lights flash continuously. With both glide-slope alerts, the flashing of the amber GND PROX switch-light his and audio repetition rate are increased as glide-slope deviation increases and radio altitude decreases. Pushing either GND PROX switch-light while the airplane is below 1000 feet (radio altitude) inhibits the Mode 5 alerts and warnings and a GS CANCEL status message is posted on EICAS. Modes 1 through 4 alerts have priority over Mode 5 alerts.

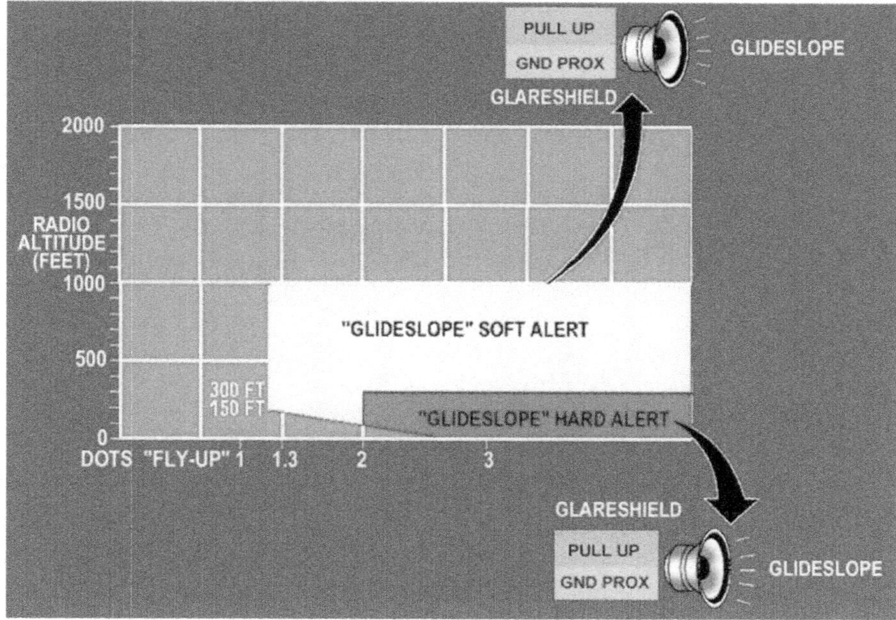

Figure 16 - 30: Mode 5

Mode 6: Descent Below Minimums/Excessive Bank Angle

Mode 6 supplies audio alerts and call-outs for descent below predefined altitudes. decision height (DH). minimums. approaching decision height and approaching minimums. If the aircraft descends below the DH, the audio message "MINIMUMS" sounds.

Altitude call-outs on the approach ("APPROACHING MINIMUMS', "500", "100", "50" etc.) are programmable. The audio call-outs are defined by the customer airline.

An optional excessive bank angle alert for this mode is available. An audio warning "BANK ANGLE" sounds when excessive bank angle is experienced The aural alert "BANK ANGLE" is given twice and is then suppressed. If the bank angle is increased by an additional 20% the aural alert will be reactivated.

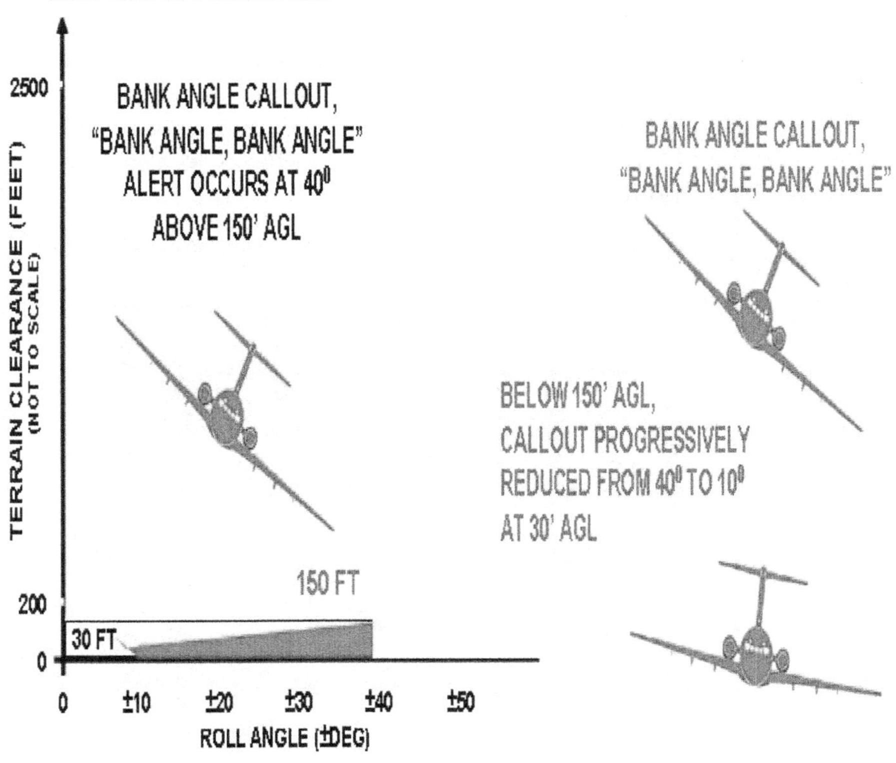

Figure 16 - 31: GPWS Mode 6

Mode 7: Windshear Detection and Recovery System

Mode 7 monitors for windshear conditions during the approach from 1500 to 10 feet radio altitude and takeoff from 10 to 1500 feet radio altitude.

Windshear Alert or Caution (increasing performance)

When the EGPWS computer detects increasing headwinds or updraft conditions that typically exist on the leading edge of a micro-burst, the following will occur:

- An amber WINDSHEAR message along with the Alpha Margin Indicator (AMI) pitch limit indicator will be displayed on both PFDs
- No escape guidance is provided
- If TOGA is selected the autopilot will disengage and the flight director command bars will be displayed for GA mode

Windshear Warning (decreasing performance)

When the EGPWS computer detects tailwind and downdraft conditions which exceed predetermined threshold limits, the following will occur:

- Audio warning "WINDSHEAR, WINOSHEAR, WINOSHEAR" sounds
- A red WINDSHEAR message will appear on both PFDs
- Alpha margin indicator (AMI) will appear on both PFDs
- Flight director escape guidance is provided automatically
- Autopilot disengages 2 seconds after windshear detected, during those 2 seconds, the autopilot will follow the windshear escape guidance

NOTE:
Windshear warnings take priority over all other audio alerts and warnings except a stall warning

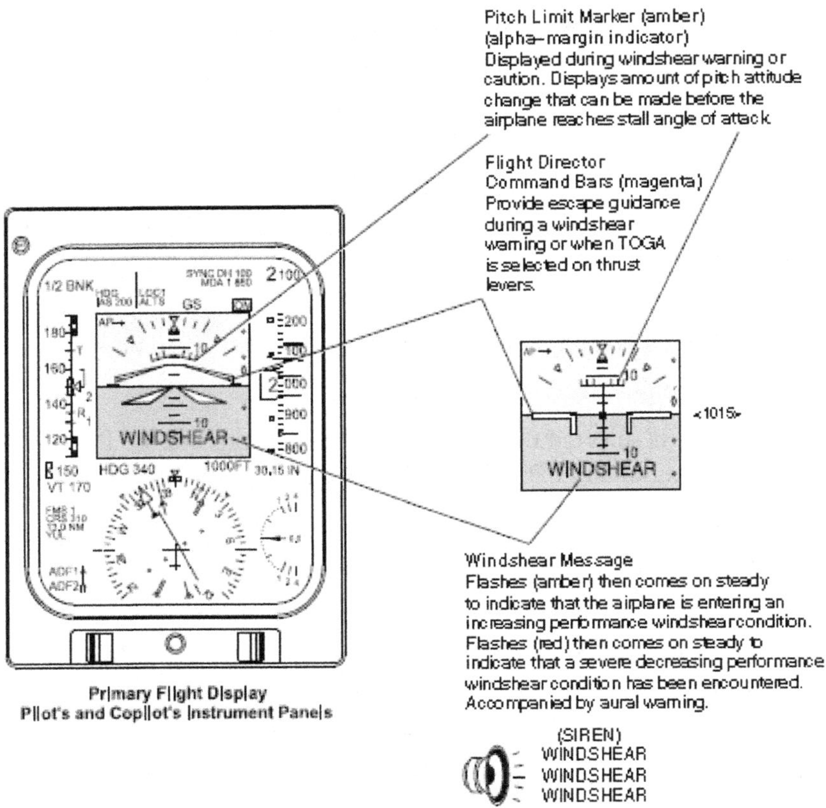

Figure 16 - 32: Windshear PFD Indications

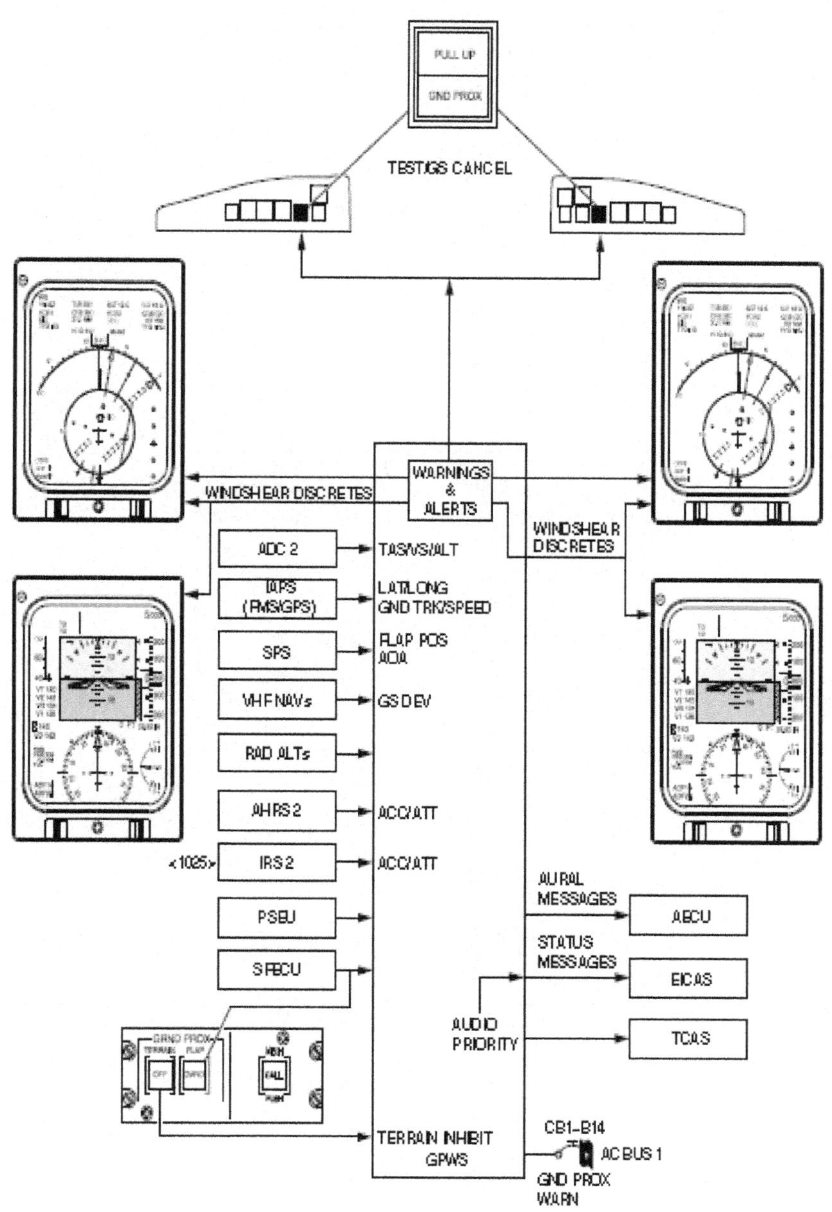

Figure 16 - 33: GPWS System Interface

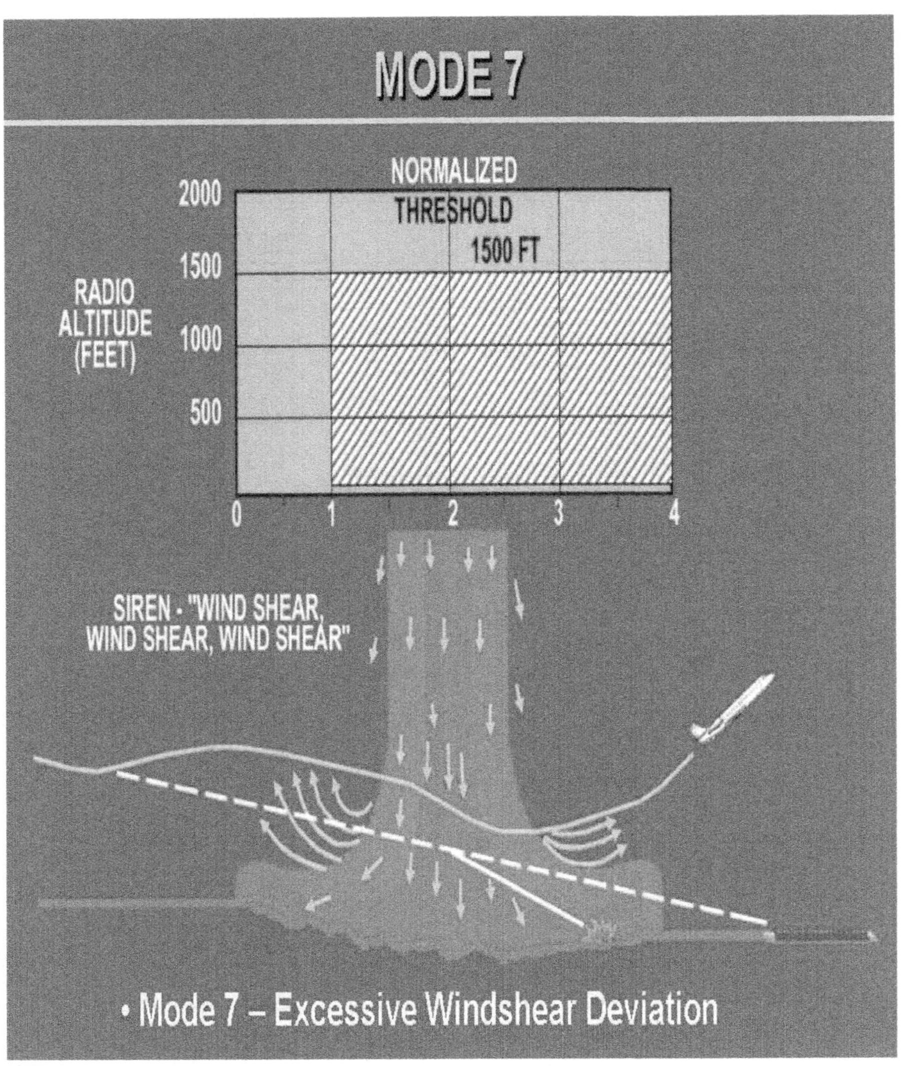

Figure 16 - 34: GPWS Mode 7

Traffic Alert and Collision Avoidance System

The traffic alert and collision avoidance system (TCAS) is an airborne system that interrogates the air traffic control transponders of nearby airplane to identify and display potential and predicted collision threats.

The TCAS surveillance range is 1-4 nautical miles (minimum) to 30 nautical miles (typical). The TCAS can track and monitor up 150 Mode S targets and can display up to 30 airplanes simultaneously. The system computes range, bearing and closure rates of other Mode S transponder equipped airplanes.

A mode S transponder provides air-to-air communications for coordinating the resolution maneuvers between TCAS equipped airplanes. The TCAS system provides no indication of traffic conflicts if the intruder airplane is without an operative Mode S transponder.

TCAS provides symbology that depicts surrounding airplane in terms of relative altitude, range, clock position, and vertical rate. The flight deck displays also provide data on closure rates. The system displays four types of traffic.

The display control panels are used to activate TCAS and to set range display. Weather radar data can be overlaid on the MFD, in TCAS mode. TCAS mode and altitude format are displayed on the top level page of the radio tuning units. Testing and setting changes are made on the TCAS main page.

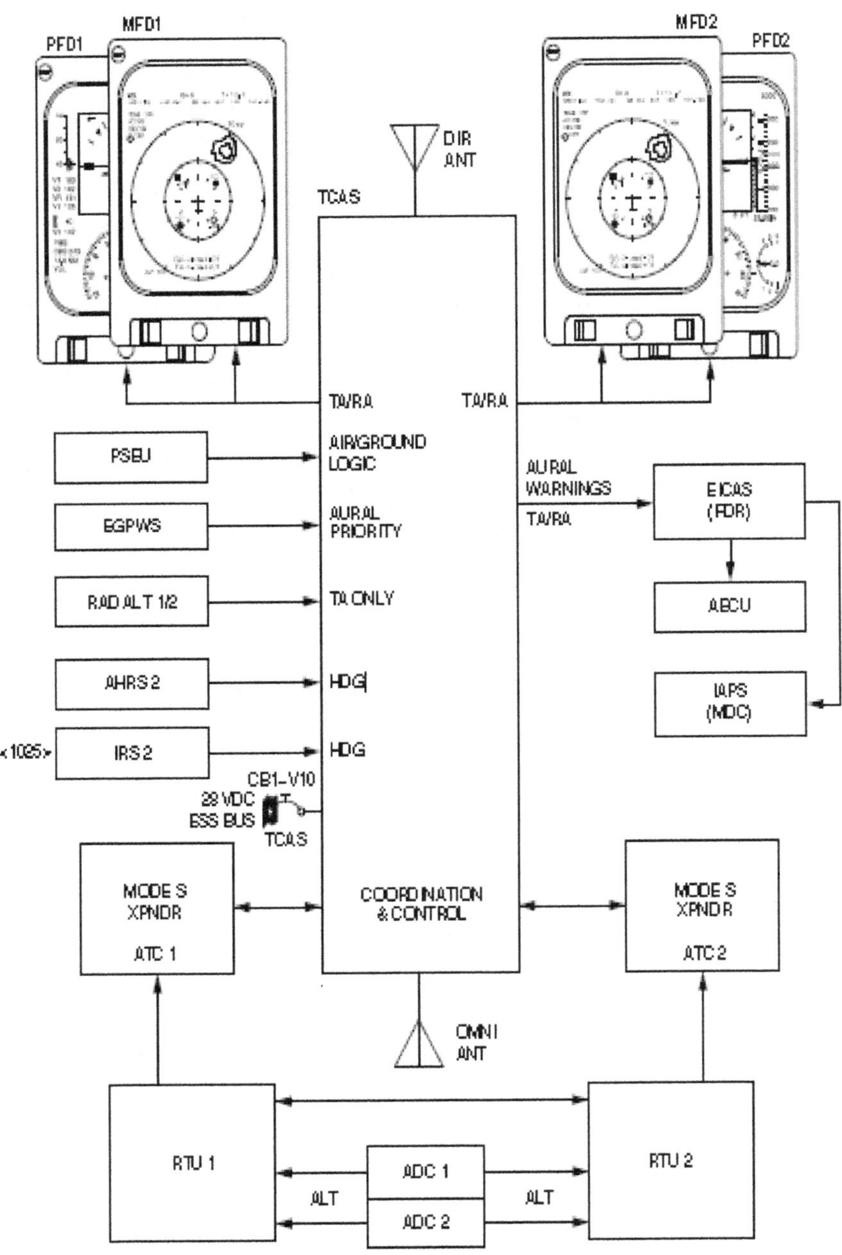

Figure 16 - 35: TCAS Interface

Components and Operation

The TCAS includes the following components:

- Transceiver - The transceiver is located in the avionics bay contains the logic for receiving, transmitting, signal processing for interfacing with the Mode S transponder, radio altimeters, RTUs, PFDs, MFDs, and the EICAS.
- Directional antenna - The directional antenna has four elements with one located at each of the cardinal 90 degree points. When the TCAS transmits interrogations, the received replies are then processed to determine azimuth and place the target on the display.
- Omni-directional antenna - The omni-directional antenna responds equally to signals from any direction and is used to send the interrogation signals in Mode C and Mode S.

Four different TCAS symbols are used on the TCAS page to represent TCAS targets. Indicators that represent the intruder's vertical speed and relative altitude data accompany each TCAS symbol.

Arrows are used to indicate the vertical speed of the target aircraft. The arrow appears only when the target aircraft is climbing or descending at more than 500 ft/min. An up arrow indicates climbing and a down arrow indicates a descending intruder. Plus or minus signs along with a number symbolize the relative altitude of the threat airplane. Plus (+) is above and minus (-) is below.

The TCAS transceiver transmits interrogations and receives replies from other airplane transponders through both antennas. The signals are processed to produce information on range, azimuth and altitude of airplane in the vicinity. With the information, the TCAS computer determines the trajectory of all airplanes in the area and then compares it to its own trajectory to determine if any conflict exists.

Other Traffic

Other traffic is defined as any non-threat aircraft traffic within the TCAS range limit. With the TCAS page selected on the MFD, 40 nm is the maximum detection range.

Selecting the TCAS page on the MFD can monitor other traffic There are no aural alerts associated with other traffic.

Proximate Traffic

Proximate traffic is any aircraft traffic within surveillance range and ± 1200 feet vertical separation but is not deemed a threat.

Selecting the TCAS page on the MFD can monitor proximate traffic. There are no aural alerts associated with proximate traffic.

Traffic and Resolution Advisories

The TCAS generates two levels of warning advisories (traffic and resolution).

Traffic Advisory

The traffic advisory (TA) is issued to indicate the relative positions of intruding airplanes that are about 35 seconds from the closest point of approach. The traffic advisory allows the flight crew an opportunity to visually locate the intruding aircraft. The advisory is always displayed on the PFDs or can be displayed on the TCAS page of the MFD if selected from the display control panel. Traffic advisory will be displayed automatically when the airplane is 1000 ± 100 feet or below, and will revert to preselected mode automatically when the airplane is above 1000 ± 100 feet.

Resolution Advisory

Resolution advisories (RA) will direct the flight crew to resolve a threat by executing a maneuver that will increase separation. This is done through the active Mode S transponder on the airplane using the Mode S data link function. This occurs when the TCAS computer predicts that the intruding airplane is within about 30 seconds from the closest point of approach.

Resolution advisories are displayed on the vertical speed indicator (VSI) portion of the PFD. The VSI shows the appropriate vertical maneuver to avoid the threat. The VSI provides vertical guidance to maintain safe vertical separation as follows:

- Corrective RAs - Fly from the red zone to the green zone
- Preventive RAs - Do not fly into the red zone

The vertical maneuver is also accompanied by TCAS voice warnings.

NOTE:

The TCAS resolution advisory programs are based on the pilot initiating the RA maneuver within approximately 5 seconds. If an additional corrective resolution advisory is issued (e.g. a reversal), the maneuver must be initiated within 2.5 seconds.

TCAS Modes

The TCAS operation includes the following Modes:

- TACS/Mode S - The ground installation receives the airplane IDENT code, range, bearing, altitude, and maneuver intent from the TCAS equipped airplanes. The ground station can also control the sensitivity of the Mode S transponders and adjust their reply probabilities.
- Mode C - In this configuration, the TCAS interrogates the other airplane in Mode C to derive range, bearing and altitude information. The TCAS will provide the resolution advisory information.
- Mode S - When the intruder airplane has a Mode S transponder installed, the TCAS can receive the airplane ID, range, bearing and altitude. In this case, avoidance maneuver coordination will be performed between the two airplanes.

TCAS Aurals

The TCAS system gives an aural RA over the aircraft flight deck audio system or an independent speaker system in addition to the RA that shows on the PFD. If the TCAS senses that the other aircraft is not responding in the planned manner, it will issue opposite corrective commands. For example, if the TCAS issues a descent but the other aircraft also descends, the TCAS will issue a corrective climb command.

The following table lists the TCAS aurals and gives a brief description of each

Advisory	Voice Message	Proper Pilot Response
Clear	*"Clear of Conflict"*	Resume normal flight, apparent conflict of airspace has been resolved.
Traffic	*"Traffic, Traffic"*	Gain visual contact with traffic. Check TCAS display for traffic bearing and range if necessary. Assess the threat and prepare to execute the evasive maneuver TCAS issues.
Preventative RA	*"Monitor Vertical Speed"*	Be alert for approaching traffic. Make sure that the VSI needle does not enter the red area of the TCAS VSI display.
Preventative RA	*"Maintain Vertical Speed Maintain"*	Maintain current vertical speed and direction. Make sure that the VSI needle does not enter the red area of the TCAS VSI display.
Preventative RA	*"Maintain Vertical Speed Crossing Maintain"*	A flight path crossing is predicted, but is being monitored by TCAS. Maintain current vertical speed and direction. Make sure that the VSI needle does not enter the red area of the VSI display.
Preventative RA	*"Adjust Vertical Speed Adjust"*	Indicates a weakening of the RA. Pilot should initiate a return to the assigned altitude.

Figure 16 - 36: TCAS Advisories (1 of 3)

Advisory	Voice Message	Proper Pilot Response
Corrective RA	*"Climb, Climb"*	Change vertical speed to 1,500 FPM climb or as indicated on the TCAS VSI display.
Corrective RA	*"Climb, Crossing Climb Climb Crossing Climb"*	Same as CLIMB except that it indicates that the flight paths will cross at some altitude.
Corrective RA	*"Descend, Descend"*	Change vertical speed to 1,500 FPM or as indicated on the TCAS VSI display.
Corrective RA	*"Descend, Crossing Descend, Descend, Crossing Descend"*	Same as DESCEND except that it indicates that the flight paths will cross at some altitude.
Corrective RA	*"Adjust Vertical Speed Adjust"*	Reduce climb vertical speed to that shown on the TCAS VSI display.
Corrective RA	*"Adjust Vertical Speed Adjust"*	Reduce descent vertical speed to that shown on the TCAS VSI display.

Figure 16 - 37: TCAS Advisories (2 of 3)

Advisory	Voice Message	Proper Pilot Response
Corrective RA	*"Increase Climb Increase Climb"*	Follows a CLIMB advisory. Increase the climb vertical speed as shown on the TCAS VSI display (typically 2,500 FPM).
Corrective RA	*"Increase Descent, Increase Descent"*	Follows a DESCEND advisory. Increase the climb vertical speed as shown on the TCAS VSI display (typically 2,500 FPM).
Corrective RA	*"Climb, Climb Now Climb, Climb Now"*	Follows a DESCEND advisory to indicate a change in vertical direction is needed. Initiate an immediate climb from the current descent to provide adequate separation.
Corrective RA	*"Descend, Descend Now, Descend, Descend Now"*	Follows a CLIMB advisory to indicate a change in vertical direction is needed. Initiate an immediate climb from the current descent to provide adequate separation.

Figure 16 - 38: TCAS Advisories (3 of 3)

PFD TCAS Indications

Resolution advisories are presented on the PFD vertical airspeed indicator. The red and green arcs are overlaid on the ring of the analogue gauge to provide vertical instruction as avoidance guidance. The pilot must maintain the vertical speed of the aircraft in the green arc of the VSI. The word TRAFFIC appears on top of the VSI scale in the following colors.

- A red TRAFFIC appears for a resolution advisory (RA)
- An amber TRAFFIC appears for a traffic advisory (TA)
- A white TCAS TEST appears when the system is tested
- A white TA ONLY appears when RA is inhibited
- A white TCAS OFF when the system is OFF

The failure messages appear on the VSI and MFD. They are:

- TCAS RA FAIL
- TCAS FAIL

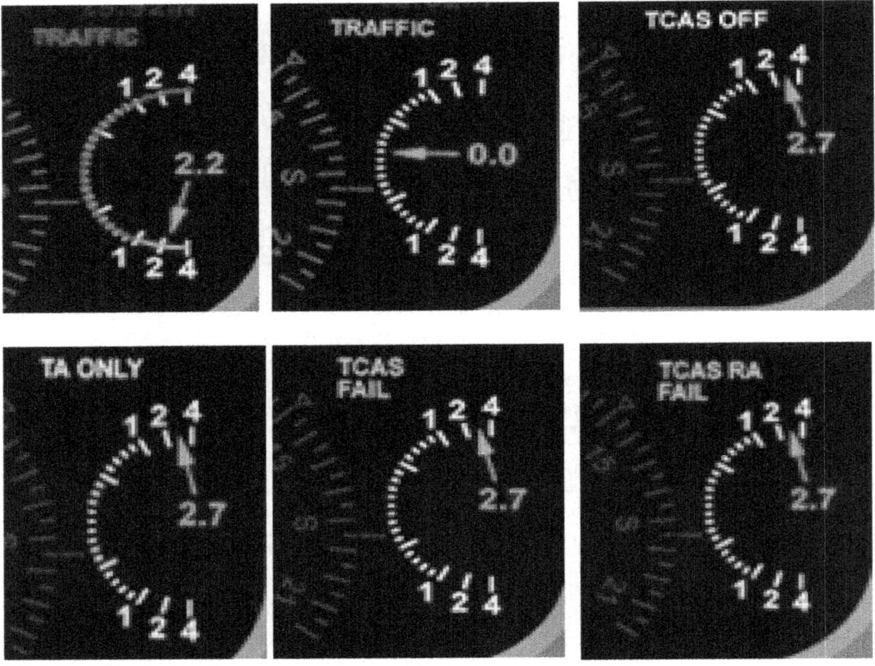

Figure 16 - 39: TCAS Indications

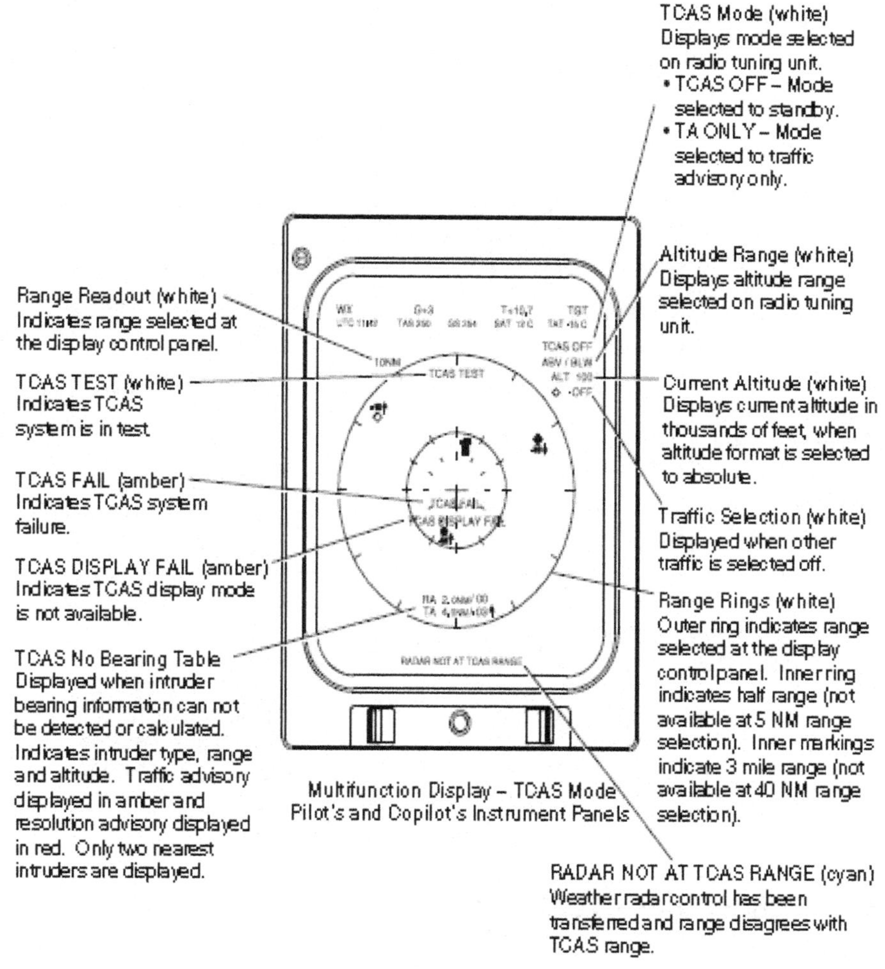

NOTES

1. Weather radar can be displayed on the MFD when in TCAS mode (range: 5,10, 20 and 40 nm).
2. TCAS can be overlaid on any map display mode.
3. During an electrical transient, TCAS display range may default to 10 nm.

Figure 16 - 40: TCAS MFD Indications

Display Control Panel

The display control panel is used to select the following TCAS functions:

- TFC pushbutton selects the TCAS page for display on the MFD
- RANGE selector knob is used to select the range on the TCAS page.

The display selections are 5, 10, 20 or 40 nm

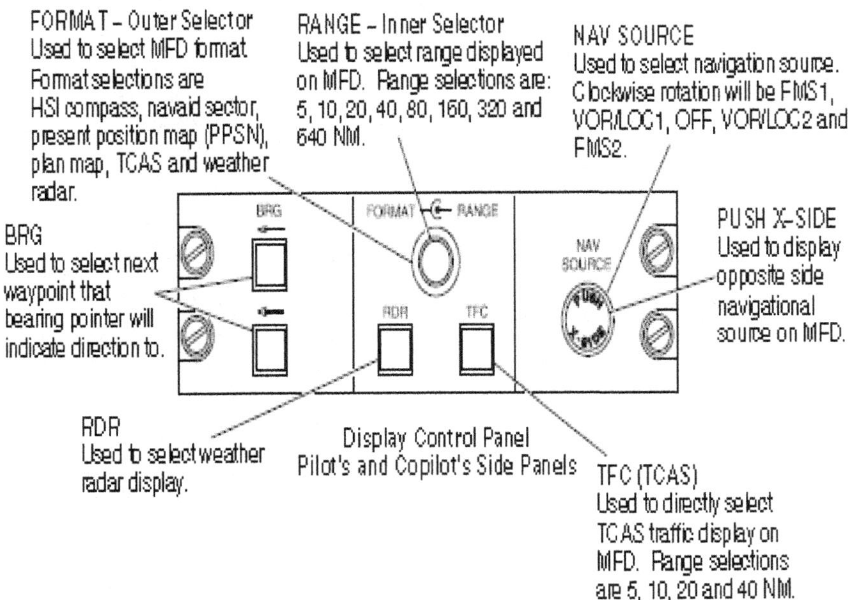

Figure 16 - 41: Display Control Panel

Radio Tuning Unit - TCAS Controls

RTU Top Page

The top page provides access to the main page settings of the TCAS system. When the main page default settings are changed, the new settings are presented in cyan on the top page.

RTU TCAS Main Page

MODE LINE KEY

The mode key is scrollable from TA/RA, STBY and TA ONLY:

- TA/RA mode, both RAs and TAs are generated (default mode)
- STBY mode, all TCAS interrogations are inhibited
- TA ONLY mode, traffic advisories only are generated

ALTITUDE FORMAT LINE KEY

- REL presents the altitude separation between the aircraft and the intruder aircraft (default mode)
- ABS presents the corrected barometric altitude for the intruder's altitude

TEST LINE KEY

When pushed the TEST line key starts the TCAS self-test. The following occurs:

- TCAS TEST is shown above the VSI

VSI scale displays the RA

- When TCAS is selected on the MFD the four colored target symbols appear

At the completion of the test, an audio message is announced "TCAS SYSTEM TEST OK" or "TCAS SYSTEM TEST FAIL".

OTHER TRAFFIC LINE KEY

When selected to OFF, OTHER TRAFFIC symbols will not be presented on the TCAS page of the MFD. (ON is default).

ALTITUDE RANGE KEYS

The three altitude range keys ABOVE, NORM and BELOW select the height and ceiling of the aircraft's surveillance airspace:

- ABOVE sets altitude range of 9900 feet above and 2700 feet below
- NORM sets altitude range of 2700 feet above and below (default)
- BELOW sets altitude range of 2700 feet above and 9900 feet below
- ABOVE/BELOW keys when selected together set the altitude range to 9900 feet above and below

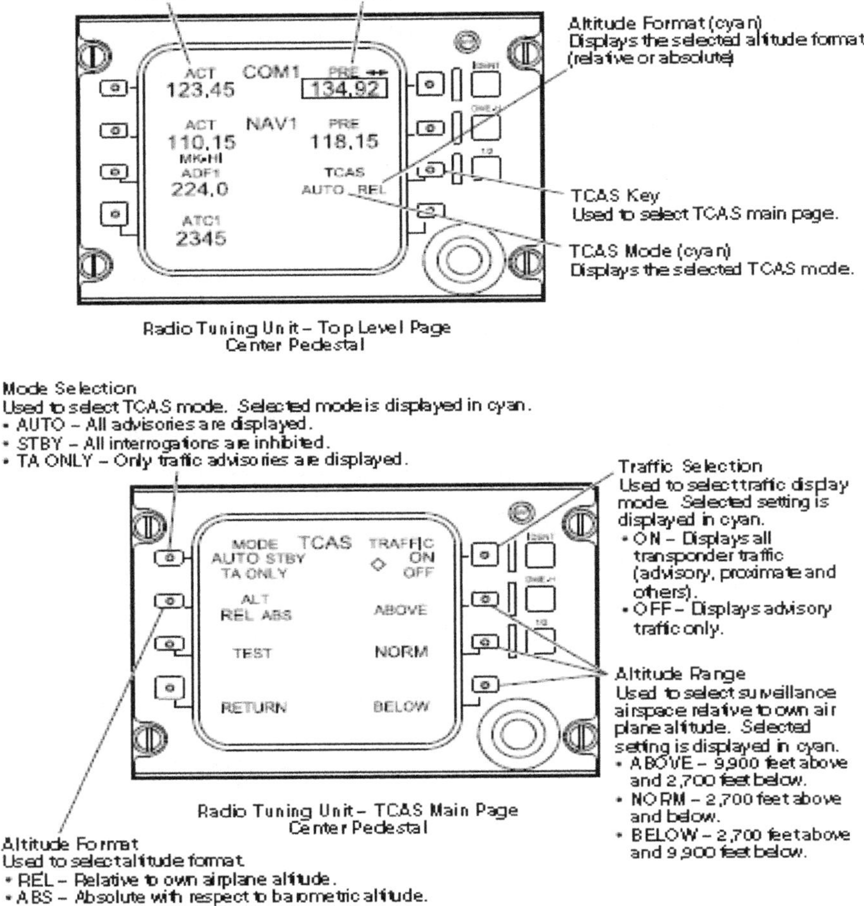

Figure 16 - 42: TCAS RTU - AC 15001 - 15276

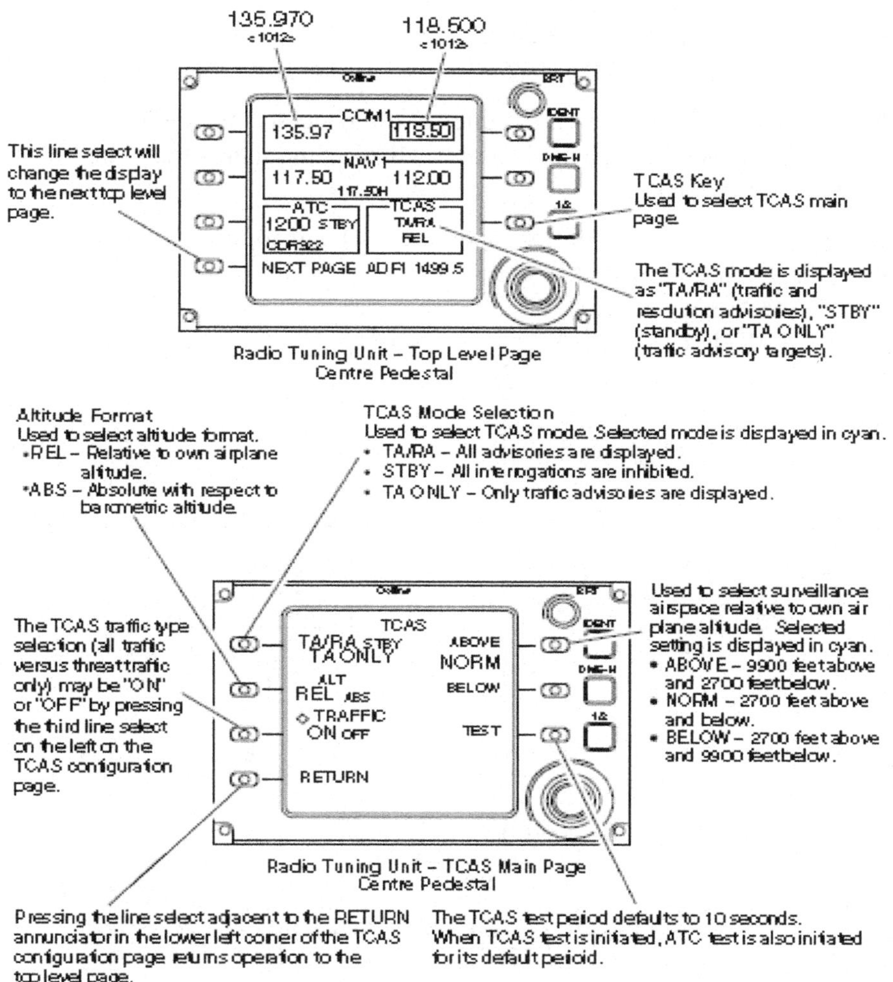

Figure 16 - 43: TCAS RTU - AC 15277 and Subsequent

Weather Radar

The Collins WXR-840 is a solid-state, low-power, X-band weather radar system. The radar is used for weather detection and ground mapping.

The standard radar installation has one control panel located on the center pedestal. Radar can be presented on both MFDs at the same time if the MFD ranges are set the same.

With the optional second control panel, the pilot and copilot can operate the radar at different ranges and antenna tilt elevations.

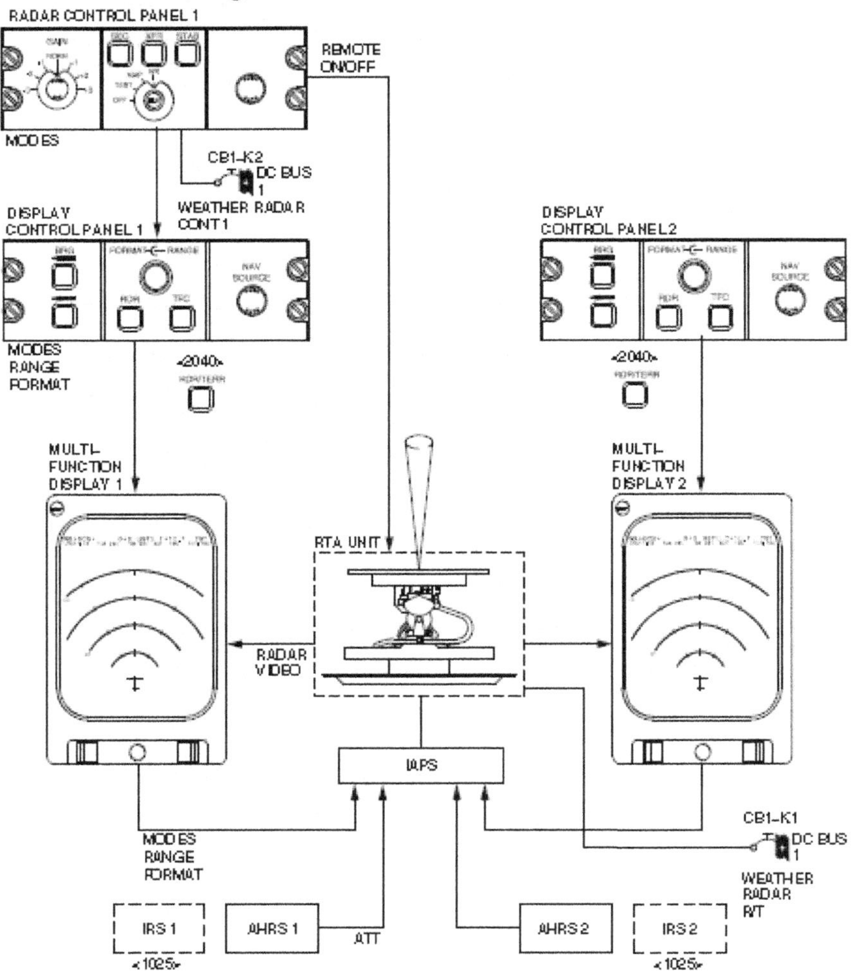

Figure 16 - 44: Weather Radar System Interface

Components and Operation

Weather Radar Control Panel

- Mode select knob
- Receiver gain and ground clutter suppression knob
- Sector scan switch
- Transfer switch
- Stabilization switch
- Tilt control knob

Mode Select Switch

The rotary mode select switch is used to select the following modes of operation.

The OFF mode inhibits the transmitter and parks the antenna facing forward. The cyan RADAR OFF message is displayed on the radar mode line on the MFD.

When TEST mode is selected, the radar system completes a self-test of the internal components and the antenna operation. The transmitter does not transmit. The MFD radar test pattern consists of six equally-spaced colored bands.

In MAP mode, the system provides a radar picture of the terrain forward of the aircraft. In MAP mode, path attenuation correction alerts (PAC) and ground clutter suppression (GCS) are disabled.

In the WX mode, the radar system operates as a weather radar. GAIN Control. The gain control knob performs two functions, gain control and ground clutter suppression.

The seven position GAIN rotary knob is used to manually adjust the color presentation of the radar picture. NORM is a calibrated setting that presents the best color presentation for most conditions. The pilot can change the gain setting to modify the radar presentation. Each setting from NORM changes the sensitivity one color level. To ensure accuracy of presentation, the GAIN knob should always be returned to NORM.

Selecting the button in the center of the GAIN knob when operating in the WX mode activates ground clutter suppression (GCS). GCS decreases the intensity of the ground returns on the radar display. By de-cluttering the presentation, precipitation returns can be assessed more accurately.

When selected, the GCS is activated for 12 seconds only. The cyan message +GCS is presented on the MFD radar mode line when the GCS is activated.

TILT and AUTOTILT Control

The TILT control knob allows the pilot to select the tilt elevation of the antenna or, by pressing the button in the center of the knob, the auto-tilt feature.

The antenna stabilization in normal operations is oriented with the horizon.

The rotary TILT knob is used to adjust the antenna elevation ± 15 degrees.

The antenna tilt angle is displayed in cyan on the right side of the MFD radar mode line. The letter 'T' is displayed before the tilt angle.

The auto-tilt button is used to automatically adjust the antenna elevation for different altitudes and ranges.

When auto-tilt is on and a different range is selected or the aircraft changes altitude, the system automatically tilts the antenna to maintain the same relationship between the beam deflection and the display.

A cyan 'A' message is displayed on the MFD radar mode line.

Antenna Stabilization

In normal operations, antenna elevation is automatically adjusted to maintain the orientation of the antenna to the horizon. The antenna stabilization signal comes from the AHRS or IRS. The STAB pushbutton engages or disables the automatic stabilization function.

If the antenna becomes unstabilized, an amber USTB message is displayed on the radar line. When the STAB button is selected, the antenna is oriented with the longitudinal axis of the aircraft and the amber message is replaced with a cyan USTB. The pilot can still manually control the tilt for the best radar presentation.

Sector Scan

There are two sector scan selections, 60 degrees each side of the nose or 30 degrees.

In the standard installation, the radar image is refreshed in both directions of the sweep.

In the dual installation, the radar presentation is refreshed on alternate sweeps. This allows the pilot and copilot to operate the radar at different ranges and tilt settings.

Transfer Control

In the standard single-radar installation, the XFR button determines which pilot has control of the radar range selection. The pilot with the white MFD range markers has control of the radar set

Amber range markers are provided at the non-controlling pilot's MFD. The non-controlling pilot must select the same range as the other pilot to have a radar presentation. If any other range is selected, the message RADAR NOT AT THIS RANGE is displayed.

For dual installations, should a control panel fail, the XFR switch is used to transfer all radar control to the other weather radar control panel. When in XFR, the radar line on the MFD become amber and an 'X' appears in front of the mode line.

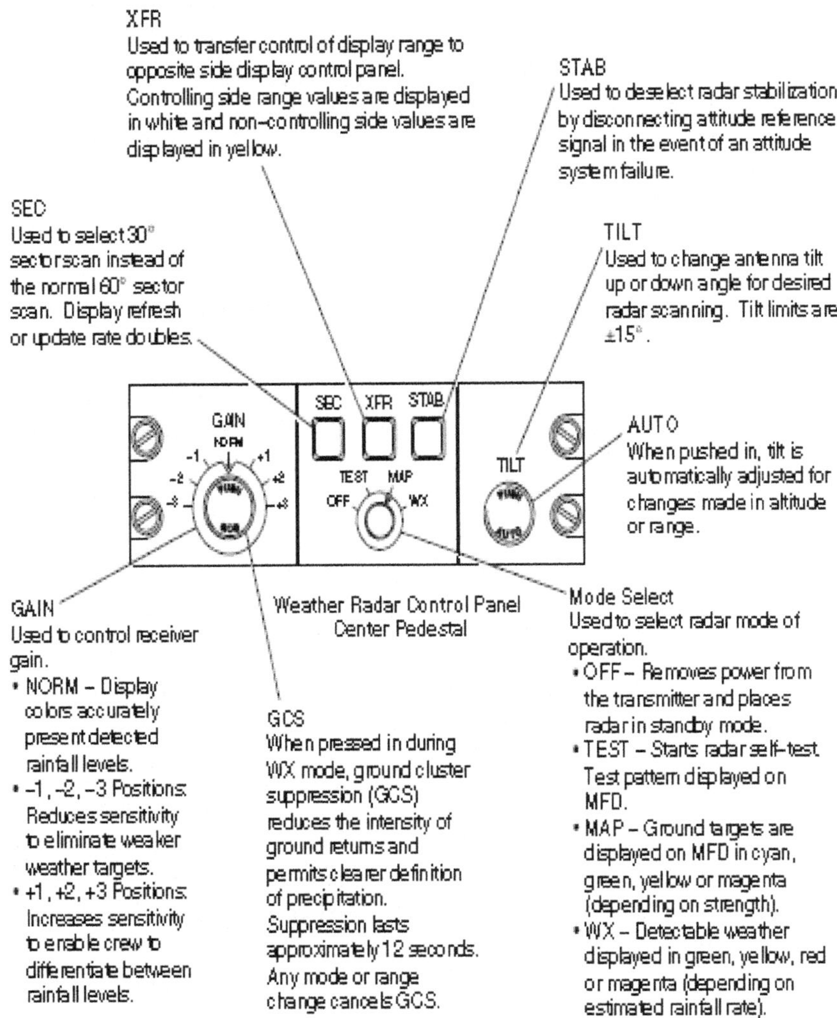

Figure 16 - 45: Weather Radar Control Panel

Display Control Panel

Radar MFD presentation is controlled at the DCP. The DCP FORMAT knob allows the pilot to select the radar page and the RDR/TERR select pushbutton allows the radar to be overlaid on other MFD formats. The DCP RANGE knob is used to select the range on the MFD.

RDR/TERR Select Pushbutton

The RDR select switch allows radar to be overlaid on NAV SECTOR and FMS MAP. A second selection of the switch removes the radar presentation.

FORMAT Knob

The DCP FORMAT knob can be used to select the MFD radar format.

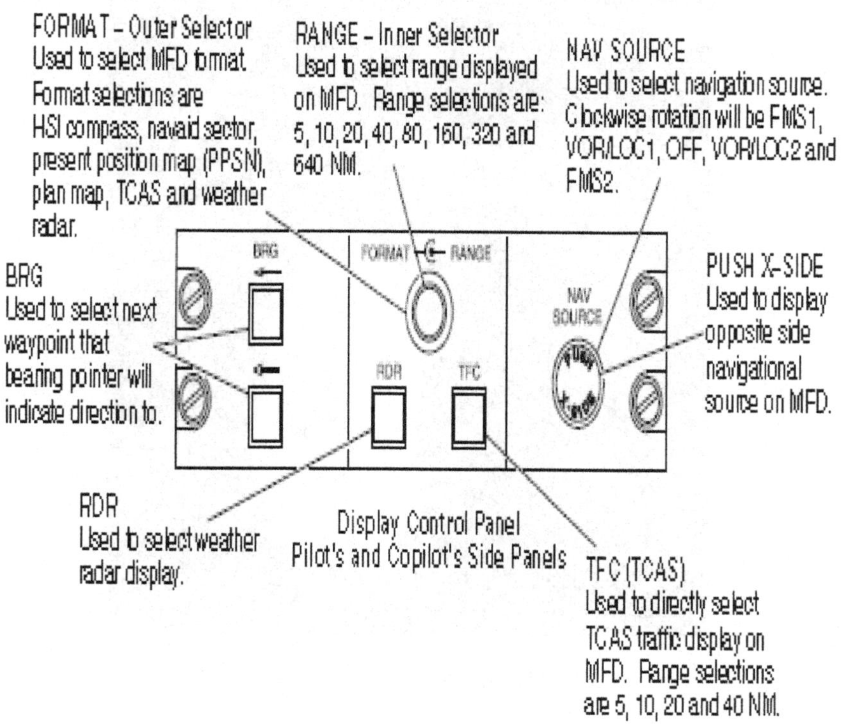

Figure 16 - 46: Display Control Panel

Path Attenuation Correction and Alert

The PAC feature compensates for the radar energy that is absorbed as the beam penetrates a precipitation cell. When the radar detects weather formations that could cause energy to be absorbed and possibly give a false return, the radar sensitivity is corrected.

If the precipitation cell is of sufficient dimension to use the full range of attenuation correction, a condition known as PAC alert occurs. The yellow PAC alert arc informs the pilot that the radar beam is being severely attenuated, and that the area of shadow behind the intervening rainfall may contain hidden and possibly significant precipitation. Avoid flight into the area between the displayed weather and the yellow PAC alert.

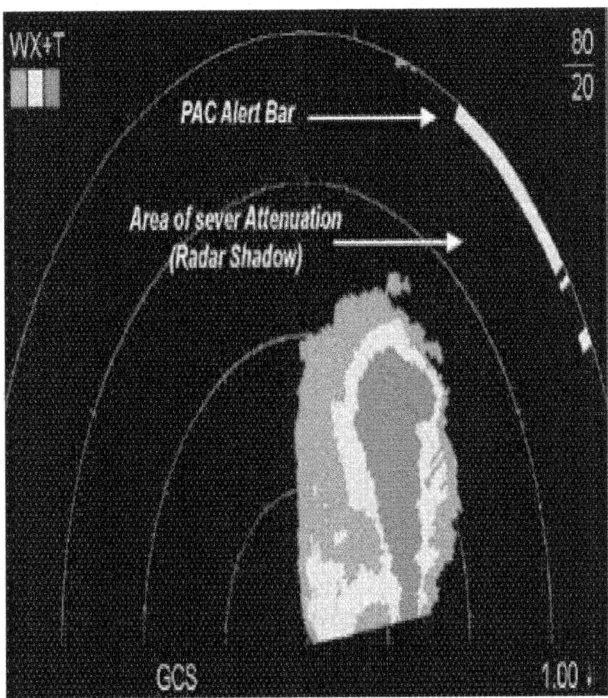

Figure 16 - 47: Pac Alert

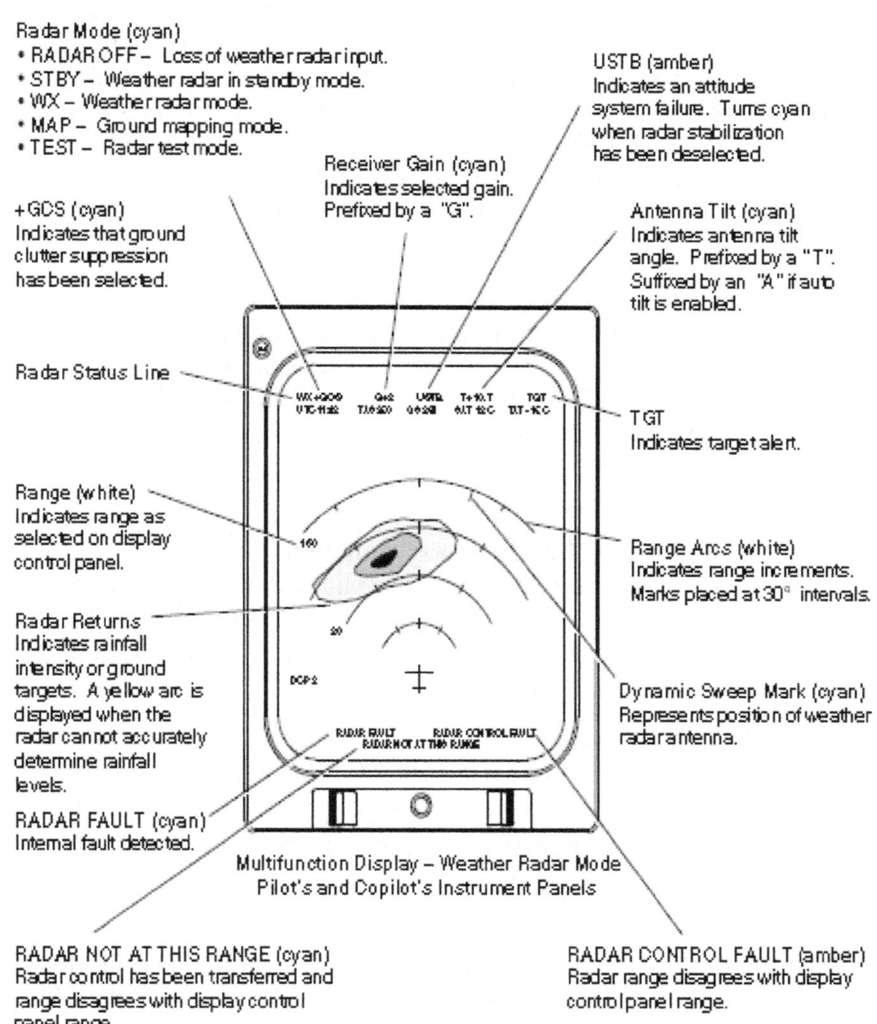

Figure 16 - 48: Weather Radar MFD Indications

Global Positioning System

The global positioning system (GPS) is a 12-channel receiver that provides the flight management system with GPS-based on-route and terminal navigation and GPS-based approach capability. With data from more than four satellites received, the GPS sensor can improve the fault tolerance and accuracy of the navigation solution by using receiver autonomous integrity monitoring (RAIM) and fault detection and exclusion (FDE) to detect satellite failures.

A minimum of four satellites with acceptable geometry or three satellites plus calibrated barometric altitude are needed to calculate a navigation solution. The GPS receiver can also provide predictive RAIM which can be used to find whether the satellite geometry at the destination airport will be sufficient enough to support an approach at the planned time of arrival. The computed position. velocity and time are input to the FMS which integrates the date in the flight plan based navigation solution

In the basic configuration, the GPS receiver, or global navigation satellite system (GNSS) sensor, processes the transmissions from multiple GPS satellites to calculate navigation solutions based on data from all satellites in view. A GNSS receiver uses the GPS satellites and ground station signals to calculate position, speed and time.

An optional GNSS receiver can detect and process corrective information from the space based augmentation system (SBAS) for improved accuracy. The advance GPS based system means a GNSS receiver operating in SBAS coverage has stand alone, satellite only, approach and navigation capability.

There are several GPS systems in orbit with only one SSAS coverage approved for aerospace navigation. The FMS provides selectable access to GNSS which includes the navigation satellites and ground systems to support specific phases of flight. All GPS/GNSS date is accessed through the FMS CDU.

The basic GNSS sensor can provide GPS non precision approach capabilities. The optional GNSS SSAS sensor provides non-precision approach and precision approach capabilities.

Components and Operation

U.S. GPS

The GPS has three main segments:

- Space segment
- Control segment (ground)
- User segment

GPS Space Segment

The space segment of GPS is made up of 24 NAVSTAR satellites. Twenty-one are operational at all times and three are spares. The satellites:

- Are in a 10,900 mile-high orbit
- Are in six orbital planes with three to four satellites in each plane
- Have a 12-hour orbit
- Are spaced to provide a minimum of four satellites in view at all times
- Have atomic clocks and transmit on two coded frequencies

GPS Control Segment

The control segment monitors the space segment and adjusts the orbits when operationally required. It includes five ground stations around the world. One is the master control station. the others are monitor stations.

The monitor stations passively track all satellites in view and acquire range data from them. This data is forwarded to the master control station. The master control station uses the data from the monitor stations to estimate satellite orbit and clock data. This information is transmitted to the satellites and down to the users as a navigation message.

Satellite Based Augmentation System

The satellite based augmentation system (SBAS) improves the GNSS position accuracy, integrity and availability with additional corrections that augment the GNSS. SBAS coverage in North America is a combination of wide area augmentation system (WAAS) satellites and a network of ground stations providing WAAS capability.

SBAS error correction compensates for ionospheric, time and satellite orbit errors. The enhanced signal accuracy allows the GNSS to used as an independent navigation system with GNSS approaches.

GPS User Segment and FMS Interface

The FMS uses the signal from the positioning satellites as a navigation sensor to upgrade its own position.

A minimum of three satellites is required for two-dimensional calculations and four satellites are required for three-dimensional calculations.

The GPS sensor can use all available navigation sensors (GPS, DME/ DME, VOR/DME and IRS if installed) and weighs the information prior to use in aircraft position calculations. VOR sensor function is automatically disabled at FMS startup. GNSS (GPS}, because of its accuracy, is normally the primary sensor.

Aircraft equipment consists of single or optional dual GPS installations. All installations are capable of receiving the US global positioning system information.

The FMS CDU provides the pilot with access to GPS data and control settings. Consult your FMS handbook for specific details.

Flight Management System

The flight management system (FMS) provides point-to-point, curved path navigation and position determination using multiple sensor inputs.

The FMS provides waypoint lateral and vertical flight plan management functions such as flight plan creation, flight plan editing and retrieval of a flight plan from a data base. The FMS has following functionalities:

- Global navigation satellite system management (GNSS)
- Temperature compensation

The FMS is the interface between the GPS and the aircraft navigation systems. The GPS is a sensor used by the flight management system. All GPS data is accessed through the FMS CDU.

Components and Operation

FMS GNSS C ONTROL Page

On GNSS equipped aircraft the CONTROL page allows the pilot to make the following basic GNSS selections:

- GNSS Control - enable or disable the use of GNSS data in calculating the FMS blended solution
- STATUS - shows status of each sensor including the difference between the FMS position and the GNSS position
- NPA RAIM - used to predict if RAIM is available for non-precision approaches and to deselect satellites that are unusable by NOTAM

Satellite Deselect

Individual satellites can be deselected from use by entering the identifier for the satellite in the FMS. This function would be used if a satellite is NOTAMed under test. When deselected, the FMS ignores the data from the satellite.

Predicted RAIM

If the GPS is going to be used for navigation, the satellite receiver autonomous integrity monitoring or RAIM can be checked for constellation accuracy. RAIM is used to assure that the GPS solution (satellite computation) meets the required accuracy. The accuracy levels are:

- 4.0 nm for oceanic/remote
- 2.0 nm for en-route
- 1.0 nm for terminal
- 0.3 nm for approach

Predicted approach RAIM for the destination is displayed on the CDU along with arrival time.

The pilot can enter the ETA in the FMS and the predicted APPR RAIM is calculated. If not available, NOT AVBL is displayed on the CDU.

The destination is automatically entered based on the active flight plan. If a GPS approach is to be made at an airport other than the destination shown on the flight plan, the flight plan must be updated.

Required Navigation Performance (RNP)

RNP is a statement of the navigation performance accuracy necessary for operation within a defined airspace. The RNP value may be specified by the pilot by entering the value on the PROGRESS page. On the FMS, position accuracy is automatically monitored for each phase of flight.

The FMS computes its expected position accuracy based on the sensors in use. The expected position accuracy (POS ACCURACY) is also displayed on the PROGRESS page. When flying in RNP airspace, the FMS position error should be less than the RNP value 95% of the time. If the expected position accuracy error is more than the RNP value, a message CHK POS is displayed on the FMS CDU and navigation by other means must be utilized.

Temperature Compensation (FMS 4.2)

FMS temperature compensation (TEMP COMP) permits the operator to automatically compensate the altitude constraints and vertical path angles associated with the approach, the approach transition, and the missed approach.

TEMP COMP ON/OFF is selectable on the default and TEMP COMP page. When TEMP COMP is active, the FMS adds a copyright symbol © to indicate that temperature compensation is active and being applied to appropriate altitude constraints. This indication will appear wherever temperature compensation is occurring. For example, the bottom CDU message line, MFD or PFD display.

FMS Advisory Vertical Navigation (VNAV)

The FMS is capable of providing vertical guidance for nearly all segments of flight including FMS approaches. The advisory VNAV feature assists the pilot in meeting the active flight plan climb, descent and altitude constraints and provides detailed instruction for top of descent planning.

Advisory VNAV is valid when the following conditions are met:

- Advisory VNAV is enabled on the PERF MENU page of the CDU
- Active flight plan with a TO waypoint exists in the FMS
- Preselected altitude, barometric altitude and airspeed/Mach are all valid
- FMS is the selected NAV source on the display control panel

Advisory VNAV information is presented on the PFD, MFD, and CDU. With NAV selected on the FCP, FMS altitude constraint readout is presented as a cyan digital readout above the altitude tape. During the climb the next altitude constraint. time and distance to go and the minimum vertical speed are presented on the appropriate displays.

During descent advisory VNAV computes the vertical path to each waypoint that has an altitude constraint providing vertical guidance relative to the pilot selected path angle (VPA). Waypoint information is presented similar to the climb with the recommended vertical speed to reach the next altitude constraint displayed as a bug on the vertical speed scale.

Advisory Descent

As the aircraft approaches the TOD point, a vertical deviation scale appears on the PFD and a snowflake is presented at the top of the scale indicating the FMS computed vertical path angle (VPA). When the aircraft overflies the TOD the snowflake tracks downward on the scale.

When the snowflake is centered the pilot selects the appropriate AFCS vertical mode and reduces thrust accordingly to start the descent. During descent when the snowflake is centered the deviation scale matches the selected VPA

For additional information on FMS features, such as advisory VNAV, refer to the Collins Flight Management System Pilot's Guide.

Nav-to-Nav Capability

When the aircraft is within 31 nm of the destination airport, the displays, autopilot and localizer receiver are all automatically preset and tuned by the FMS, once selection of a localizer-based approach is made. The FMS tunes the Localizer frequency, selects it at the preset navigation source and sets the inbound preset second course.

When the aircraft is in a position to capture the localizer and upon selection of the flight control system approach mode, The flight control computer automatically changes the NAV source from FMS to the localizer. Vertical navigation guidance is provided up to the point of glide-slope capture for an ILS approach to the final approach fix (FAF) for a localizer-only or localizer back course approach.

Other FMS Selections and Indications

The FMS provides updated CDU information on the following DEPARTURE/ARRIVAL page selections:

- NDB/DME selection changed to N/D
- VOR/DME selection changed to V/D
- Addition of ARR DATA line select key
- Entering NOB waypoint designator has been updated, e.g. UFXNB is now entered as UFX

FMS POS Error Indication

The FMS PFD navigation block indication has been updated. The GPS navigation error white MSG has been replaced with a loss of integrity (LOI) indication. LOI is displayed in white or amber, depending on the phase of flight.

Alternate Approach-LNAV

The flight director vertical speed mode can be used with minimums corrected to the charted LNAV approach.

During the final approach segment, if the required navigation accuracy to complete the approach is lost, a yellow NO APPR message is displayed on the PFD.

Loss of Integrity Alerting

A yellow loss of integrity (LOI) message is issued during en-route operation should the GNSS HPL value exceed the current HAL. This is annunciated by an amber message on the PFD. MFD and CDU message line.

The GNSS loss of position accuracy will remove the current white message (EN ROUTE or LPV) and display one of the following PFD amber status annunciations:

- LOI: Loss of integrity en-route or oceanic
- LOI TERM: Loss of integrity in terminal mode
- NO APPR: Displayed during final approach

Inertial Reference System (IRS)

The Inertial Reference System (IRS) provides inertial outputs of attitude, magnetic heading, angular rates, linear acceleration and present position information. Some of the information is used for display on the EFIS, and to provide navigation information to the flight management system. The IRS is aligned on the ground to establish a starting reference point. After alignment, the system is ready to navigate by tracking and updating the present position, attitude and heading through motion sensing provided by the inertial instruments of the system.

The IRS is a dual system with two Inertial Reference Units (IRU) and a dual Mode Select Unit (MSU). Each IRU receives information from the same side air data system. The IRS utilizes inertial principals to sense, compute and supply the information that follows:

- Attitude (roll, pitch, yaw)
- Angular rates (roll, pitch, yaw)
- Linear accelerations (longitudinal, lateral, normal)
- True heading
- Magnetic heading
- Present position
- Track (true and magnetic)

The IRU measures inertial motion sensed by the inertial instruments and computes attitude and heading data. This information is processed and sent to the integrated avionics processor system which interfaces with the flight control computers and flight management computers. These signals are also routed to the TCAS, EGPWS, fuel system, stall protection system, flight data recorder and data concentrator units. The MSU provides pilot selection of the IRS modes.

The IRS provides attitude and heading information to the electronic flight instruments. Attitude is displayed on the Attitude Direction Indicator (ADI) of the primary flight displays and heading is displayed on the Horizontal Situation Indicator (HSI) portions of the displays. Heading is selected to magnetic or true using the flight management system.

The IRS normally operates in navigation mode. In navigation mode, it is not possible to update the IRS position, however, it is possible to perform a rapid realignment while on the ground.

Attitude mode is a reversionary mode, used when the IRU has detected an inertial failure or inaccuracies of the navigation operation in flight. Attitude mode does not provide position data. In attitude mode, the heading may drift and must be corrected using the Flight Management System (FMS). If the FMS is not available, the EICAS control panel can be used to make heading corrections. Attitude mode is annunciated on the EICAS status page.

NOTE:
Crew may experience minor control column oscillation during flight with the autopilot engaged, at a frequency of approximately 2-6 Hz, a displacement of approximately 0.25 - 0.50 in., and the observed N1 is in the 30-45% range, this should not be considered as unusual. It should be noted that the control column oscillations, which may vary through the 30-45% N1 range, will disappear as N1% is increased.

NOTE:
The column oscillations are directly related to airplane vibrations (in the 44-48 Hz range) while engine N1 was in the 30-45% range. Since the IRS sampling frequency is 50 Hz, the resonant frequency of approximately 44-48 Hz observed at the IRS mounting structure induces an aliasing effect in the IRS, resulting in a minor pitch command output to the Flight Control Computers (FCCs) of approximately 2-6 Hz. On most airplanes this condition is imperceptible

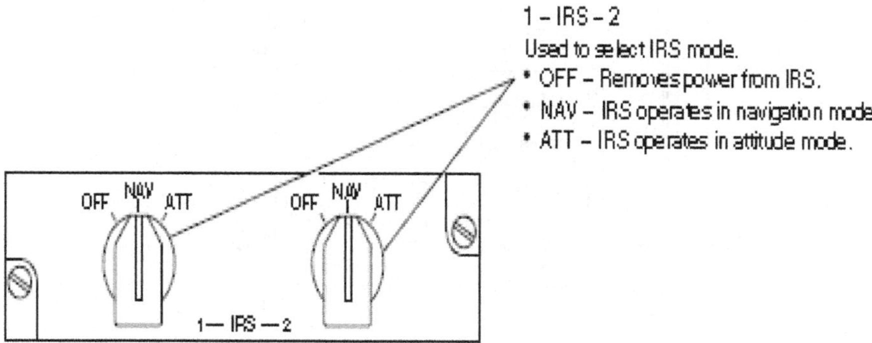

Figure 16 - 49: IRS Mode Control Panel

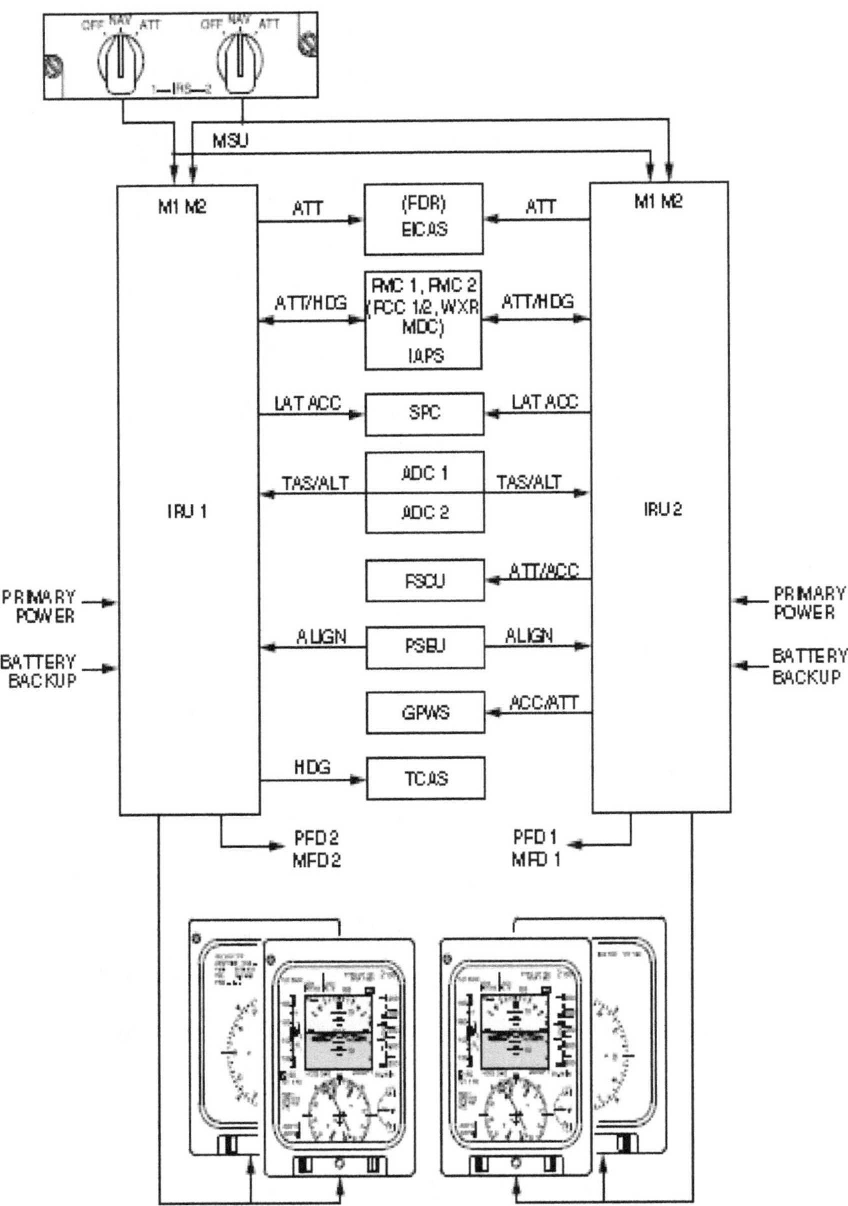

Figure 16 - 50: IRS Interface

Initialization and Alignment

IRS initialization takes about seven minutes at normal temperature. The IRS requires that the initial position be entered using the flight management system (FMS). The primary flight displays present a flashing initialization alignment message during initialization. Upon successful alignment, the IRS will automatically sequence into navigation mode. Attitude alignment takes one minute provided the airplane is stationary on the ground or in straight and level flight. When switching from navigation to attitude mode, alignment takes 34 seconds.

NOTE:
In the event of a failure of the FMS, the IRS initial position can be entered by selecting the EICAS Menu page. On the MENU page, SET POS is selected and then your present position (latitude and longitude) is entered.

EICAS Messages

Caution Message	Cause
ADS-B OUT FAIL	Indicates ADS-B Out function failure of the active ATC or invalid data of GPS and/or GPS failure.
XPDR FAIL	Indicates failure of the active transponder.
Status Message	**Cause**
ADS-B OUT 1 FAIL	Indicates failure of the ADS-B section of transponder 1 that does not affect the MODE-S operation. Displayed only when ATC 2 is selected and serviceable or, when ATC SEL is on STBY.
ADS-B OUT 2 FAIL	Indicates failure of the ADS-B section of transponder 2 that does not affect the MODE-S operation. Displayed only when ATC 1 is selected and serviceable or, when ATC SEL is on STBY.
GPWS FAIL	Failure in the basic ground proximity warning modes.
GS CANCEL	Glideslope mode 5 alerts inhibited.
IRS 1(2) DC FAIL	Comes on to indicate that IRS back-up power has failed.
IRS 1(2) IN ATT	Indicates that IRS is operating in attitude mode.
IRS 1(2) OVERTEMP	Indicates that an over temperature condition exists.
TERRAIN FAIL	Failure in the terrain map display.
TERRAIN NOT AVAIL	Terrain map display is not available due to position inaccuracy.
TERRAIN OFF	Terrain map display has been selected and the terrain functions have been inhibited.
WINDSHEAR FAIL	Failure in the windshear detection system
XPDR 1 INOP	Indicates transponder 1 failure
XPDR 2 INOP	Indicates transponder 2 failure

System Schematic

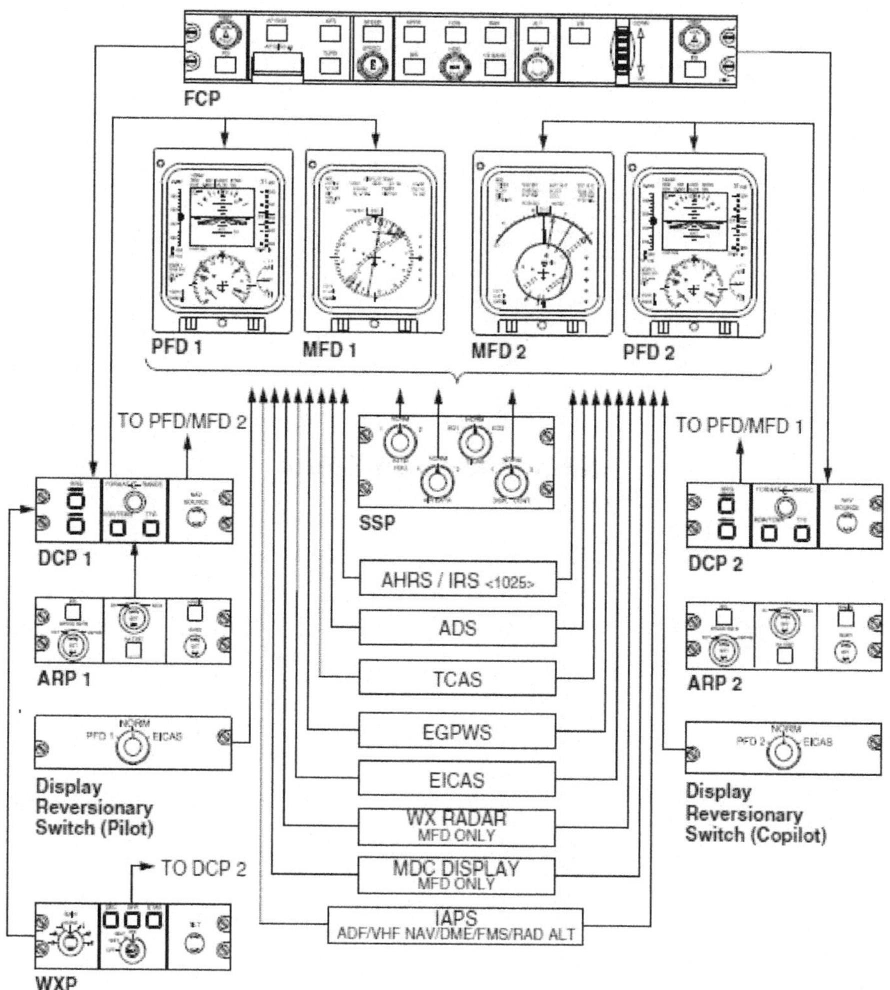

Figure 16 - 51: EFIS Interface Block Diagram

Intentionally Left Blank

Chapter 17: Oxygen and Emergency Equipment

Intentionally Left Blank

Chapter 17: Oxygen and Emergency Equipment

General

This chapter describes the systems and equipment which are essential to the safety of the passengers and crew during a fire, rapid decompression, ditching and emergency evacuation. The emergency equipment consists of the following:

- Oxygen equipment
- Fire fighting equipment
- Emergency locater transmitter (ELT)
- Over water emergency equipment
- Flight deck and cabin emergency equipment

Placards containing symbols are used to indicate the location of the emergency equipment

Oxygen Equipment

The oxygen systems supply oxygen to the flight crew and passengers in emergencies such as depressurization, decompression, smoke, fumes, first aid and during certain aircraft operations.The aircraft oxygen consists of two independent oxygen systems. One system supplies stored oxygen to the flight deck crew and the other supplies generated oxygen to the passengers and flight attendants. In addition, portable oxygen cylinders are provided in specific areas in the passenger cabin.

Crew Oxygen System

The crew oxygen system consists of an oxygen cylinder, a ground servicing panel and three crew oxygen masks with mask stowage containers.

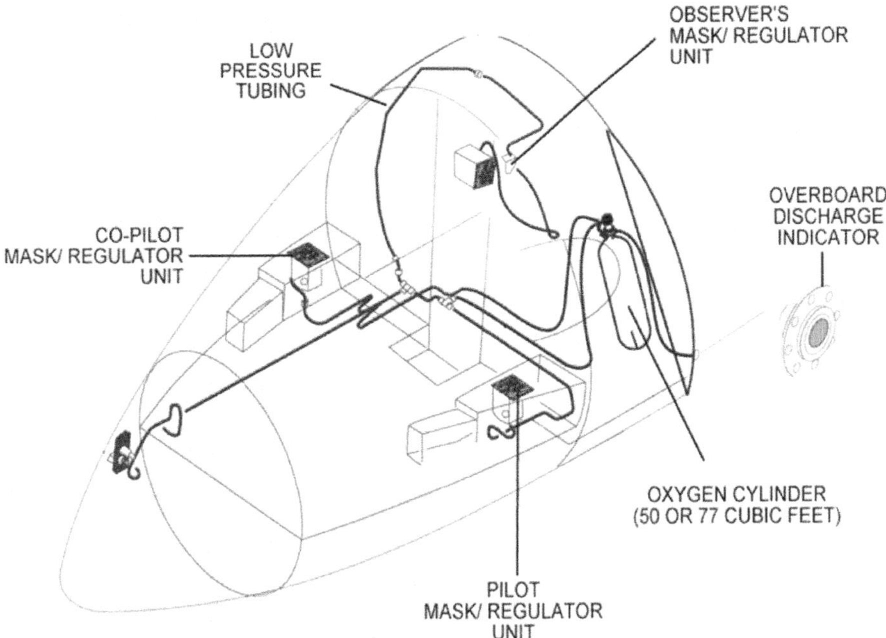

Figure 17 - 1: Crew Oxygen Overview

Crew Oxygen Cylinder

The crew oxygen cylinder contains 50 cubic feet (1419 liters) [77 cubic feet (2180 liters) optional) of oxygen and is located in an enclosure behind the entrance storage compartment. Normal cylinder charge pressure is 1850 psi at 21 °C (70°F). The enclosure is well ventilated with a permanent flow of ECS air to the under floor avionics compartment. The air is then dumped overboard through the outflow valve.

The cylinder assembly consists of a manual (lever type) shutoff valve, regulator, pressure gauge, pressure transducer, pressure switch, and a pressure relief valve.

The oxygen cylinder outlet is monitored by a pressure transducer. If the outlet pressure decreases below 1410 psi (960 psi optional), the EICAS will display an OXY LO PRESS caution message on the primary page.

Output pressure is regulated to between 60 and 85 psi. If the output exceeds 94 psi, a low pressure relief valve opens venting the oxygen. The cylinder is protected from overpressure by a frangible high pressure relief valve. If the cylinder pressure reaches 2500 to 2775 psi, the valve ruptures and the oxygen is vented overboard through the high pressure discharge indicator on the left side of the forward fuselage.

The pressure switch monitors the output pressure from the regulator. If the pressure decreases below 45 psi, an OXY LO PRESS caution message will be displayed on the EICAS primary page.

NOTE:
If the OXY LO PRESS caution message is displayed, the crew should refer to the dispatch requirement charts.

When the contents of the oxygen cylinder is vented through the high pressure discharge indicator, a green snap disc dislodges, presenting a visual indication that the oxygen cylinder contents have been vented.

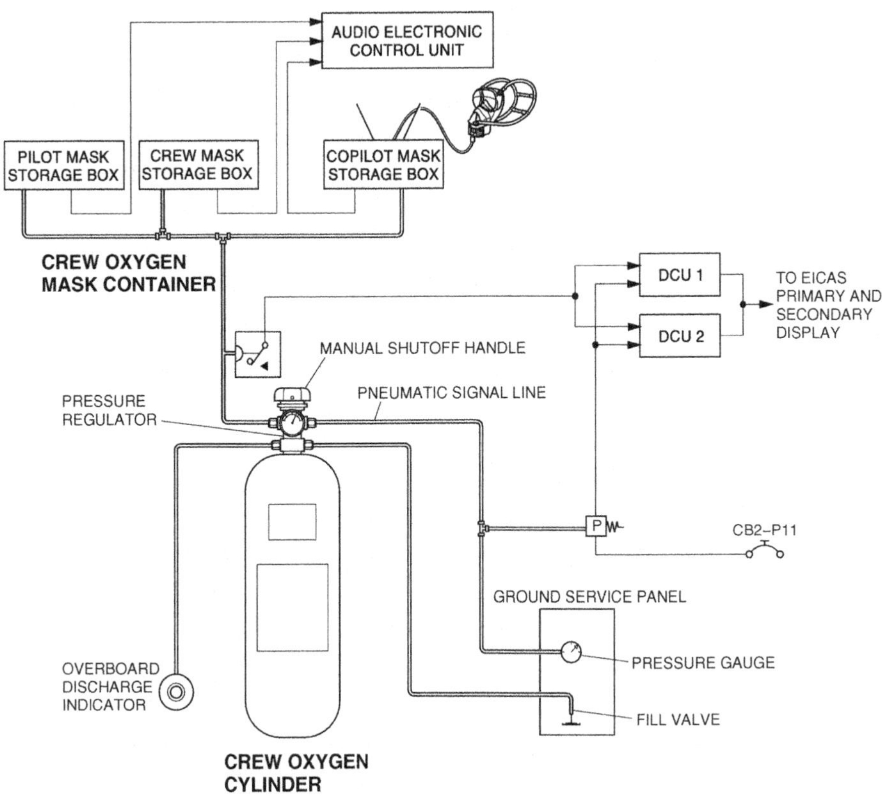

Figure 17 - 2: Crew Oxygen Cylinder

Ground Service Panel

The oxygen servicing panel is located on the right side of the forward fuselage. The service panel contains a fill port, a pressure servicing chart and a pressure gauge. Check valves in the fill and supply lines, prevent loss of oxygen when the cylinder is removed or when the cylinder replenishment source is disconnected.

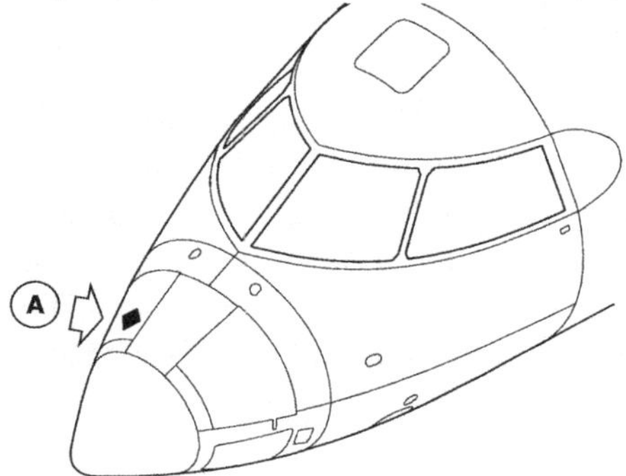

OXY. CYL. SERVICING CHARGE CYL. AT RATE NOT TO EXCEED 200 PSI/MIN TO "FULL" PRESSURE	
FULL PRESS. PSI	AMBIENT TEMP. °C
1990	38
1900	27
1805	16
1710	5
1620	–7
1530	–18
1435	–29
1340	–40

MAX. FILL PRESSURE VERSES TEMPERATURE CORRECTION CHART

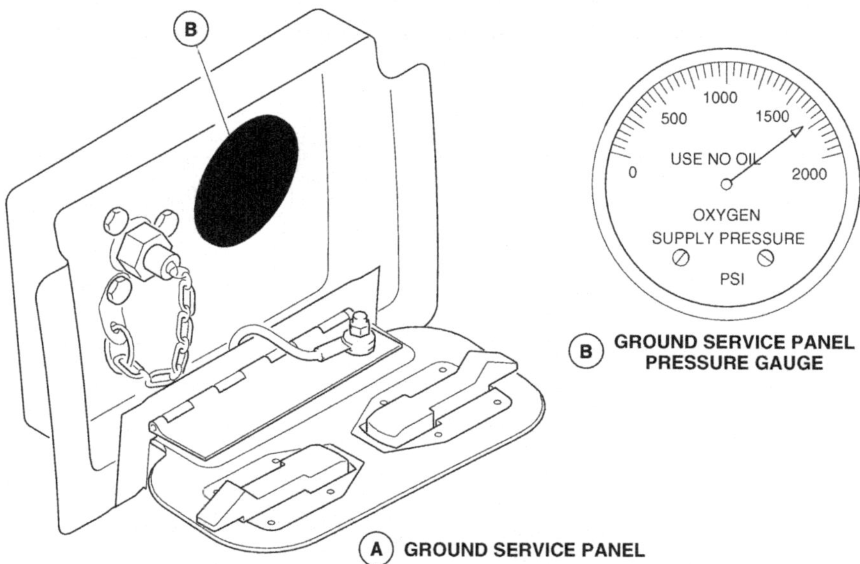

Figure 17 - 3: Ground Service Panel

Mask Stowage Containers

The pilot and copilot oxygen mask stowage containers are located in the side consoles; the third container is in the upper section of the circuit breaker panel behind the pilot.

Each container has a test lever, a yellow flow blinker, an oxygen ON flag, a set of quick-release doors and connectors for the oxygen and communication lines.

The flow blinker displays a yellow cross when oxygen is flowing. The blinker is black when there is no flow. When the mask is in use, a white ON flag comes into view on the left door of the container to indicate that the oxygen shutoff valve is open. When the use of the mask is no longer required, closing the container doors and pressing the PRESS TO TEST AND RESET lever stops oxygen flow to the mask and removes the white ON flag.

WARNING

When supplemental oxygen is in use, do not close the quick-release container doors. Inadvertent selection of the PRESS TO TEST AND RESET lever will stop the flow of oxygen to the mask.

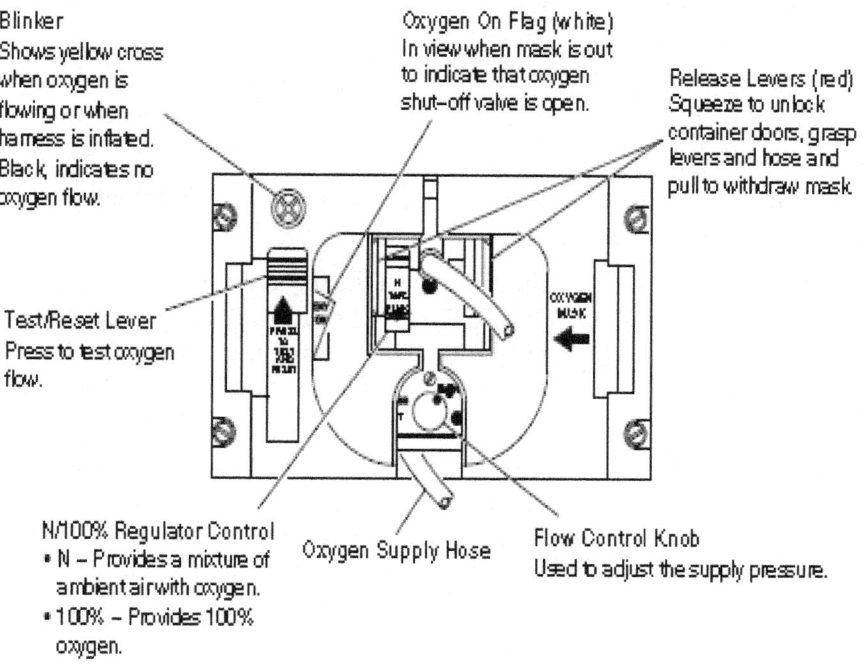

Figure 17 - 4: Mask Stowage Container

Crew Oxygen Masks

The crew oxygen masks are located in stowage containers. One for the pilot, one for the copilot and one for the observer position. The crew mask is a full face type mask. The mask includes an oxygen regulator, a pneumatically-controlled inflatable harness, a flow-control knob, a mixture control lever and a microphone. To release the mask from the stowage container. the operator squeezes the red release levers and holds them.

This action opens the quick-release doors, frees the mask and inflates the harness. The operator then dons the mask. The red levers are then released, which deflates the harness, causing the mask to install correctly on the operator's head.

Oxygen is supplied to the mask regulator at about 78 psi. The regulator control (N/100% positions) allows the user to select a mixture of oxygen and ambient air or pure (100%) oxygen.

- When the regulator control is set to the N position, a mixture of ambient air and pressurized oxygen is supplied to the mask on demand.
- With the control set to the 100% position, pure oxygen is supplied to the mask on demand.

The flow control knob is used to adjust the oxygen flow. If the knob is turned clockwise to the EMERGENCY position. the mask is supplied a constant flow of 100% oxygen at a positive pressure. To test the oxygen flow, press the flow control knob. This will momentarily supply oxygen to the mask.

When cabin altitude is more than 30,000 feet (9144 meters). the mask supplies pure oxygen regardless of the N/100% switch position. To remove the mask. the red release levers on the mask are squeezed, which inflates the harness to allow the mask to be removed from the operator's head

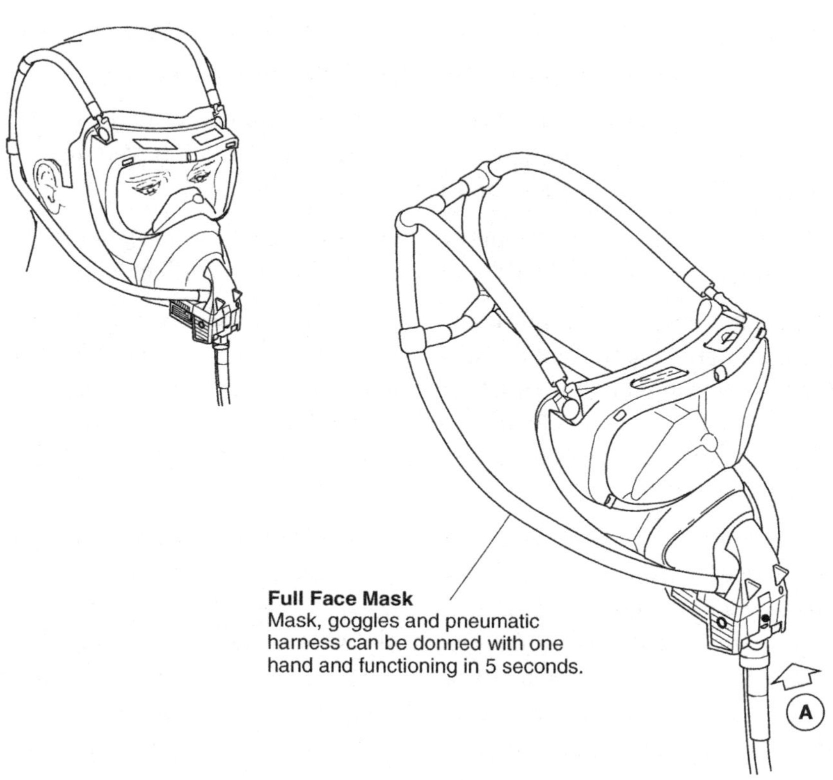

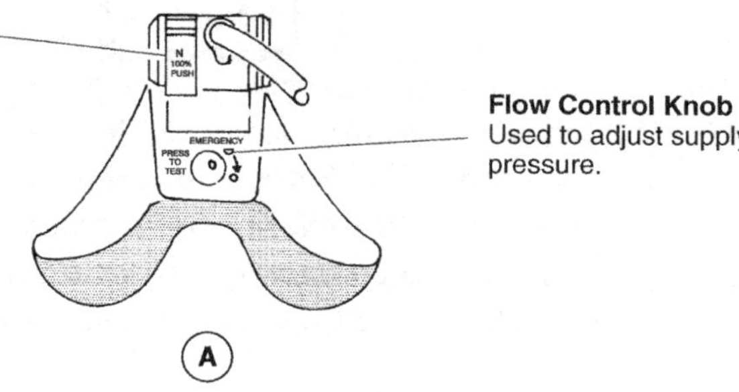

Figure 17 - 5: Full Face Mask

Passenger Oxygen System

The passenger oxygen system provides chemically generated oxygen for all cabin occupants in the event of cabin depressurization.

The oxygen generators and oxygen masks are installed as part of the passenger service unit (PSU) panel and are available at all passenger seats, in the lavatory and at the flight attendant stations.

All oxygen compartment doors will open to present the oxygen masks automatically if the cabin altitude reaches 14,000 ± 300 feet.

If the automatic system fails to open the doors, or if necessary to override the automatic system, the flight crew can operate the (guarded) PASS OXY switch-light on the overhead panel to open the oxygen doors in the PSUs. As a back-up to electrically opening the doors, each individual oxygen compartment door can be opened manually through a release hole in the door.

When the oxygen compartment doors are open, the passengers will pull the oxygen mask to their face, which pulls a lanyard connected to the firing pin of the chemical oxygen generator. This initiates the flow of oxygen to the passenger's oxygen mask. A flow indicator in the supply tube will show green when oxygen is flowing. The reservoir bags on the passenger oxygen masks begin to fill with oxygen. The chemical oxygen generator supplies approximately 13 minutes of oxygen to each mask.

WARNING

The oxygen generator surface temperature may reach 260°C (500°F) when generating oxygen. Do not touch or attempt to remove generator. Burn injury can result. If an active generator is inadvertently removed from the compartment, the generator must be placed in a metal container such as a lavatory or galley sink. The generator's heat will scorch other materials or fabrics.

NOTE:

Odor similar to scorched cloth may be created b y activation of a generator. The odor does not affect the purity of the oxygen and there is no fire hazard.

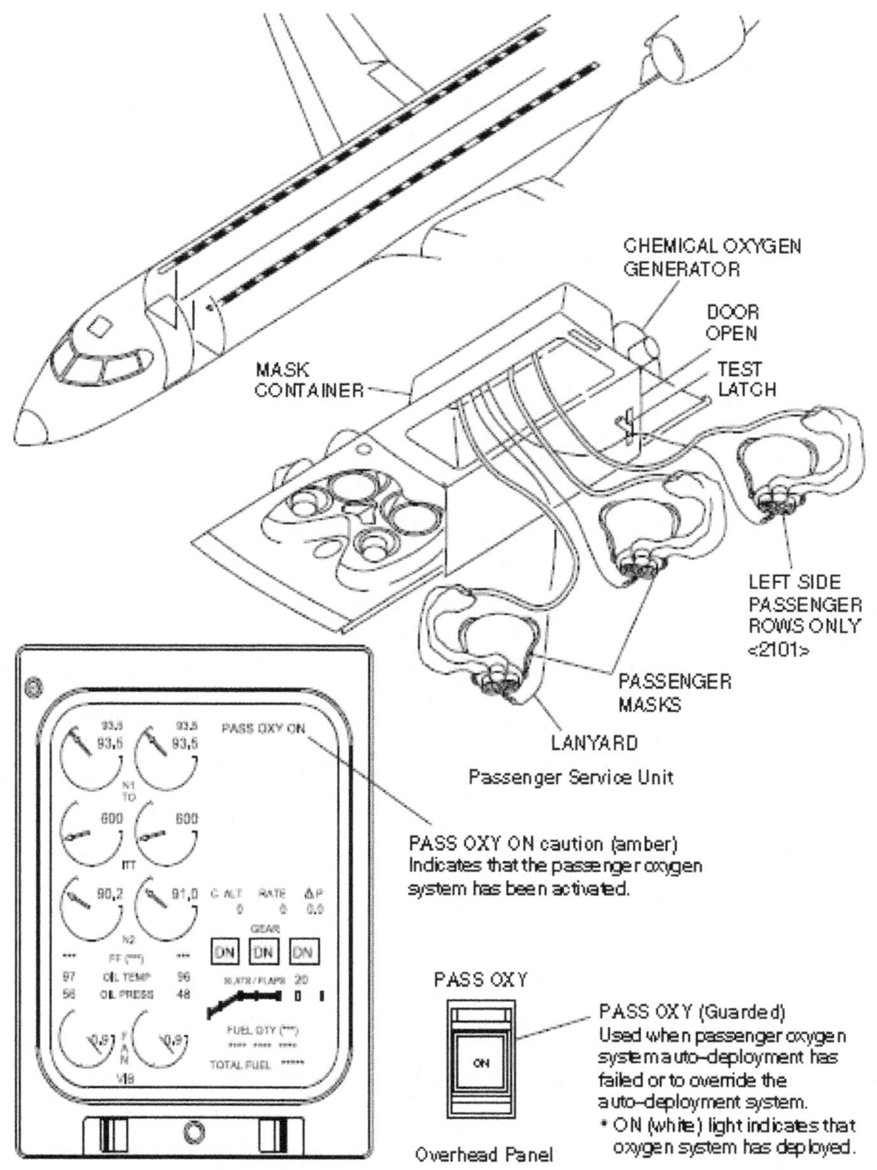

Figure 17 - 6: Passenger Oxygen System

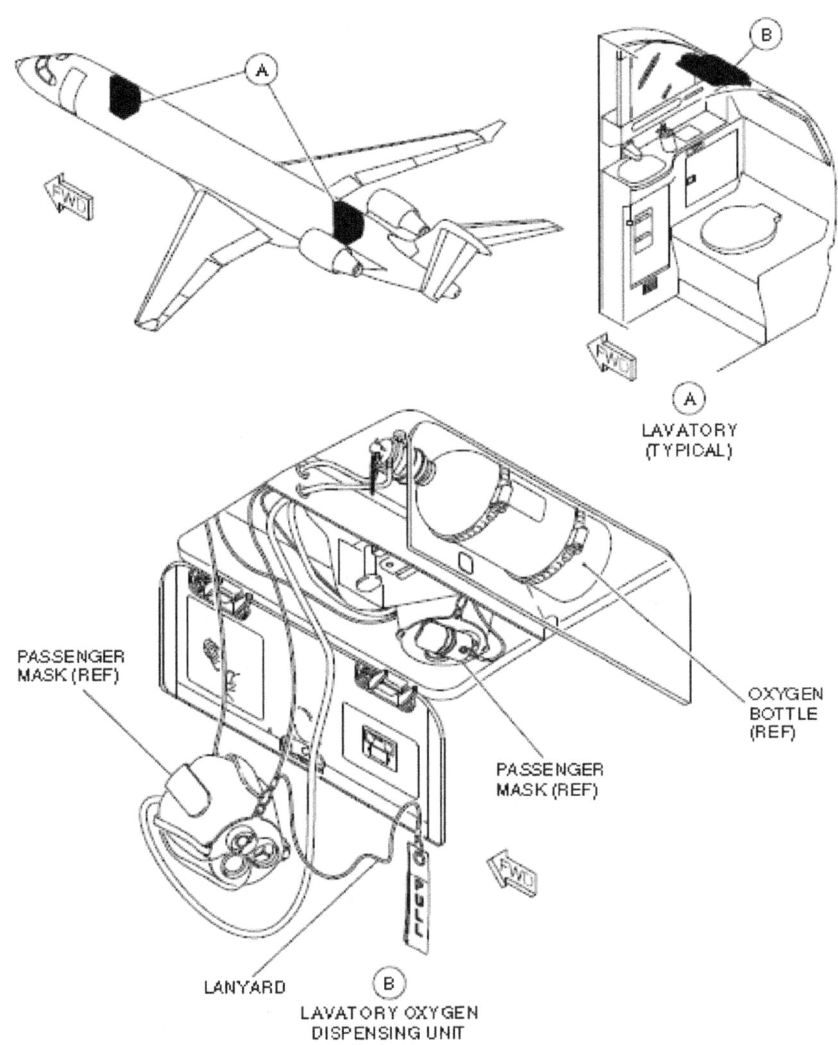

Figure 17 - 7: Lavatory Oxygen Bottle

Portable Oxygen System

The portable oxygen system is available to supply oxygen to the crew or the passengers during an emergency.

Portable oxygen cylinders are provided, as protective breathing units, to be used for protection against smoke and harmful gases or as therapeutic oxygen for first aid purposes.

Two portable oxygen cylinders, with disposable masks, are located near each flight attendant station. The portable oxygen cylinders allow the flight attendants to move about the passenger cabin during an emergency. Each cylinder has two regulator outlets which are color coded and preset to provide appropriate flow rates. An instruction decal located on the cylinder provides clear, easy-to-read operating instructions.

The contents gauge on each portable oxygen cylinder indicates from 0 to 2000 psi with a red band between 1800 to 2000 psi. When the gauge needle is in the red band this indicates that the cylinder is fully charged.

WARNING

Take precautions to ensure that oxygen cylinders do not come into contact with oil, grease, or other contaminants during handling. An explosion and fire could result if this happens.

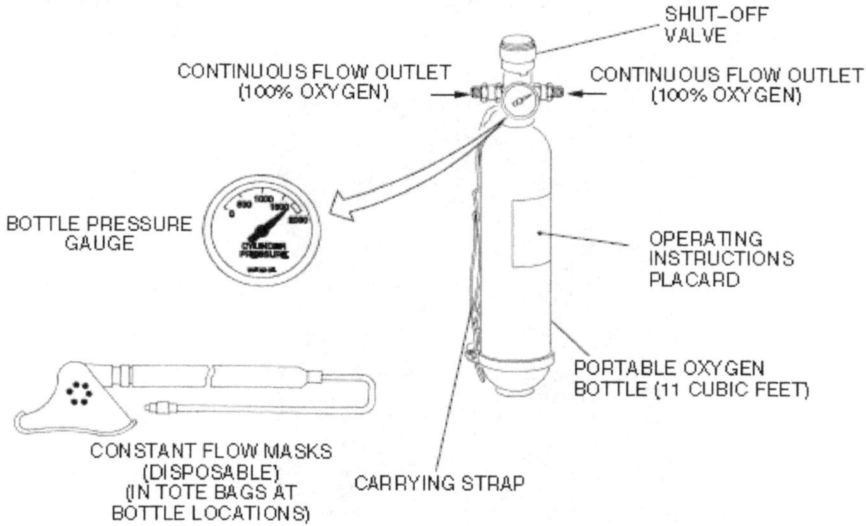

Figure 17 - 8: Portable Oxygen System

Fire Fighting Equipment

Fire protection gloves, portable fire extinguishers, and protective breathing equipment are provided to fight a fire occurring inside the flight deck or passenger cabin.

Three hand-operated fire extinguishers containing Halon 1211 are provided. Halon 1211 is effective on electrical, oil and fuel fires, and is suitable for use in cold weather.

The location of the fire extinguishers are:

- One (1) extinguisher in the flight deck, behind the copilot's seat
- One (1) extinguisher in the entrance stowage compartment
- One (1) extinguisher in the left storage bin forward of the emergency exit

Effective discharge time of a 3-1/2 pound Halon extinguisher is 10 to 12 seconds. Ventilate the compartment promptly after successfully extinguishing the fire to reduce gases produced by the fire and Halon.

WARNING

If a Halon fire extinguisher is to be discharged in the flight deck, all flight crew must wear oxygen masks with EMERGENCY selected (100% oxygen). Crew exposure to high levels of Halon vapors may result in dizziness, impaired coordination, and reduced mental sharpness.

Figure 17 - 9: Portable Halon Fire Extinguisher

Portable Water Fire Extinguisher

An portable water fire extinguisher is located in the cabin on the right aft bulkhead. The water extinguisher is useful for extinguishing fires fueled by wood, paper and fabric. It does not have Halon's hazardous effect on people.

WARNING

Do not use water extinguisher on electrical or flammable liquid fires.

Figure 17 - 10: Portable Water Extinguisher

Protective Breathing Equipment

The protective breathing equipment consists of three (3) protective breathing equipment (PBEs). The PBEs are self-contained smoke hoods that are supplied oxygen with either an on-demand oxygen regeneration system or with on-demand oxygen cylinders. The PBEs prevent injury to crew members from smoke inhalation. Each PBE is in a vacuum-sealed bag, and is kept in a storage container with tamper-proof seal. One PBE is installed in the flight deck, one is in the forward storage compartment and one is on the left aft passenger cabin bulkhead.

Figure 17 - 11: Flight Deck PBE

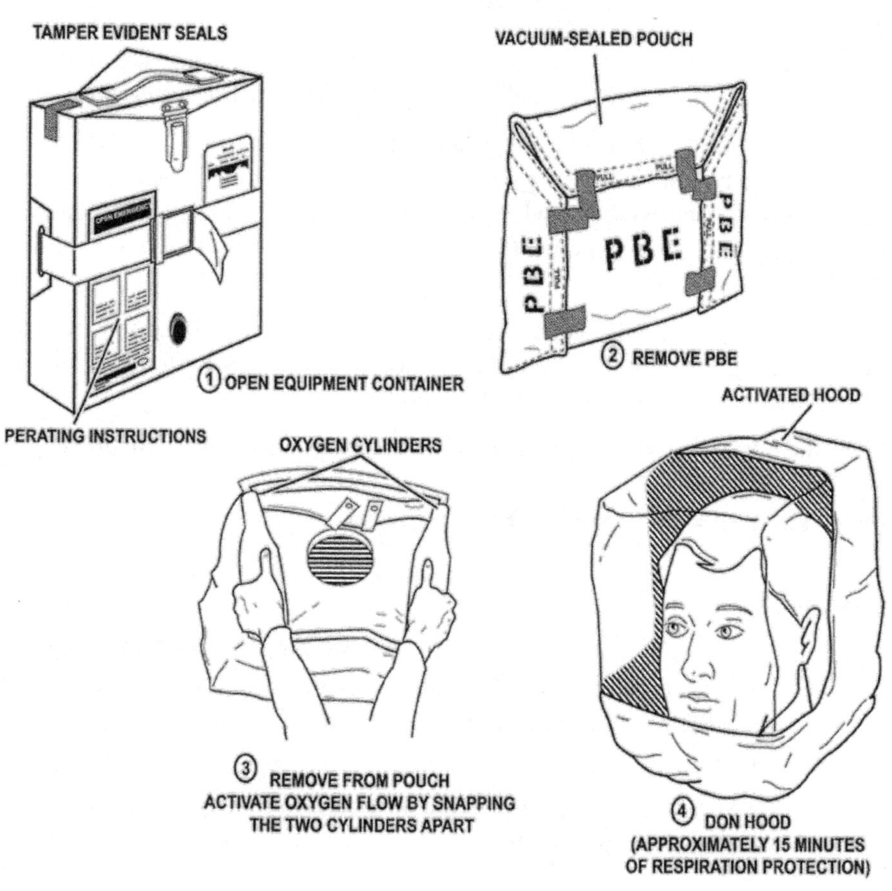

Figure 17 - 12: PBE

Emergency Locater Transmitter

The satellite capable emergency locater transmitter (ELT) is located in the aft equipment bay and is automatically activated during an aircraft crash.

The ELT transmits a standard swept tone on 121.5, 243.0 and 406.0 MHz for satellites. The two position EL T switch is located on the flight deck overhead panel and is labeled ARM/RESET and ON. The switch is used to test, arm and reset the unit. During normal flight operations, the EL T switch is in the ARM/RESET position.

The ELT can be manually activated by selecting the ELT switch to ON. To reset the unit after it has been activated automatically, the switch is selected to ON, then back to the ARM/RESET position.

When the EL T is transmitting, the caution message ELTON is displayed.

Over Water Emergency Equipment

Life Vests

An inflatable life vest is provided for each member of the flight crew. One (1) life vest is stowed under each pilot seat, one (1) life vest is stowed adjacent to the 3rd crew seat and one (1) is adjacent to each flight attendants seat.

Each life vest includes a manual and an oral inflation system, a locater light, and a system for automatic battery plug removal during life vest deployment.

Optional inflatable life vests can be installed under each passenger seat.

Additional adult and infant life vests can be stored in the left aft overhead stowage bin.

Each passenger seat cushion serves as a flotation device.

Flight Deck and Cabin Emergency Equipment

Emergency equipment that is located in flight deck and cabin includes:

- Standard D-cell type (or optional rechargeable) flashlights
- Crash ax

- Fire extinguishers
- Protective breathing equipment
- Crew life vests
- Escape rope and safety lines
- First aid kits
- Megaphone

Flashlights

There are five (5) hand-held flashlights installed in the aircraft. Each flashlight is battery operated and is stored in a bracket for easy access and removal. They are located behind the pilot's and copilot's seats, in the forward and aft flight attendant's stowage compartments and in the observers seat area.

The aircraft can be fitted with optional rechargeable flashlights that are powered using an internal Ni-Cad battery pack. When a rechargeable flashlight is removed from the retaining bracket, it is activated manually, via a slide switch on the flashlight body.

WARNING

Failure to turn the rechargeable flashlight off prior to reinstalling it in the retaining bracket, will result in the overheating of the unit and damage to the assembly.

Charging of the flashlight starts automatically when the flashlight is re-inserted into the retaining bracket. A constant illuminated LED, near head of the flashlight, indicates the battery pack is charging.

Figure 17 - 13: Emergency Flashlight

Crash Ax

The crash ax is mounted on the lower flight deck bulkhead behind the copilot. The crash ax handle is insulated to protect against electrical shock.

The crash ax has two cutting surfaces, the pointed surface is used to initiate puncture of the surface, the larger blade is used to enlarge the opening.

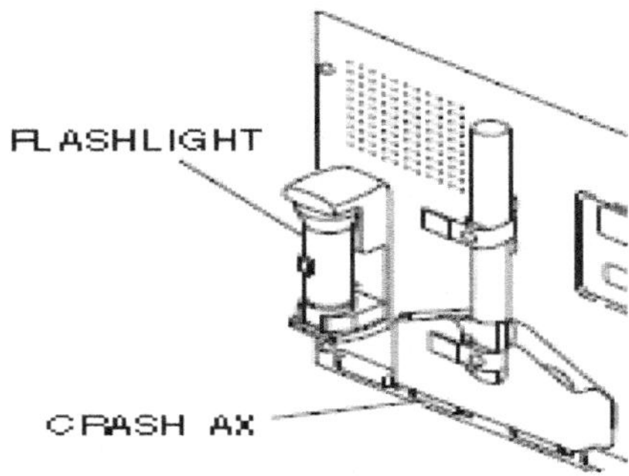

Figure 17 - 14: Crash Ax

Escape Rope (Flight Deck)

The escape rope is installed in the ceiling above the copilot's seat. It has a cover that is secured with a velcro strap. The rope is used as an aid for crew members to evacuate through the flight deck overhead escape hatch.

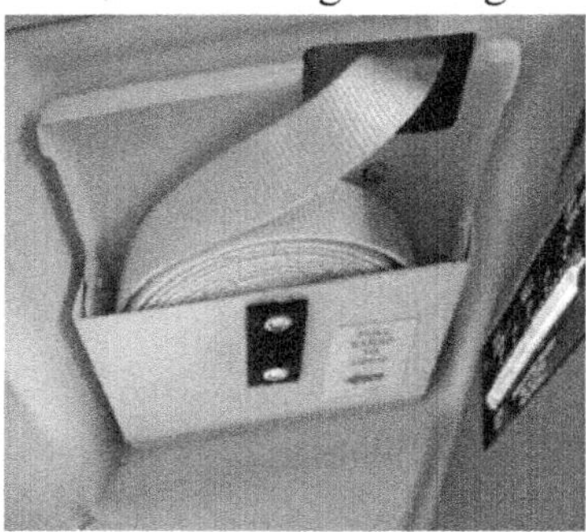

Figure 17 - 15: Escape Rope

Safety Lines (Over wing Exits)

A safety line is provided at each over wing exit to assist in evacuation during a ditching.

First Aid Kits

Two (2) first aid kits are installed in the aircraft. One (1) first aid kit is located in the forward entrance compartment and one (1) is located in the right aft overhead storage bin.

Megaphone

A battery-powered megaphone is located in the right aft overhead storage bin.

EICAS Indications

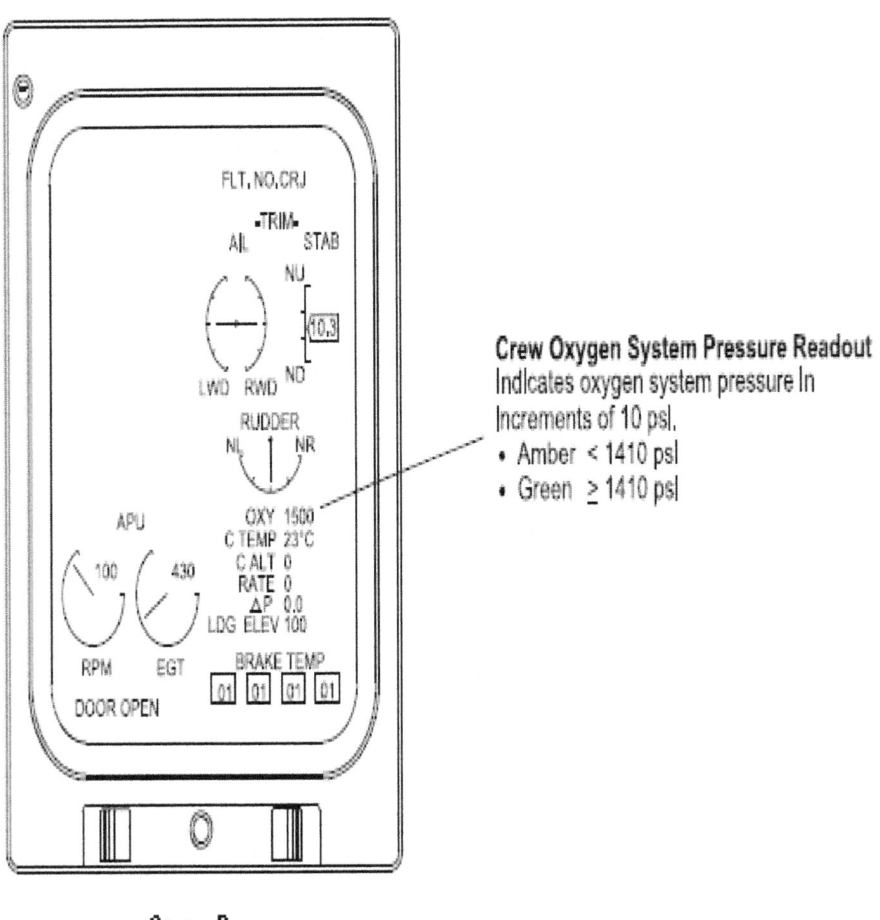

Figure 17 - 16: ED 2 Indications

EICAS Messages

Master Caution	Cause
ELT ON	Indicates the ELT has been activated
OXY LO PRESS	Crew oxygen bottle pressure less than 1410 psi.
PASS OXY ON	Oxygen system has deployed, accompanied by an aural alert

System Schematic

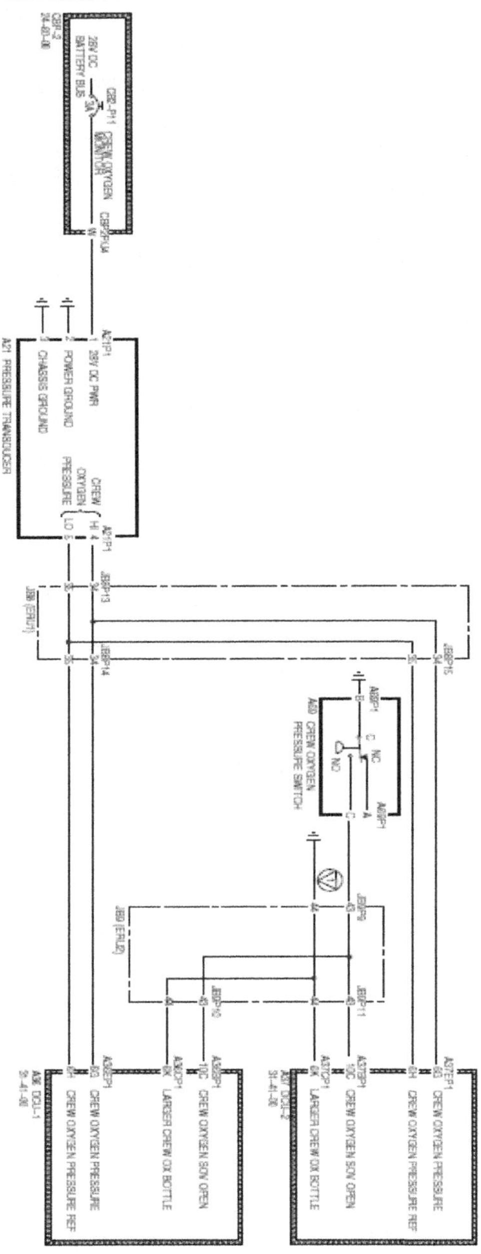

Figure 17 - 17: Crew Oxygen Schematic

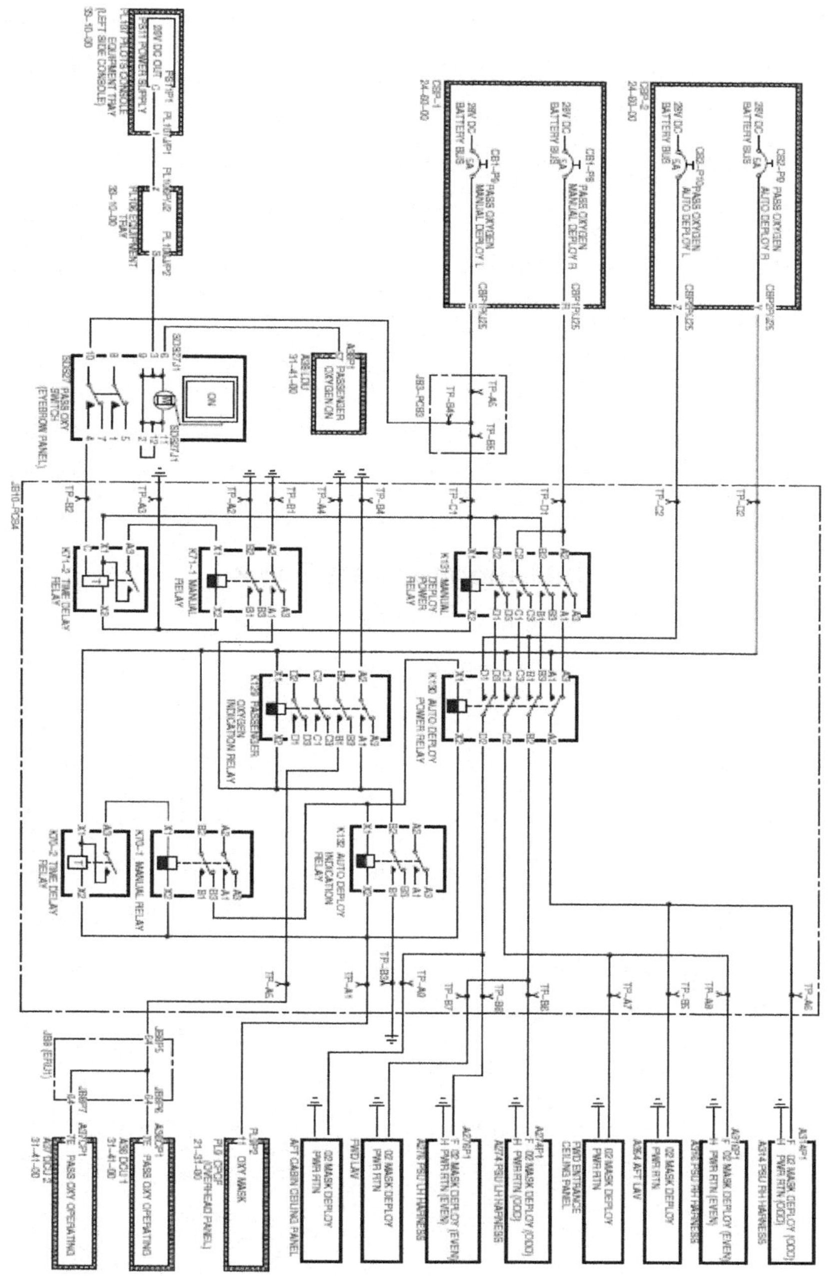

Figure 17 - 18: Passenger Oxygen Schematic

Intentionally Left Blank

Chapter 18: Pneumatics

Intentionally Left Blank

Chapter 18: Pneumatics

General

The pneumatic system receives pressurized air from either the 6th or 10th stage of the left and right engine compressors, the APU, or from an external ground air source. Pneumatically-operated aircraft systems include engine starting, cowl and wing anti-icing and air-conditioning pressurization.

Bleed air management is fully automated; however, manual control of the system is provided for non-normal operations.

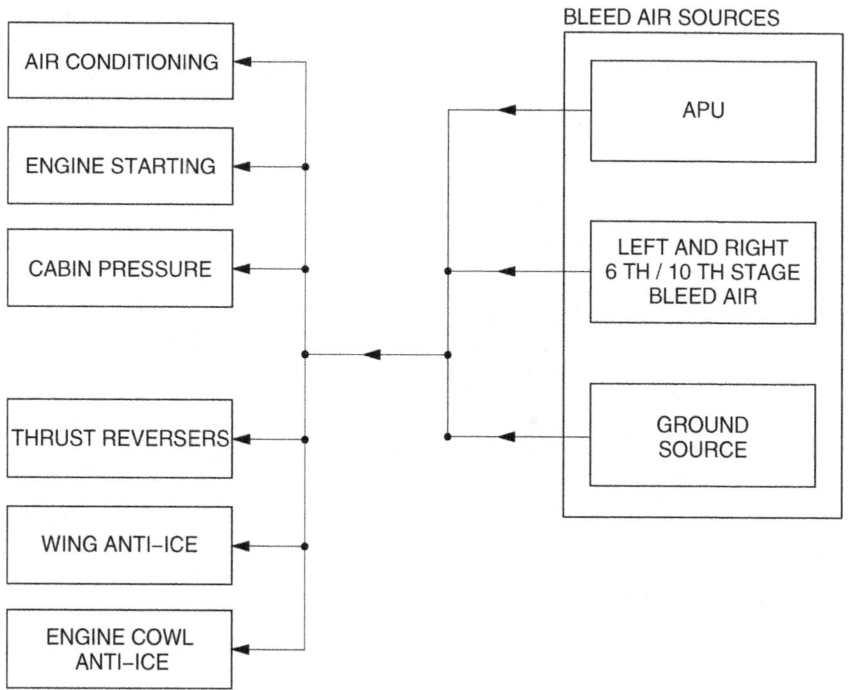

Figure 18 - 1: Bleed Air System - Block Schematic

Bleed Air Distribution

A common bleed air manifold distributes the bleed air for the pneumatic aircraft services.

Bleed air management, which consists of bleed source selection, valve operation and air distribution, is automatically controlled by two air-conditioning system controllers (ACSC 1 and ACSC 2).

The pilot can manually control the pneumatic system from the BLEED AIR control panel.

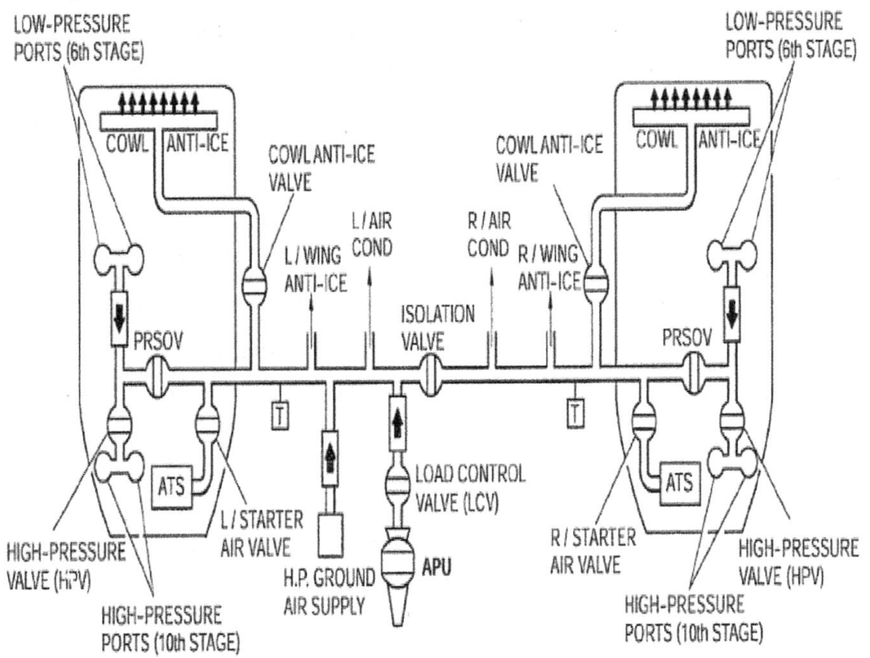

Figure 18 - 2: Pneumatic System

Components and Operation

Bleed Air Manifold

The bleed air manifold is located in the engine pylons and aft equipment bay and is divided into two sections by the isolation valve. The left side of the manifold is pressurized from an external air source, the APU or the left engine. The right engine normally pressurizes the right side of the manifold.

External Ground Air Connector

An external ground air connector is located on the fuselage aft of the aft cargo bay door. This connector allows external air to be supplied to the bleed air manifold.

The external ground air connector and duct are not depicted on either the ECS or ANTI-ICE synoptic pages. However, when external air is connected, the pressure can be read as a digital readout on the ECS synoptic page.

One-Way Check Valves

Four check valves are installed in various locations in the bleed air ducting.

An intermediate pressure check valve is installed in each engine's 6th stage bleed air duct. The check valve blocks the output of 6th stage bleed air to the manifold when the high-pressure valve (HPV) is open and prevents the reverse flow of 10th stage bleed air to the lower pressure 6th stage duct.

Check valves are installed at the external ground air connector and in the APU bleed air duct. The check valve at the external ground air connector prevents the loss of manifold air to the atmosphere. The check valve in the APU duct prevents the reverse flow of manifold air to the APU.

The check valves are not represented on the ECS or ANTI-ICE synoptic pages.

Engine Bleed Valves

Engine bleed air pressure increases proportionally with N2 compressor speed. The pressure regulating function of the engine bleed valve limits the engine's discharge pressure to 45 ± 3 psi. When N2 compressor speed is low and the engine bleed air pressure is less than the regulated value, the valve is fully open.

The bleed valves are located on the engines (one per engine) and are pneumatically operated and electrically controlled by their associated ACSC.

The bleed air valves are fail-safe closed when:

- The engines are shut down
- The ACSC control signal is removed
- The associated ENG FIRE PUSH switch-light is selected

The valves can also be manually selected closed by the pilot at the BLEED AIR control panel.

To enhance dispatch reliability, a manual override lever on the valve housing allows the valve to be locked in the closed position.

High-Pressure Valves

Normally, the 6th stage of the engine compressor supplies air to operate the pneumatically-powered aircraft systems. When system demands exceed the capability of the 6th stage, the high pressure valve (HPV) is commanded open to supply higher-pressure 10th stage bleed air to the manifold. The HPV is controlled automatically by the ACSC.

The HPV is closed when the associated:

- PRSOV is closed
- Upstream bleed pressure is less than 13 psi
- Bleed air duct leak (automatically commanded closed by AILC)
- Control signal is removed (ACSC or pilot manually commanded)
- ENG FIRE PUSH switch-light pushed

To enhance dispatch reliability, the valve can be manually closed. The manual override lever is on the HPV casing.

There are no EICAS indications of HPV operation.

APU Load Control Valve

The APU electronic control unit (ECU) controls the operation of the APU and determines when the APU is ready for bleed air loading.

The APU load control valve (LCV) is modulated to control the amount of bleed air taken from the APU. The ECU controls the opening, closing and modulation of the LCV. When the aircraft's pneumatic system requires APU bleed air, the ACSC signals the ECU to open the LCV. When the LCV is open, APU bleed air is delivered to the left side of the bleed air manifold and the white APU LCV OPEN status message is displayed.

The APU LCV is fail-safe closed. APU shutdown or the loss of the ECU or ACSC signal closes the valve.

Isolation Valve

The isolation valve (ISOL) provides interconnection and isolation capabilities between the left and right sides of the bleed air manifold.

The ACSC controlled 28 VDC powered ISOL valve cannot be motored opened when any engine is supplying bleed air to the manifold or cowl and/or wing anti-ice is selected ON.

In AUTO mode the ISOL valve:

- Automatically opens when both engine PRSOVs are closed and the APU is supplying bleed air to the manifold (LCV open)
- Cycles open and closed during cross bleed engine starts or right engine start using APU bleed or external ground air
- Cycles open or closed during the auto transfer of bleeds between the APU and engines

The pilot can manually select the ISOL valve open or close by switch selection at the BLEED AIR control panel. Manual control does not override the system's protective features nor restrict automatic operation of the isolation valve during engine start.

To enhance dispatch reliability, the ISOL valve can be locked in the closed position. The manual override lever is on the valve housing.

Bleed Air Management

Two air-conditioning system controllers (ACSC 1 and ACSC 2) manage the distribution of bleed air and control the air-conditioning system. The ACSCs monitor the duct pressures and position of all bleed and anti-icing valves and use the information to program the:

- High-pressure valves (HPVs)
- Engine bleed valves
- Isolation valve
- APU LCV (via the APU ECU)

ACSC 1 controls the left HPV and left engine bleed valve; ACSC 2 controls the right HPV and right engine bleed valve. Normally, the active channel of ACSC 1 controls the APU LCV and isolation valve. If ACSC 1 fails, control of the APU LCV and ISOL valve are transferred automatically to ACSC 2.

Each controller consists of two channels. One channel of each controller actively manages bleed air distribution for its applicable side while the second channel maintains a standby/monitoring mode. Active channel assignment is automatically switched on a daily basis. There are no EICAS indications of active ACSC channel assignment.

Engine Bleed Source Switching

The ACSC looks at the following parameters in determining HPV operation:

- Synthesized 6th stage duct pressure
- Manifold pressure (pack inlet pressure)
- Air-conditioning system demands
- Anti-icing system demands

Air is usually bled from the low pressure ports of the engine's N2 compressor (6th stage). When the engine N2 rpm is low or pneumatic demands are high, the ACSC commands the HPV to open to allow the higher-pressure 10th stage air to pressurize the manifold.

When engine rpm increases and the 6th stage compressor output increases to a satisfactory level, the HPV is commanded closed by the ACSC.

Engine Bleed Air Pressure Regulation

Engine bleed air pressure increases proportionally with N2 compressor speed and the pressure regulating valve limits engine bleed air pressure to 45 ± 3 psi. If a pressure regulating valve fails and manifold pressure exceeds 60 psi, the ACSC commands the applicable bleed valve closed and the EICAS caution message L or R ENG BLEED appears.

APU and Engine Bleed Air Priority - AUTO mode

Normally, the bleed air management is operated in AUTO mode. On the ground, with both the engines and APU running, the APU always has priority in supplying the bleed air for the aircraft pneumatic systems.

When the APU pneumatic ready-to-load (RTL) signal is sent, the ACSC commands the APU ECU to open the APU LCV. After the APU LCV opens, the isolation valve (ISOL) is automatically opened by the ACSC to pressurize both sides of the manifold

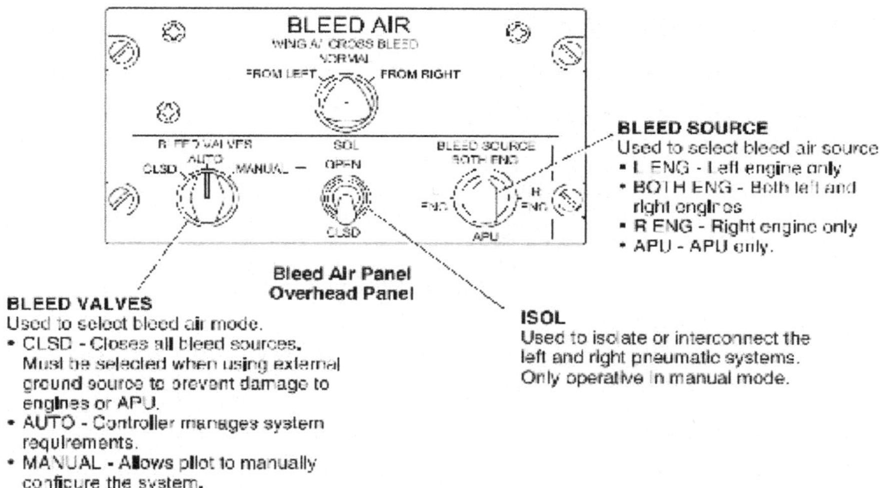

Figure 18 - 3: BLEED AIR Control Panel - AUTO Selection

APU LCV Interlock Protection

The ACSC logic provides two levels of interlock protection for the APU LCV. The control logic will not allow the engines and APU to supply the manifold at the same time. The ACSC commands the engine bleed valves to close before opening the APU LCV.

In AUTO, the control logic also prevents the APU from supplying bleed air to the anti-ice systems. If the APU is supplying bleed air to the manifold and either cowl or wing anti-icing is selected ON, the APU LCV is automatically closed and the engine bleed valves are commanded open.

If the APU is manually selected as the bleed source and either cowl or wing anti-ice is selected ON, the:

- APU LCV remains open
- Wing anti-icing valves remain closed
- BLEED MISCONFIG caution message is displayed

Engine Start Sequences

During engine starts on the ground, engaging the ENG ST ART switch-light automatically sequences the valves to allow bleed air to the air turbine starter (ATS).

When the ENG START switch-light is pressed, the:

- Isolation valve opens (if previously closed)
- Both air-conditioning pack valves close (if previously open)
- Associated start valve opens
- Engine air turbine starter (ATS) engages

When the engine achieves a starter cutout rpm of 50%, the ATS disengages, the start valve closes and the air-conditioning pack valves and isolation valve return to their previous operating setting.

Bleed Switching - Single Engine to APU

During most phases of flight, the engines supply the bleed air for system operations and the pneumatic system is managed automatically.

During single engine operation, if an engine relight is attempted, the operating engine will supply the air for the cross bleed start. The APU, if operating at this time, will only become a bleed air source for engine starter operation when the BLEED AIR panel is configured as follows:

- BLEED VALVES switch to MANUAL
- BLEED SOURCE switch to APU

In-flight - APU/Engine Bleed Switching Sequence

After takeoff with the APU operating. landing gear up and flaps 20 or less, bleed air loading is transferred automatically to the engines when the thrust levers are retarded from the takeoff detent. The transfer of bleeds from the APU to the engines is totally automatic and sequential. The transfer process takes approximately 20 seconds to complete.

On approach, if the APU is available for bleed air loading, the bleed loading transfers from the engines to the APU when the flaps are greater than 20 or the landing gear is down. The transfer process is automatic and sequential and takes approximately 20 seconds to complete.

Anti-icing - APU Engine Bleed Switching Sequence

The ACSC will not allow the APU to supply bleed air to the wing anti-icing systems. In AUTO mode with the APU as the bleed source, selection of cowl or wing anti-icing immediately closes the APU LCV and transfers the bleed loading to the engines.

In MANUAL mode with t he APU as the bleed source, the automatic bleed transfer from APU to engines is not available. When cowl and wing anti-icing is selected ON, the:

- APU LCV stays open
- Engine bleed valves remain closed
- BLEED MISCONFIG caution message appears
- Wing anti-ice valves remain closed
- Cowl anti-ice valves open

APU LCV Manual Operation

In MANUAL mode the following parameters must be met before the APU LCV will open:

- APU must be operating and ready to accept bleed air loading
- BLEED SOURCE switch must be selected to APU
- BLEED VALVES switch must be selected to MANUAL

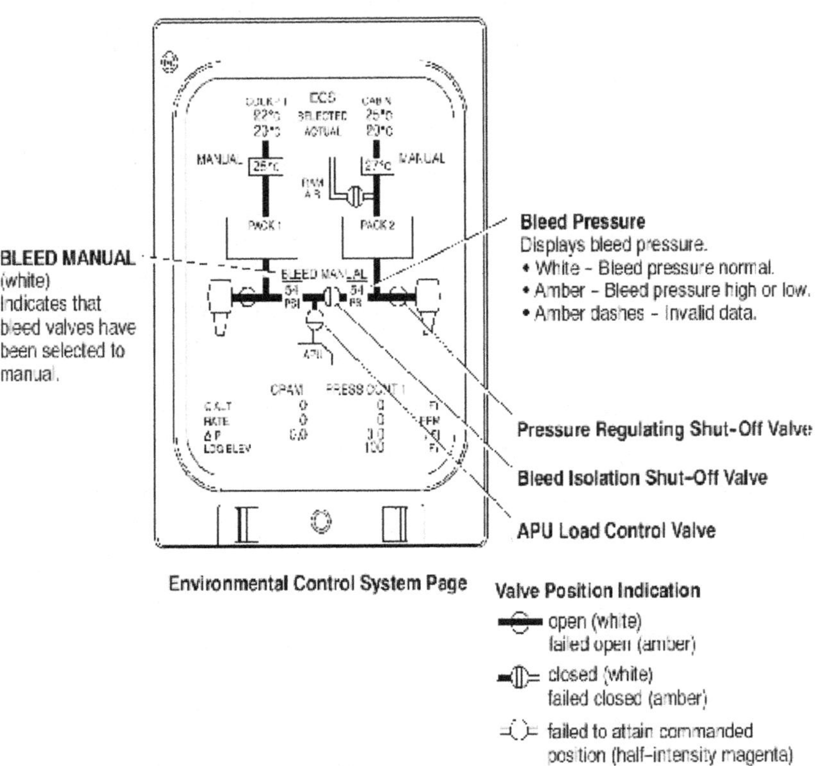

Figure 18 - 4: Bleed Air System Environmental Control Synoptic Page

Bleed Air Leak Detection System

The bleed air leak detection system monitors the pneumatic and anti-icing ducting for high temperatures associated with bleed air leakage. EICAS messages and system control are provided by an anti-ice and leak detection controller (AILC).

The bleed air leak detection consists of continuous sensing loops routed in parallel along the pneumatic ducting. The system is divided into five zones and each zone can be isolated by means of a shutoff valve. The five zones are left and right bleed zones, left and right cowl anti-ice zones and wing anti-ice zone. The wing anti-ice zone is subdivided into four loops. They are left and right fuselage loops and left and right wing loops. The supply ducting is encased in a protective cover. If a leak occurs, holes in the protective cover direct the hot bleed air towards the sensing loops.

The dual-sensing loops are used to ensure dispatch reliability and to minimize system false warnings. To prevent false indications, both loops must detect a leak before an EICAS message is posted. The leak detection sensing loops consist of two wires mounted coaxially inside a flexible metal tube. The ends of each sensing loop are connected to the controller. When hot air escapes from a leak in the ducting it is sensed by the controller which posts an EICAS message identifying the leakage zone. The duct is then isolated by closing the appropriate shutoff valve.

For normal wing anti-icing, hot bleed air from the supply ducting is released through piccolo tubes to heat the wing leading edges. The dual loops and a skin temperature sensor located in the wing leading edge are used to detect failures in the wing anti-ice ducting.

The cowl anti-ice ducts, located in the engine pylons, consist of inner and outer ducts. Bleed air for anti-icing travels through the inner duct. The area between the inner and outer duct is monitored by a pressure transducer. In the event of a failure or crack of the inner duct, the pressure transducer will sense the air pressure change and send a signal to the AILC to post an EICAS warning message.

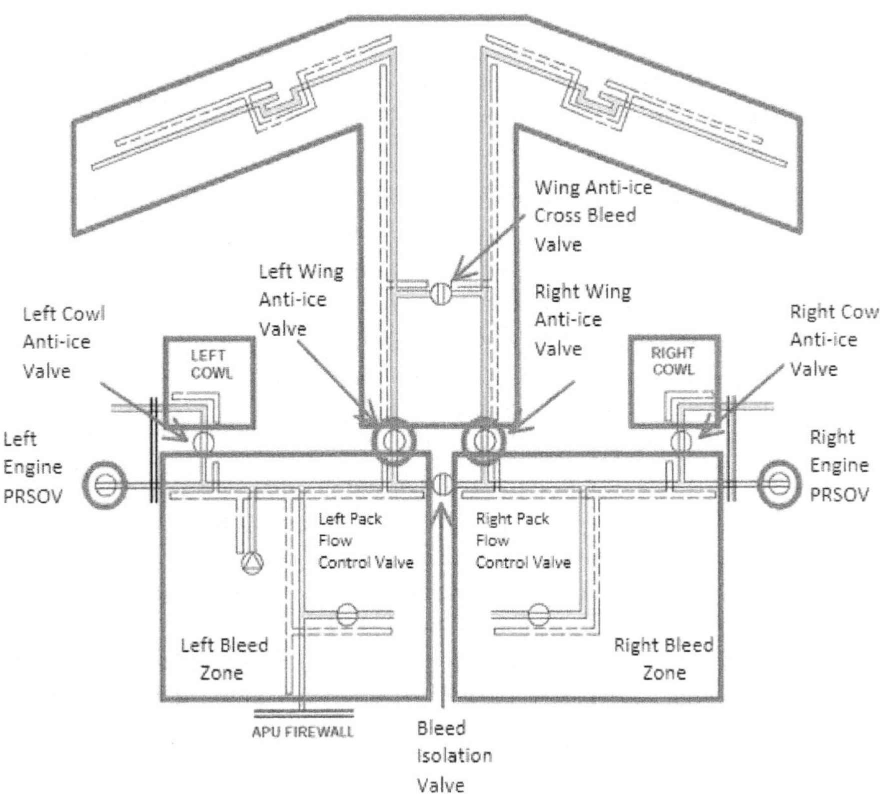

Figure 18 - 5: Bleed Air Leak Detection

Components and Operation

Dual-Sensing Loops

Dual-sensing loops are used to monitor:

- Bleed air ducts in the engine pylons and aft equipment bay
- Anti-ice ducts in the aft equipment bay, center fuselage (including the cross bleed duct) and leading edges of the inboard wings and slats

Both loops must detect the high temperatures associated with a leak before the applicable bleed air duct or anti-ice duct warning is generated.

The dual- loop system provides two significant advantages over a single-loop system. Dual-loops minimize false leak indications and allow the aircraft to be dispatched under certain conditions with one loop faulted.

Bleed and Center Fuselage Anti-icing Ducts

The bleed air ducts are constructed of stainless steel and are insulated and encased by a protective outer cover. A series of holes are drilled at predetermined locations in the duct's outer cover and dual-sensing loops are installed adjacent to these holes. If a bleed or anti-ice duct leaks, the holes direct the escaping hot air onto the sensing loops.

Wing Anti-icing Ducts

In the wing leading edges, dual-sensing loops and dual-temperature sensors are used to detect the failure of a wing anti-ice duct (piccolo tube) or duct connector. The detectors can discriminate between the overheat condition that is caused by a leaking duct and the normal high temperatures that are associated with wing anti-icing.

The extreme outer portion of slat 3 is not protected by dual-sensing loops. Instead, a dual-temperature sensor is used to detect failure of the piccolo tube.

Cowl Anti-icing Duct and Leak Detection System

The cowl anti-icing valves are located in the aft equipment bay and connected to their associated nacelle by an anti-ice duct. In the pylon, a double-walled duct is used. The pressurized hot bleed air travels through the inner portion of the pylon duct. The outer portion is not pressurized and is monitored by a pressure transducer.

If the inner duct ruptures, the escaping pressurized bleed air is contained inside the outer duct and the transducer detects the increase in outer duct pressure. The transducer signals the DCU and the associated L or R COWL A/I DUCT warning message is generated.

Anti-ice and Leak Detection Controller

A dual-channel anti-ice and leak detection controller (AILC) continuously monitors for failure of the bleed air and anti-icing ducts and integrity of both channels of the detection system. AILC channel A monitors loop A, while channel B monitors loop B. Both loops must sense the high temperatures associated with a bleed air leak before a leak detection warning message is posted.

If a leak occurs downstream of the wing anti-icing valve (aft equipment bay, center fuselage or wings), the AILC will automatically close both wing anti-icing valves and display the ANTI-ICE DUCT warning message.

If the leakage occurs in the bleed ducting before the wing or cowl anti-icing valves, the appropriate L or R BLEED DUCT warning message is displayed. The applicable engine bleed valve automatically closes to isolate the failed duct.

Leak Detection Testing

The leak detection system is continuously tested.

Failure of both detection channels in any detection zone is indicated by the applicable L or R BLEED LOOP or ANTI-ICE LOOP caution message. Failure of a single-channel (loop) is indicated by the DUCT MON FAULT or WING A/I FAULT status message.

EICAS Messages

Master Warning	Aural	Cause
ANTI-ICE DUCT	"Anti-ice duct"	Illuminates when bleed air leak is detected in left and/or right fuselage or wing anti-ice ducts.
L or R BLEED DUCT	"Bleed air duct"	Indicates a bleed air leak has been detected by the anti-ice and leak detection controller in the respective engine bleed ducting.
L or R COWL A/I DUCT	"Anti-ice duct"	Indicates that the respective cowl anti-ice duct pressure is above 17 psig or a bleed air leak has been detected by the anti-ice and leak detection controller.

Master Caution	Cause
ANTI-ICE DUCT	Indicates bleed shutdown after bleed leak detected and system not selected off.
ANTI-ICE LOOP	Indicates loss of both wing anti-ice leak detection loops during power-up test.
BLEED MISCONFIG	Indicates an impossible bleed configuration in manual mode
ISOL FAIL	Indicates that bleed isolation shut-off valve failed to attain selected position.
L or R BLEED DUCT	Indicates respective bleed shutdown after bleed leak detected and system has not been reconfigured.
L (R) BLEED LOOP	Indicates loss of both left or right engine bleed air leak detection loops during power-up test.
L (R) COWL LOOP	Indicates loss of both left or right cowl bleed air leak detection loops during power-up test.

L or R ENG BLEED	Indicates failure of high pressure valve, pressure regulating shut-off valve or controller.
Status Message	**Cause**
BLEED CLOSED	Indicates that all bleeds are closed.
BLEED MANUAL	Indicates that the bleed system is in the manual mode.
DUCT MON FAULT	Indicates a loss of redundancy in bleed leak detection system.
ISOL CLOSED	Indicates that bleed isolation shut-off valve is fully closed.
ISOL OPEN	Indicates that bleed isolation shut-off valve is fully open.
L or R ENG BLEED CLSD	Indicates that respective engine bleed is not selected with respective high pressure valve and pressure regulating shut-off valve closed.
L or R ENG BLEED SNSR	Indicates respective pack inlet pressure sensor has failed.

System Schematic

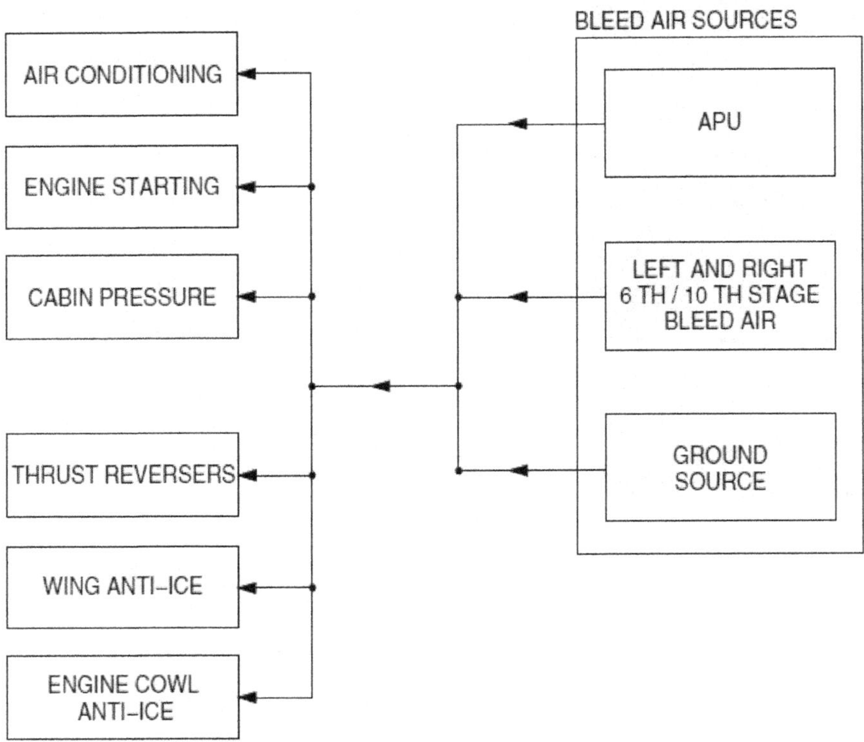

Figure 18 - 6: Bleed Air System - Block Schematic

Chapter 19: Water and Waste

Intentionally Left Blank

Chapter 19: Water and Waste

General

The water and waste system stores and supplies potable water to the galley(s) and lavatorie(s). and removes waste (drain) water from the sinks. The system also collects and stores waste from the lavatory toilet(s).

The water and waste system consists of the following:

- Potable/wash water system
- Lavatory waste disposal system

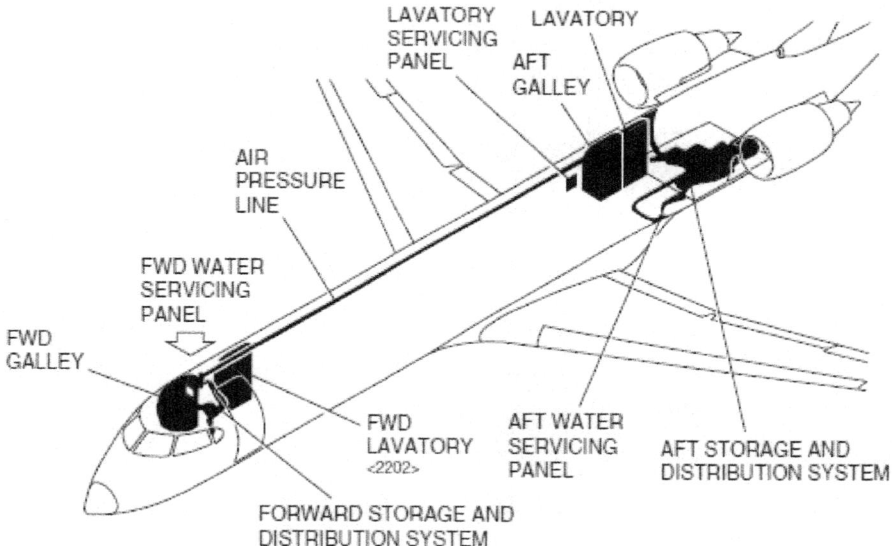

Figure 19 - 1: Water and Waste System

Potable/Wash Water System

The potable/wash water system (PWS) stores, supplies and controls the flow of water to the galley(s) and lavatorie(s). The PWS drains waste water to the waste disposal system, and ensures safe drainage of waste water from the waste disposal system out of the aircraft. The system controls temperature to prevent freezing of the drain masts, drain lines, supply lines and water tanks.

The potable water system is pressurized by regulated bleed air, provided from either the ECS system, a compressor or a ground unit.

The control panel is located at the forward galley (G1).

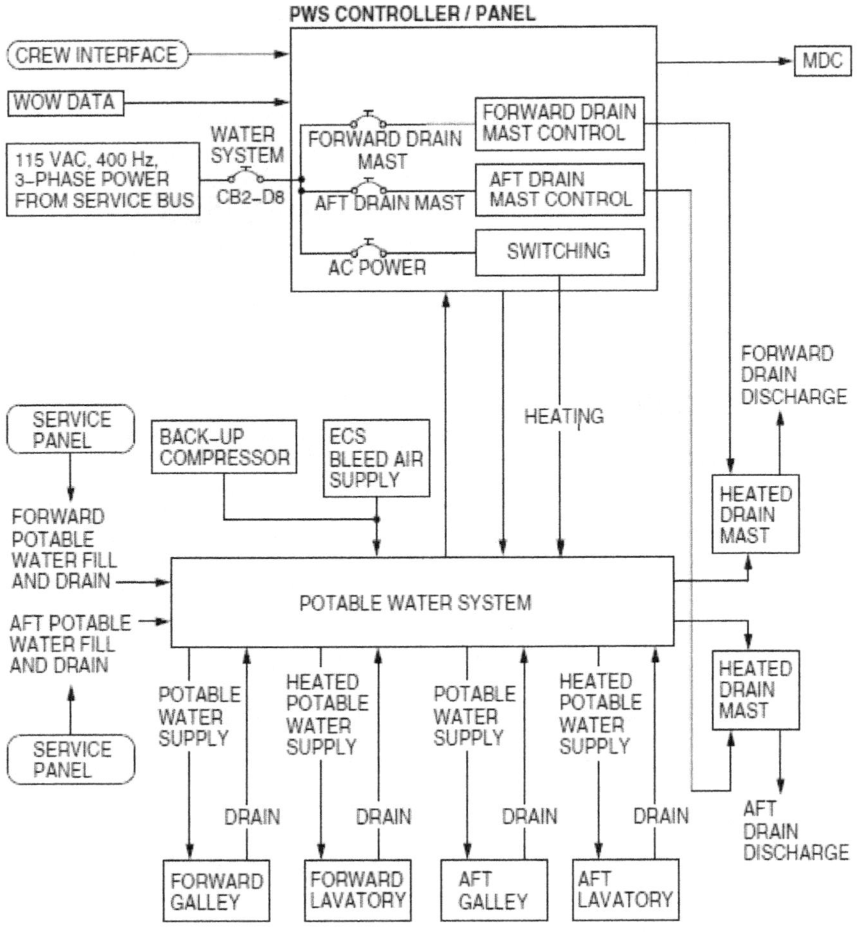

Figure 19 - 2: Potable Water System

Lavatory Waste Disposal System

The lavatory waste disposal system stores waste from the lavatory toilets. Removal of waste material is carried out at the lavatory service door on the right aft side of the fuselage.

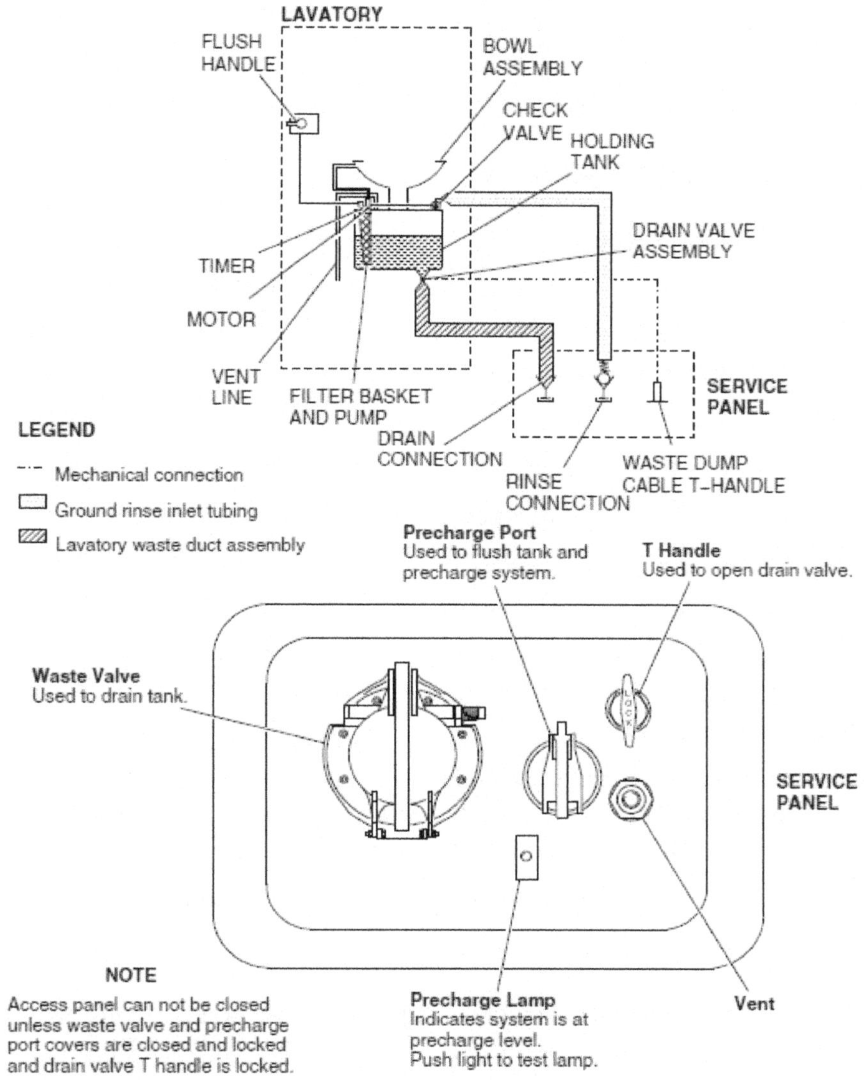

Figure 19 - 3: Lavatory Waste System

Potable/Wash Water System

The potable/wash water system (PWS) stores and supplies potable water to lavatory sinks and galley coffee makers. Two storage and distribution subsystems are installed, one forward and one aft. Water does not flow between the forward and aft systems, however the systems share a common controller and air pressurization system.

The controller supplies power to the forward and aft PWS. Software in the controller manages all of the functions of the forward and aft PWS.

Components and Operation

Potable Water Storage

Potable water is stored in forward and aft tanks and includes the following components:

- Two water storage tanks (fwd and aft)
- Water level sensors
- Lavatory water heater
- Fill (supply) and drain lines
- Drain valve
- Drain mast
- External service panels (fwd and aft)

Water Storage Tanks

The forward water storage tank is located in the forward galley sink compartment. The tank is insulated with a blanket to minimize noise. and does not require a heater due to its location in the cabin.

The aft tank is installed under the floor in the aft tail section. An external heater blanket is installed on the tank. and the tank and heater are insulated.

The tanks are replenished with an external service cart through an exterior service panel.

Water Level Sensors

The storage tanks have five water level sensors to correspond with the control panel indications. Aircraft level and attitude have little effect on the accuracy of the indications. The FULL/EMPTY indicator light at the forward galley (G 1) provides tank-level status.

Lavatory Water Heater

The lavatory water heater is located directly under the lavatory sink and warms the potable water for the lavatory faucet. The water heaters are automatically powered by the potable water system control unit.

There is a manual reset switch on top of the heater. in case of overheat. It will only accept reset once temp falls.

Drain Mast

The drain mast exhausts waste and potable water from the water system. The forward and aft water systems have separate drain masts. The drain masts mount externally on the underside of the fuselage. The chosen locations ensure no hazard to the aircraft results from overboard discharge from the drain masts in-flight. The water drains and drain masts are heated to prevent them from freezing. The drain mast heaters are always operational.

When a DRAIN FAULT indication on the PWS controller occurs, the affected system (fwd or aft) is not to be used.

Fill (Supply) and Drain Lines

The water tank fill (supply) and drain lines are heated. Shrouds are installed on water lines to direct water leakage to non-water sensitive areas of the aircraft.

Drain Valve

The drain valve reduces pressurization leak noise in the galley drain line.

External Service Panels

The service panel includes the following components:

- Water fill connection
- Tank drain/fill handle

The aft service panel has a fitting to allow ground pressurization of the system.

The forward service panel is on the right side. The aft service panel is on the left side of the aircraft under a panel in the aft wing root fairing. The adapter contains an open screen to prevent large debris from entering the water system.

The fill/drain valve controls water flow for fill, flight and drain conditions. In the FILL position, the fill port is connected to the supply lines and storage tank, and the overflow line is routed to the drain mast. If the tank is overfilled, water will flow out through the drain mast overflow outlet.

In the FLIGHT position, the overflow lines and supply lines are blocked from the drain mast, and the fill port at the service panel is open straight to the drain mast. The handle must be in the FLIGHT position to close the service door.

In the DRAIN position, the overflow and tank supply lines are open to the drain mast and all water can exit the PWS. This permits the tank to be drained to replace the water supply or to empty the tank when the aircraft is parked. When the handle is operated, the water is discharged to the ground.

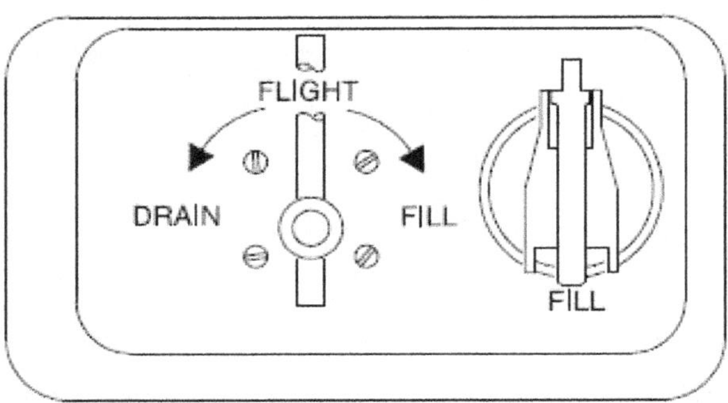

Figure 19 - 4: Potable Water Servicing Panel

Potable/Wash Water System Air Supply

The potable water system is pressurized by air. The primary source of pressurized air for the water and waste system is regulated aircraft bleed air from the engines or APU. The air pressurization system taps into the APU/engine bleed air system downstream of the left ECS pre-cooler. The pressure regulator/shutoff valve of the bleed air system normally regulates the bleed air pressure. The PWS pressure regulator reduces downstream pressure even further.

The secondary source of pressurized air is an air compressor, used when the primary air pressure is insufficient. The controller/panel measures line pressure to operate the air compressor.

The system may also be pressurized on the ground through a fitting in the aft service panel.

Controller and Control Panel

The PWS controller/panel is an integral unit located in the forward galley (G1) and is used as the primary operational interface.

The controller maintains temperature and prevents freezing of the liquid at the drain masts, drain lines, supply lines and water tanks for the lavatory and galley, as well as controls pressurization of the system.

There are two sets of water level lights on the control panel. One set is for the forward water tank level, the other is for the aft water tank level. The water level indications are FULL,3/4. 1/2, 1/4, and EMPTY. Each indicator will illuminate green when the sensed water is at or above its indicated level.

The diagnostic light on the panel indicates that non-critical maintenance needs to be performed on the PWS. This light illuminates (white) should a fault occur in the PWS that may reduce its functionality. A FWD or AFT DRAIN FAULT light will illuminate (red) when a drain mast failure is detected. In this case, the affected system (fwd or aft) is not to be used. When the water goes below the empty sensor, the empty light changes from green to amber, while all other level indicators remain off.

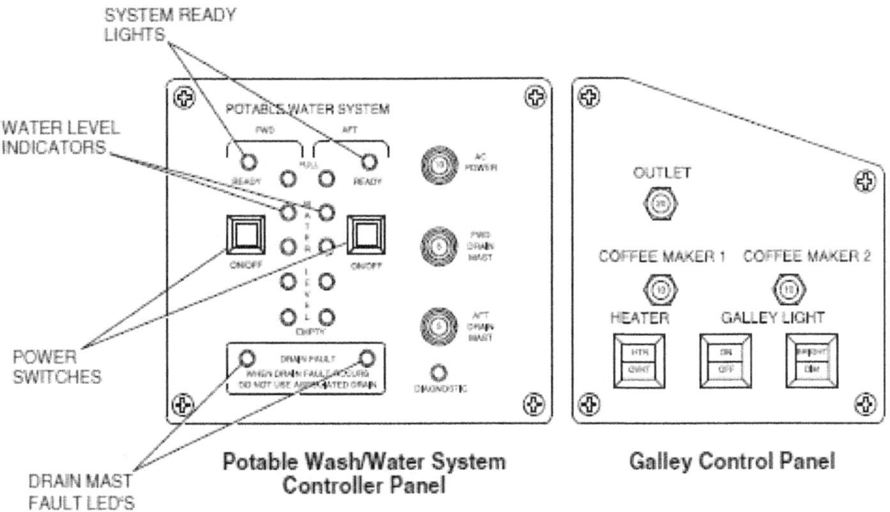

Figure 19 - 5: Potable Wash/Water System & Galley Control Panel

Waste Disposal System

The waste disposal system collects and stores waste from the lavatory toilet(s). The toilet is self-contained with an integral seat bowl and holding tank. The toilet unit is filled with clean water and drained of waste water through the service panel.

Components and Operation

Toilet Operation

When the toilet flush handle is pushed, the flush cycle is initiated. Flushing fluid is drawn through a filter by an electrical pump, and discharged into the toilet bowl and waste material is carried into the tank assembly.

The filter basket is contained inside the motor pump filter assembly. The waste and flushing fluid inside the holding tank are pulled by suction through the filter basket. The filtered liquid portion from the filter basket is used as the flushing fluid.

Drain Valve Assembly

The drain valve assembly lets the servicing crew remove the waste materials from the holding tank. The valve is manually operated through the waste dump cable T-handle, located on the service panel. When the drain valve assembly is opened, the waste materials in the holding tank are drained overboard through the lavatory waste duct assembly.

The valve is locked in the open position when the T-handle on the service panel is turned counterclockwise, pulled, and turned clockwise. The T-handle must be in its initial lock and closed position to close the lavatory service door.

After the removal of the waste materials from the holding tank. a rinsing agent and the flushing fluid are sent through the ground rinse inlet tubing.

Servicing

The servicing of the lavatory waste system is accomplished through the service panel. There is one for the forward lavatory, and one for the aft lavatory. The service panel includes a waste drain outlet, a rinse/fill connection point. a waste tank drain handle, and a tank level indicator.

A service vehicle drain line is connected to the drain outlet, the drain valve assembly for the holding tank is opened through the waste dump cable T -handle on the service panel. The waste material, contained in the holding tank is drained into the servicing vehicle. The T-handle must be placed in the closed and locked position in order to close the service panel door.

During the cleaning of the holding tank, rinse water from the servicing vehicle goes in through the rinse/fill connection. The rinse water goes into the holding tank through the check valve and through a rotating spray nozzle. From the spray nozzle, the rinse water cleans the filter basket and the inside walls of the holding tank. After draining and rinsing, the holding tank must be pre-charged (filled) with a dye, deodorant, chemical and water solution. This solution is used as the flushing fluid.

Upon draining the waste, the recharging line from the servicing cart is connected to the recharge port. The waste tank is then flushed and filled with a fresh quantity of chemical solution. The T-handle must be turned toward the right and pushed in to close the drain for filling.

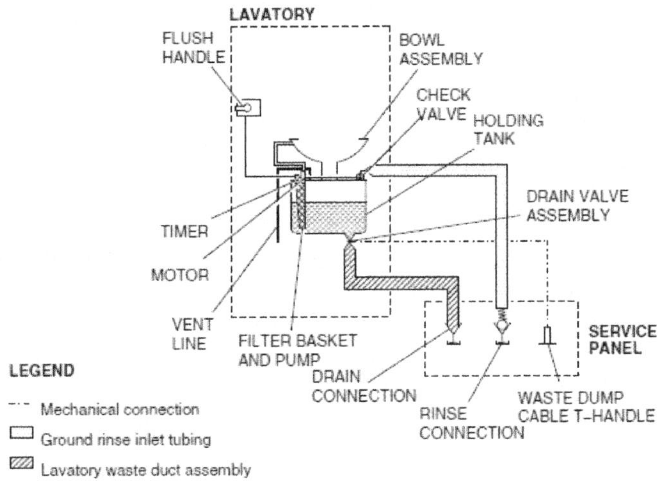

Figure 19 - 6: Lavatory Waste System

System Schematic

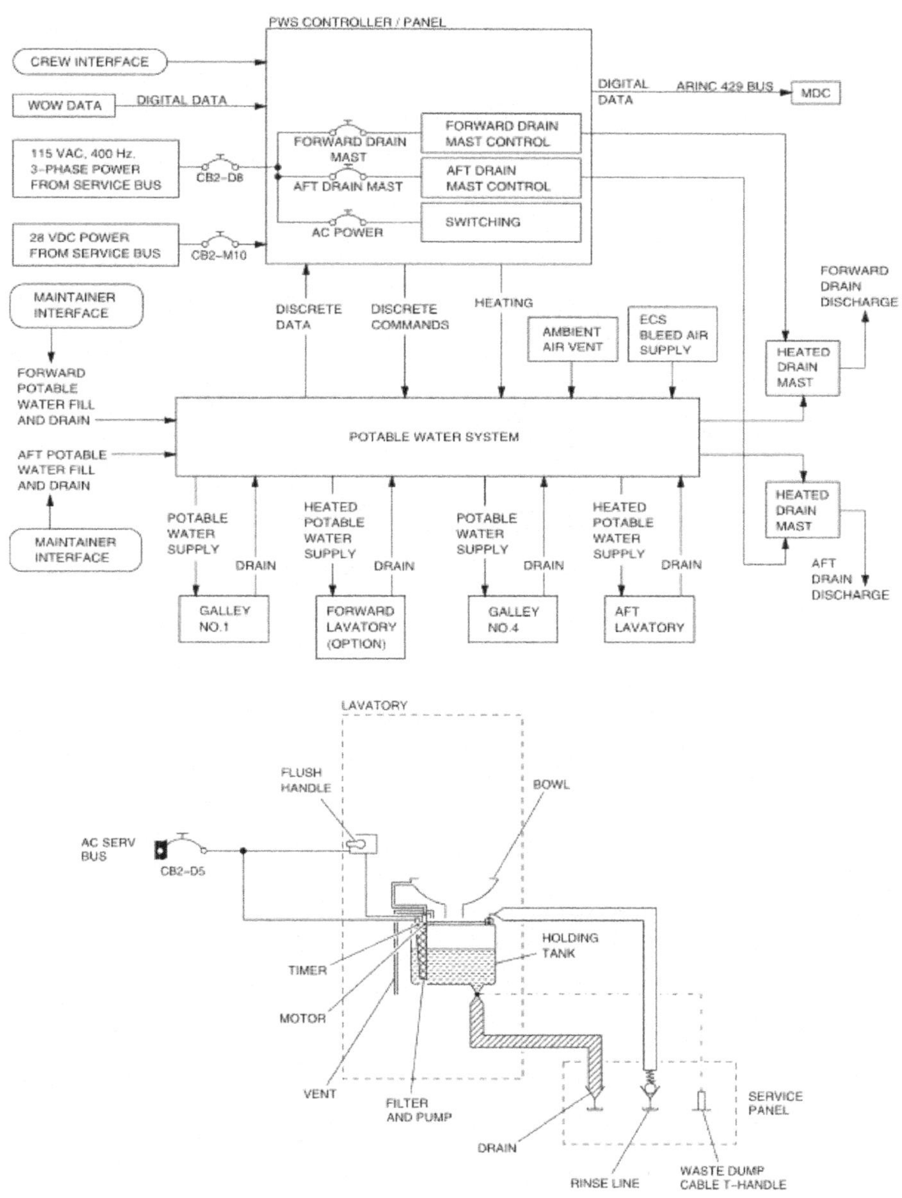

Figure 19 - 7: Water & Waste Schematic

Intentionally Left Blank

Chapter 20: Airborne Auxiliary Power

Chapter 20: Airborne Auxiliary Power

General

The auxiliary power unit (APU) is a fully automated gas turbine with automatic fault detection, shutdown. In the event of an APU fire on the ground, automatic fire extinguishing protection.

The APU's primary function is to operate the gearbox mounted 40 kVA generator. Its secondary function is to supply bleed air for air-conditioning (ECS) and engine starting. Under high demand the APU's electrical loads take priority over pneumatic loads.

The maximum operating altitude of the APU is 41,000 feet; with APU starting up to 37,000 feet. The 40 kVA APU generator may be utilized up to 41,000 feet. APU bleed air extraction is limited to 25,000 feet and below.

An electronic control unit (ECU) controls and monitors all facets of APU operation, including the recording of operating hours and start cycles. The DC-powered ECU sets up the appropriate fuel acceleration and EGT temperature schedules and relays appropriate operating data to the pilots through the EICAS displays.

Two switch-lights on the APU control panel in the flight compartment initiate APU start and shutdown. Operational control of the APU electric and bleed air systems is performed at the ELECTRICAL POWER, BLEED AIR and AIR-CONDITIONING panels.

The APU is mounted in the airplane tail cone and is accessed via two clamshell doors located on the fuselage underside. Airflow through the APU's intake is split into two directions to provide airflow to the intake plenum (the compressor) and to the APU oil/air heat exchanger. The air is heated by the oil/air heat exchanger and is then discharged into the compartment to create ventilation of the enclosed space. The ventilating air is then exhausted overboard through the APU exhaust eductor. The tailpipe's high-velocity exhaust gas creates the venturi action needed to induce airflow through the eductor.

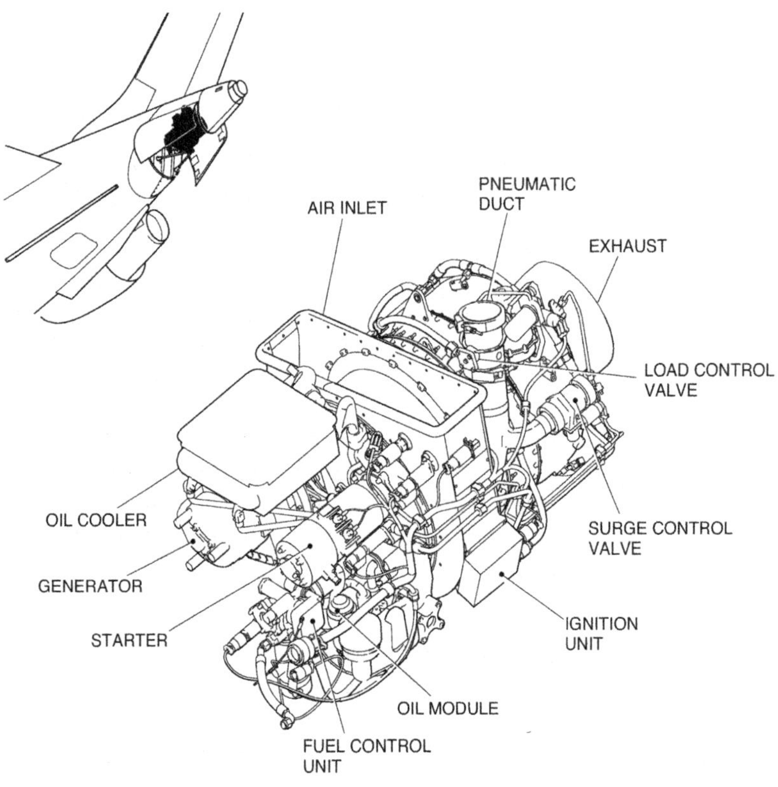

Figure 20 - 1: Auxiliary Power Unit (APU)

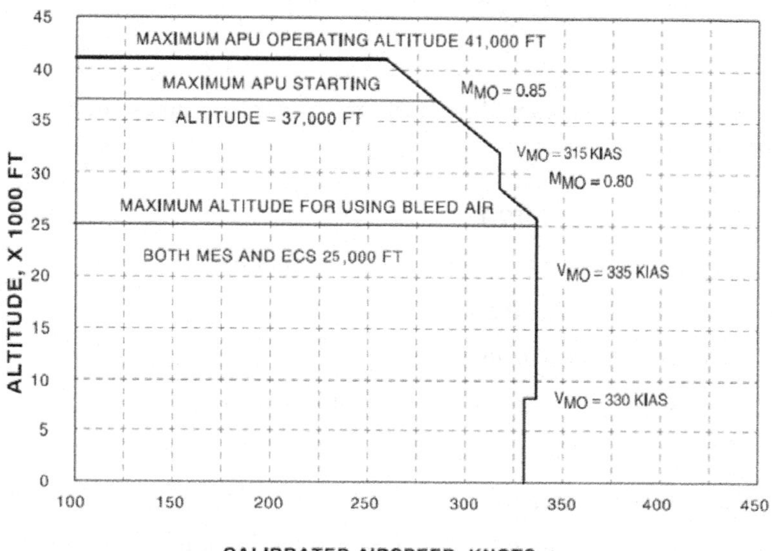

Figure 20 - 2: APU Altitude and Airspeed Envelope

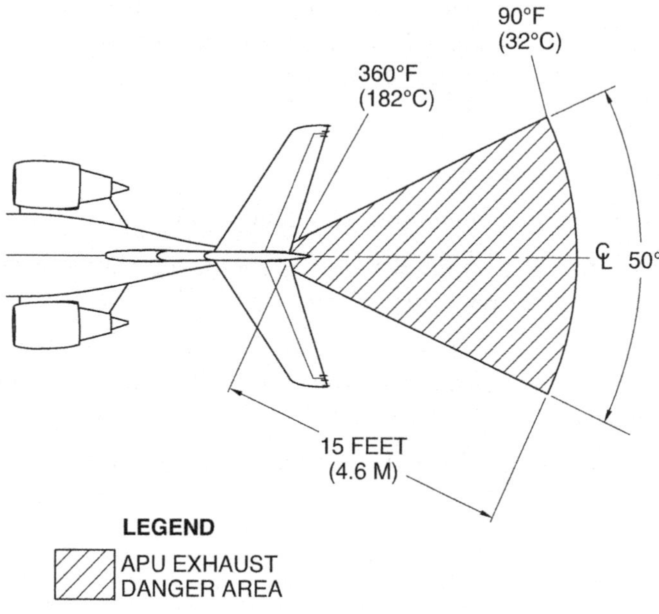

Figure 20 - 3: Exhaust Danger Area

APU Construction

Description

The APU is a fully-automated gas turbine with its own integrated oil, ignition and start systems.

Components and Operation

The APU consists of the following major components:

- Single-stage centrifugal compressor
- Two-stage axial flow turbine
- Accessory gearbox with an integral oil reservoir
- Annular reverse flow combustor
- Exhaust duct and an eductor

Engine power is developed through compression of ambient air by a single-stage centrifugal compressor. The compressed air, when mixed with fuel and ignited in the annular reverse flow combustor, drives a two-stage axial flow turbine. The rotating shaft of the turbine rotor drives the compressor, the gearbox and the accessories needed to support APU operation. The driven accessories include:

- Speed sensor
- Oil pump and lubrication module
- Fuel control unit and high-pressure fuel pump

The APU gearbox also serves as the sump for the lubrication system and incorporates the oil gravity fill cap and quantity sight-glass.

The DC-powered starter motor is used to drive the accessory gearbox which in turn accelerates the gas turbine to a self sustaining rpm during the start sequence.

A 40 kVA generator is mounted on and driven by the gearbox.

Air Intake and Exhaust

Description

The APU air intake is located on the upper right hand side of the rear fuselage just above the APU compartment. The intake door opens outward and into the air-stream. The variable position intake door is controlled by the ECU. Door position is dependent on weight-on-wheels, APU rpm and Mach number.

APU exhaust gases are expelled through an exhaust pipe and eductor in the tail of the fuselage.

Components and Operation

Air Intake Door Control

The ECU controls the position of the intake door via a DC powered actuator.

Should the door actuator fail, maintenance personnel can mechanically lock the door in the open or closed position. This provision allows the aircraft to be dispatched with an unserviceable door actuator.

CAUTION:
If the APU door position is unknown, airspeed is restricted to 220 KIAS, or the APU must remain in operation.

NOTE:
The critical rotation speed is between 4% and 30% rpm. RPM below 4% does not produce enough heat to cause bearing damage, while RPM above 30% ensures sufficient cooling oil is available to ensure proper bearing lubrication.

Intake Door Operation

The APU door is electrically actuated and scheduled open by the ECU. On the ground, the door is commanded fully open upon selection of the PWR/FUEL switchlight. In-flight the door opens with the START/STOP switchlight.

During an in-flight start, the APU intake door position varies with APU rpm. The door opening is limited at low rpm to off-load the APU compressor. At APU rpm greater than 50%, the intake door is fully open.

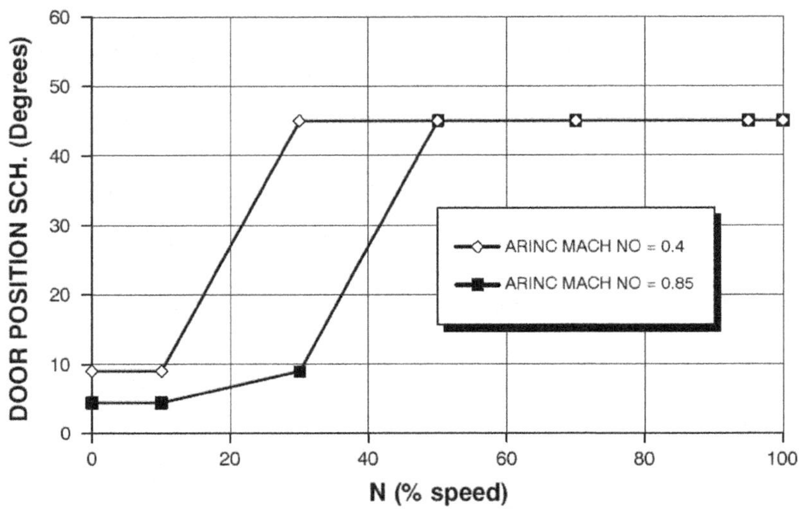

N (% speed)	ARINC MACH NO = 0.4 DOOR POSITION SCH. (Degrees)	ARINC MACH NO = 0.85 DOOR POSITION SCH (Degrees)
0	9.0	4.5
10	9.0	4.5
30	45.0	9.0
50	45.0	45.0
70	45.0	45.0
95	45.0	45.0
100	45.0	45.0

APU DOOR POSITION SCHEDULE

Figure 20 - 4: APU Door Position Chart

Intake Door Indications

The APU door position is displayed on the EICAS status page.

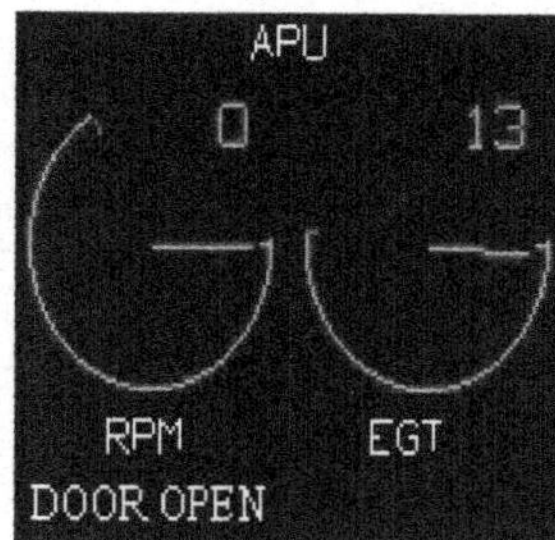

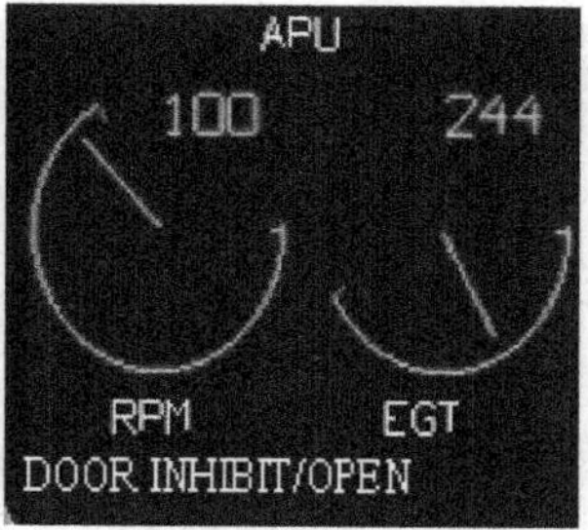

Figure 20 - 5: APU Door Indications

Exhaust

The APU exhaust gases are discharged overboard through an APU tailpipe. The tailpipe is covered by an acoustically insulated exhaust baffle, and includes an exhaust eductor.

Exhaust Gas Temperature Indications

APU exhaust gas temperature (EGT) is indicated on the EICAS status page. The EGT indication uses an indicating arc that expands and contracts with changing APU loads.

EGT is influenced by:

- Ambient air temperature
- Air density
- APU generator loading
- Bleed air demands

When the APU is supplying bleed air, a portion of the air that was used to cool the APU is diverted to the bleed air manifold and an increase in EGT results.

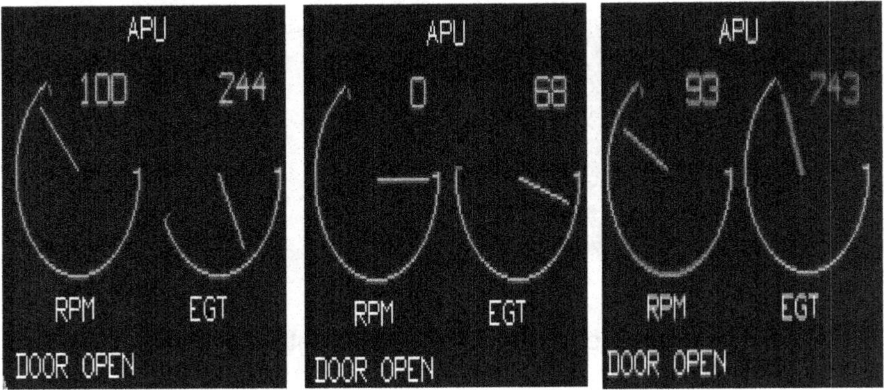

Figure 20 - 6: APU EGT Indications

Lubrication System

Description

The APU is provided with a self-contained lubrication system. During ground operation the loss of oil pressure or high oil temperature results in automatic shutdown. In-flight either fault results in the presentation of an associated caution message but the APU will continue to operate.The APU oil state is also monitored via the EICAS maintenance diagnostic computer and a magnetic chip collector provides a visual indication of metal particles in the lubrication system.

Components and Operation

The oil pump directs filtered oil under pressure to lubricate the APU bearings, gearbox and generator bearings.

Oil is cooled in the APU air/oil heat exchanger, with airflow provided by the APU intake duct. Intake air is split at the APU intake duct, allowing airflow to pass directly across the cooling fins of the heat exchanger. This cooling flow is induced by ram forces during flight and extracted by the exhaust eductor during flight and ground operations. Air heated by the oil cooler is then discharged into the APU enclosure and is used to ventilate the APU compartment.

Fuel Distribution

Description

During start, the fuel system schedules fuel to support combustion and provide for smooth acceleration of the engine to 100% rpm. When the APU is on-speed, fuel flow is modulated to meet the demands of varying pneumatic and electrical loads.

The APU receives fuel from the left collector tank via a dedicated APU fuel distribution manifold. The fuel system incorporates the APU pump, an APU fuel feed shutoff valve (SOV), a fuel shutoff solenoid, fuel system filters, a high pressure fuel pump, and a fuel control unit (FCU).

Components and Operations

APU Fuel Pump

The APU pump draws fuel from the left collector tank, and delivers fuel under pressure to the APU high pressure pump and fuel control unit Should the APU pump fail, an internal bypass valve opens, allowing the accessory gearbox mounted high pressure pump to draw the fuel, permitting continuous operation.

Fuel Feed Shutoff Valve

The APU fuel feed shutoff valve (SOV) is used to interrupt the supply of fuel to the APU. The SOV is activated open by pushing in the PWR/ FUEL switchlight on the APU control panel. It is closed automatically by the ECU when an APU fire is detected, or manually commanded closed when the PWR/FUEL or APU FIRE PUSH switchlight is pressed.

Fuel Control Unit and Fuel Shutoff Solenoid

The APU fuel control unit (FCU) meters fuel under command of the ECU. During the start sequence, the fuel shutoff solenoid, located between the FCU and the fuel nozzles, is activated open by the ECU at 5% rpm. Fuel is supplied in accordance with a schedule to accelerate the

APU throughout the start sequence. When the APU is stabilized at 100% rpm, fuel scheduling is influenced by generator/bleed air demands and deviation from either normal onspeed or EGT parameters.

During normal or automatic APU shutdown, a simulated over-speed signal is generated to close the fuel solenoid SOV. Closing the SOV interrupts the supply of fuel and the APU shuts down.

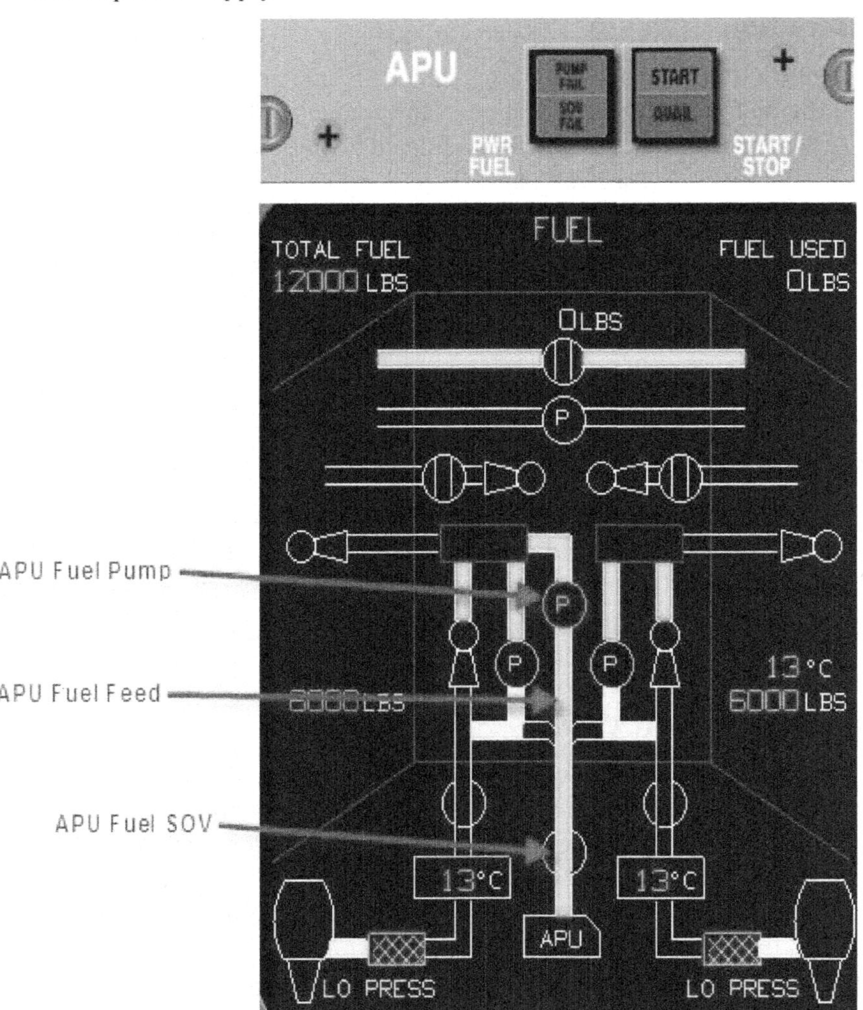

Figure 20 - 7: FUEL Synoptic Page - APU Fuel Feed

Start and Ignition

Description

The starting and ignition systems operate automatically during the APU start sequence.

Components and Operations

APU Start System

The APU start system consists of a DC starter motor and a series of relays that are controlled by the ECU. The electrical power source is:

- ECU - battery bus (BATT BUS)
- Starter - APU battery direct bus (APU BATT DIR BUS)

During the start sequence the ECU schedules starter disengagement as a function of altitude. For example, at sea level, the starter disengages at approximately 46% rpm. During in-flight starts starter cutout may occur between 46% (below 8,000 feet) and 60% (37,000 feet) rpm, as determined by the ECU. The status message APU START is displayed on EICAS when the starter is engaged.

In flight, an immediate restart (APU starter reenactment) can be initiated upon APU rpm roll down below 12%.

NOTE:
A two-minute delay must be observed between cranking attempts to allow for cooling of starter and starter contactor and for APU drainage.

NOTE:
The APU starter motor duty cycle is limited to no more than 3 starts or start attempts in 1 hour.

High-Energy Ignition

The APU ignition system consists of a high-energy ignition unit and a single igniter plug. The ignition system is energized during the APU start cycle and is de-energized when the APU start sequence is complete.

Should a flame-out occur during operation the ignition unit will automatically energize the igniter through the "auto relight" function of the ECU.

Figure 20 - 8: APU Control Panel

APU Start Sequences

On Ground

NOTE:
Both the APU and main battery are required to start and operate the APU. The minimum battery voltage for both is 22 volts DC.

NOTE:
If the APU is rotating above 5% RPM (on ground) or windmilling above 12% (in-flight) the ECU will inhibit start.

When the APU PWR/FUEL switchlight is selected, the:

- APU SOV OPEN status message is displayed (APU ECU and APU fuel pump activated)
- APU IN BITE status message is displayed until intake door is fully open
- APU RPM and EGT gauges are displayed
- APU DOOR OPEN message displayed

When the APU START/STOP switchlight is pressed, the ECU:

- Energizes the starter motor, APU START status message appears on the status page and the START light on the switchlight illuminates
- Fuel and ignition on at 5% rpm, APU start and acceleration
- 46 - 60% rpm (altitude dependent), ECU de-energizes the APU starter motor, APU START status message is removed and the switchlight START indication extinguishes
- At 95% rpm, ignition de-energized, APU SOV OPEN message removed
- At 99% rpm + 2 seconds, the green AVAIL switchlight illuminates

NOTE:
Illumination of the green AVAIL switchlight on the APU control panel indicates that the APU is available for electrical loading.

In-flight

NOTE:
Maximum in-flight starting altitude of the APU is 37,000 feet.

The in-flight start sequence is essentially the same as the ground start sequence except that the APU DOOR is not opened until the START/STOP switchlight is selected. This change in switchlight logic prevents potential windmilling of the APU.

During an in-flight start, the APU intake door position varies with APU rpm. The door opening is limited at low working rpm to off load the APU compressor. At APU rpm greater than 50%, the door is fully open.

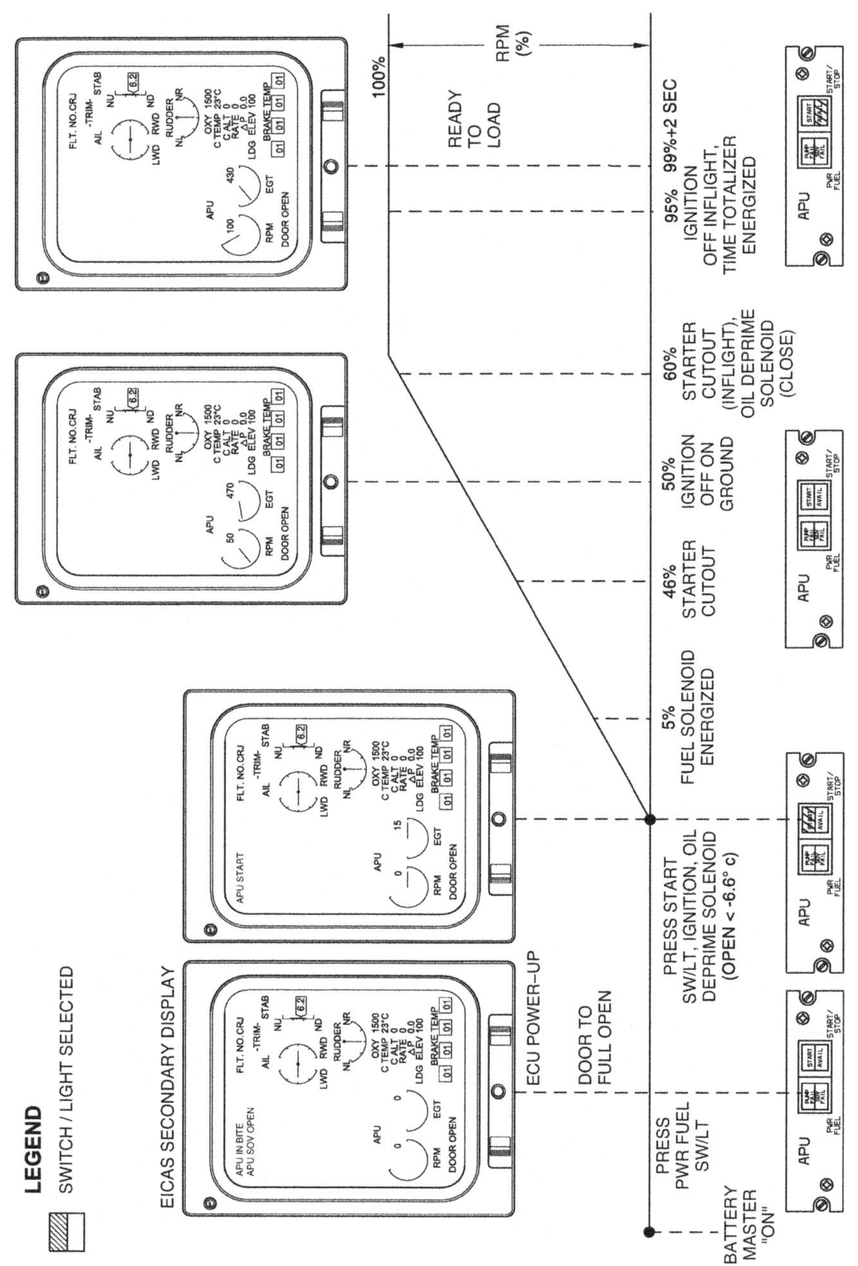

Figure 20 - 9: APU Start Sequence

APU Shutdown Sequence

Ground and In-flight

When the APU ST ART/STOP switch light is pressed:

- Green AVAIL switchlight extinguishes (APU GEN tripped OFF)
- APU fuel shutoff solenoid closes
- APU RPM and EGT decrease
- APU intake door closes when rpm decreases below 25% rpm
- EICAS parameter DOOR CLSD displayed

When the APU PWR/FUEL switchlight is pressed in, the:

- APU RPM and EGT gauges are removed from the status page after approximately 10 seconds

Protective Shutdowns

The ECU will shut down the APU (on the ground or in flight) if any of the following faults occur:

- Over-speed – APU speed exceeded 106%.
- Loss of over-speed protection – A combination of speed sensors or over-speed circuits failed.
- Loss of speed sensor signals – Both speed sensor channels failed.
- APU door failed to open within 30 seconds of command.
- APU door was open then closed without command, while the APU was operating.
- ECU internal failure.
- No APU rotation – During start, speed did not reach 5% within specified time requirement (12 seconds for warm oil; 50 seconds for cold oil).
- No APU light-off – APU light-off was not detected within specified time requirement.
- Slow start – Starting time period exceeded.
- No acceleration – Acceleration during start was less than 0.05% per second for 15 seconds.

- Speed fall-back – The APU speed dropped below 50% after starter cutout.
- Loss of DC power – Battery power lost for more than 200 milliseconds.
- APU fire/emergency – APU FIRE PUSH switch or one of the emergency shutdown switches was selected.
- Loss of air inlet door position sensor signal – Failure of air inlet door position sensor.

The ECU will shut down the APU (on ground) if any of the following faults occur:

- Over-temperature – APU EGT exceeded schedule limits.
- Low oil pressure (LOP) – Low oil pressure existed for 15 seconds with the APU operating.
- Oil pressure switch failed – Could not detect a low oil pressure condition.
- High oil temperature – Oil temperature exceeded 300° F with the APU operating.
- Reverse flow – APU inlet temperature exceeded 350° F for 5 seconds with the APU operating and LCV open.
- Under-speed – APU was operating and speed dropped below 80% for 5 seconds.
- Loss of EGT sensors – Both EGT sensor channels failed.
- APU oil filter in an impending bypass condition.

Pneumatic Supply

Description

APU compressor bleed air is used to provide pneumatics for engine starting and air-conditioning. A one-way check valve and the control logic of the air-conditioning system controllers (ACSCs) prevents bleed air from the manifold air from reverse flowing to the APU.

Components and Operation

The APU pneumatic control system consists of two components, the load control valve (LCV) and the surge control valve (SCV).

Load Control Valve Modulation/Bleed Air Loads

The pneumatic load on the APU is controlled by the ECU in response to commands sent from the ACSCs. When the LCV is opened, a portion of the APU's compressor discharge air is diverted to the bleed air manifold. The APU's EGT increases in accordance with the bleed air demand. Should EGT reach the limit, the ECU will reduce bleed flow by modulating the LCV toward the closed position, to prevent EGT over temperature.

Load Control Valve

The APU load control valve (LCV) is controlled by ECU and ACSC in automatic mode. In manual mode the valve is controlled by two switches on the BLEED AIR control panel. Manual mode has priority over the automatic function.

Surge Control Valve

The APU surge control valve (SCV) is automatically controlled by the ECU. The SCV is commanded open when the APU is supplying electrical loads at flight altitudes above 17,000 feet. This function reduces APU surge potential when operating in an unloaded bleed air condition.

Bleed Air Protection

For additional information on APU bleed air protection refer to Chapter 18 - Pneumatics

Bleed Air Availability

When in the essential (flight) mode, the ECU waits for 2 seconds after 99% rpm before sending out a pneumatic ready-to-load (RTL) signal. This signal is generated by the APU ECU, and is sent to air-conditioning system controller (ACSC) via the ARINC data bus. This pneumatic RTL signal is removed if the APU LCV fails to OPEN.

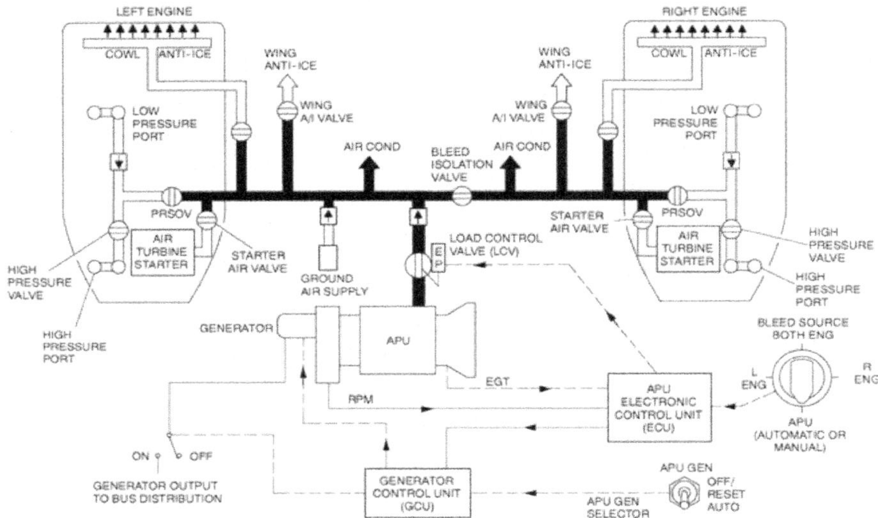

Figure 20 - 10: Bleed Air Pneumatics

Bleed Air Limitations

APU bleed air extraction is not permitted above 25,000 feet.

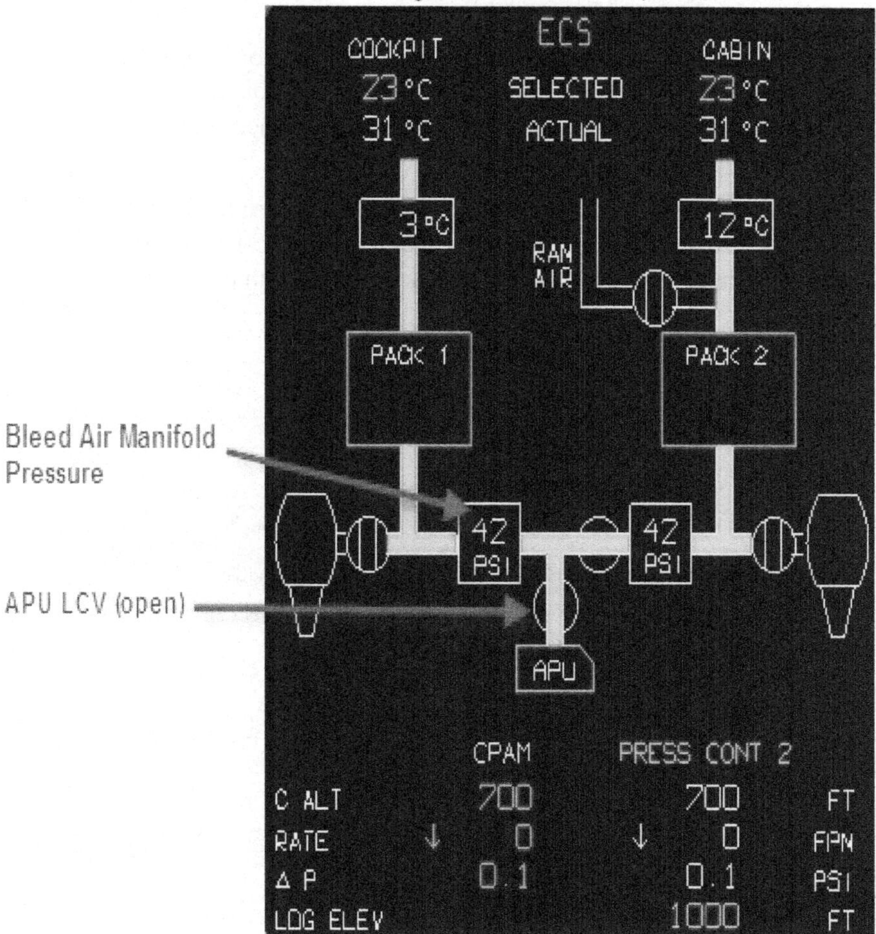

Figure 20 - 11: ECS Synoptic Page - APU Indications

Controls and Indications

Description

The APU control panel provides control and system indications for the APU.

During ground operations, remote shutoff switches located outside of the flight deck can be used to shut down the APU during a ground emergency.

AC electrical power provided by the APU generator is controlled from the ELECTRICAL POWER control panel. Bleed air services are controlled from the BLEED AIR control panel when the system is operated in manual mode.

APU system operation and fault presentation is provided on the EICAS primary and status pages. The FUEL, ECS and AC ELECTRICAL synoptic pages are used to monitor the APU system and services.

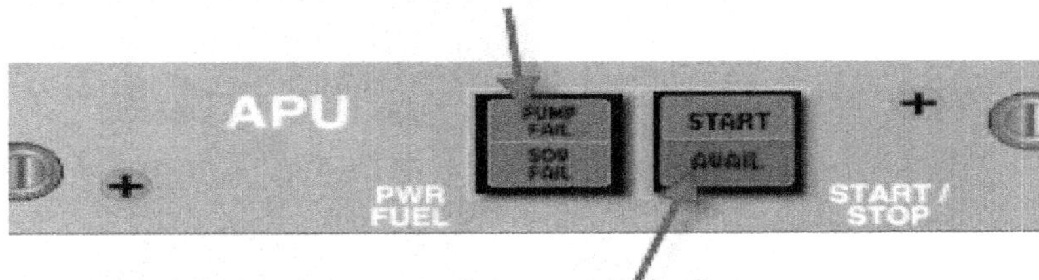

Figure 20 - 12: APU Control Panel

APU Remote Shutoff Switches

The aircraft is equipped with two APU remote shutoff switches. One switch is located on the external services panel. The second (pushbutton type) shutoff switch is located inside the APU compartment.

BLEED AIR Control Panel

The BLEED AIR control panel is used to manually select the APU as the bleed air source. The APU LCV position is displayed on EICAS as a status message and pictorially on the ECS synoptic page.

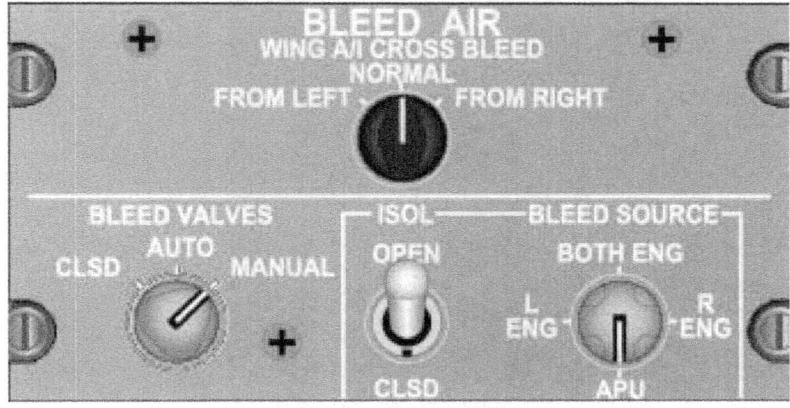

APU Manually Selected as Bleed Source

Figure 20 - 13: APU Manually Selected as Bleed Source

ELECTRICAL POWER Control Panel

The ELECTRICAL POWER control panel is used to control the operation of the APU generator.

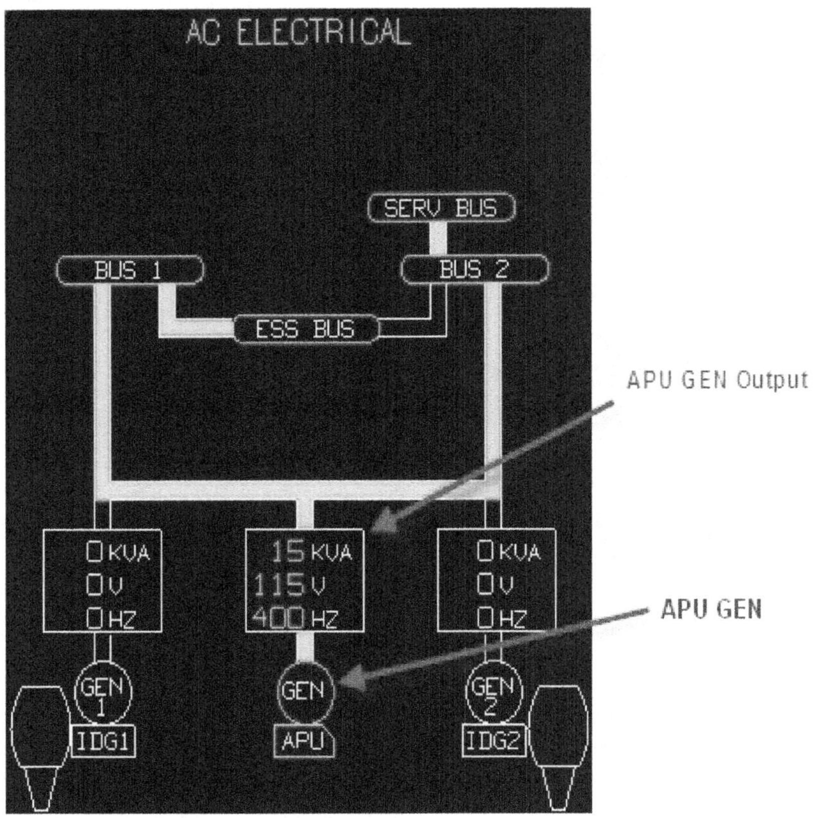

Figure 20 - 14: ELECTRICAL POWER Control Panel

EICAS Status Page

Flight compartment indications of intake door position are displayed in the lower left-hand corner of the EICAS status page.

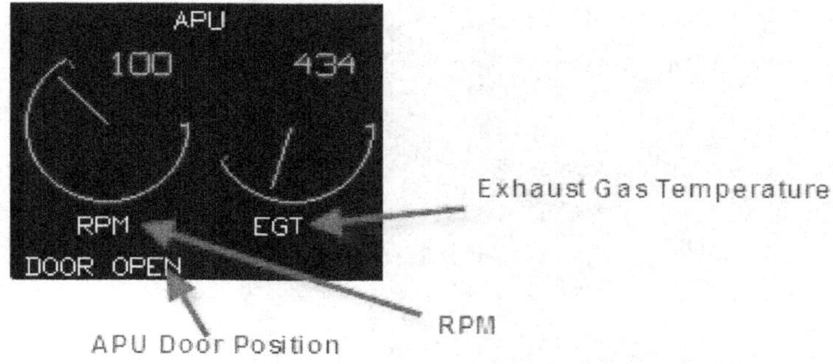

Figure 20 - 15: Status Page - APU Indications

ECS Synoptic Page

The ECS synoptic page can be used to monitor the APU LCV.

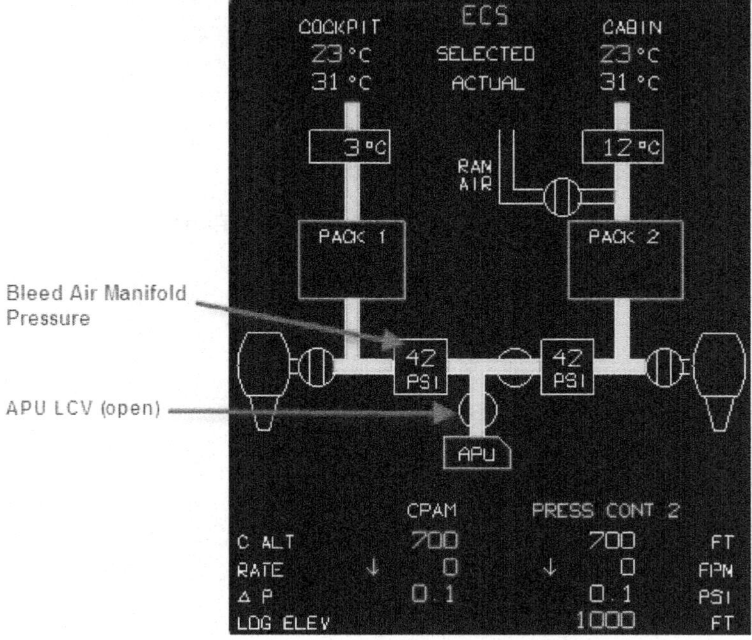

Figure 20 - 16: ECS Synoptic Page

AC ELECTRICAL Synoptic Page

The AC ELECTRICAL synoptic page can be used to monitor the APU generator.

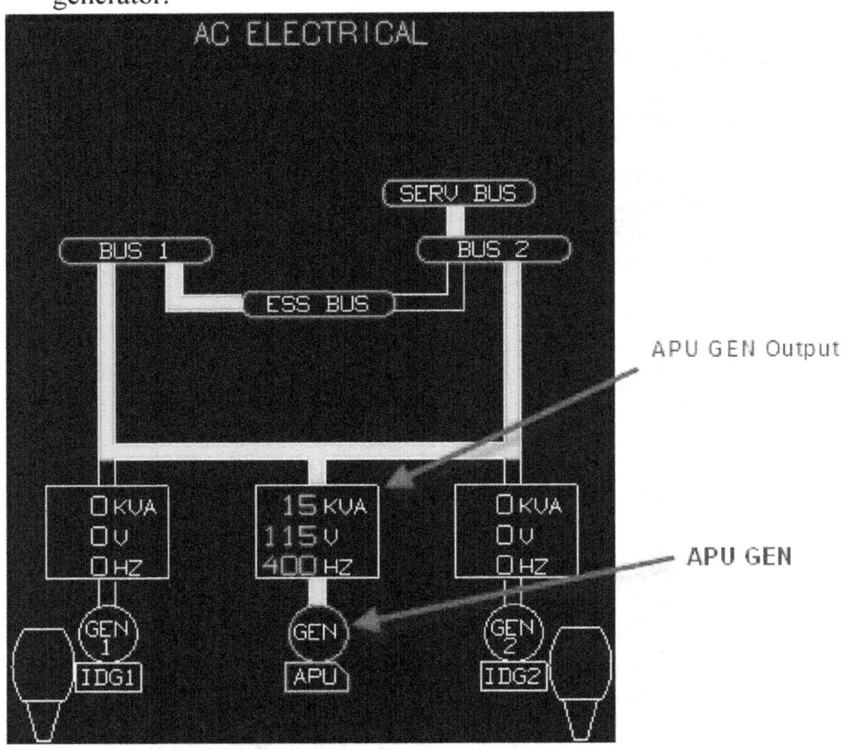

Figure 20 - 17: AC ELECTRICAL Synoptic Page

FUEL Synoptic Page

APU pump and fuel SOV operation can be monitored on the FUEL synoptic page.

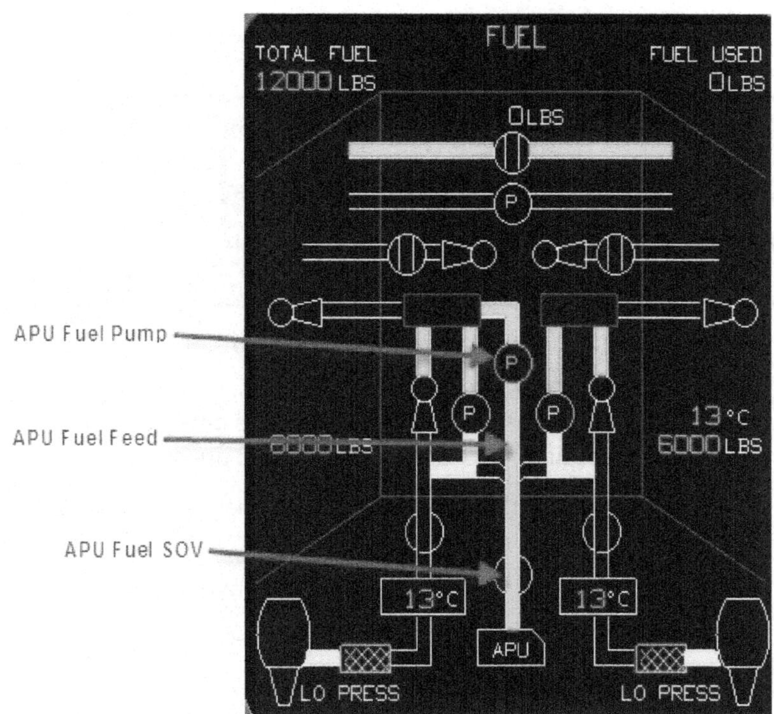

Figure 20 - 18: FUEL Synoptic Page

Controls and EICAS Messages

APU Control Panel

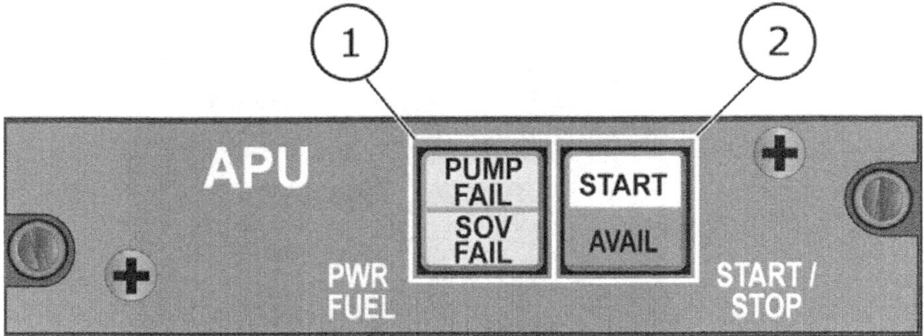

Figure 20 - 19: APU Control Panel

1. **PWR FUEL Switchlight** – When pressed in, cross-flow/APU pump is energized, APU IN BITE, APU gauges energized, door scheduled to open, and APU fuel shut-off valve opens.
   - PUMP FAIL light illuminates to indicate that cross-flow/APU pump has failed.
   - SOV FAIL light illuminates to indicate the APU fuel feed SOV has failed.

2. **APU START/STOP Switchlight**

   When pressed in, starter motor is energized and START light (white) illuminates.

   - At 50% rpm, START light extinguishes.
   - At 99% rpm, and 2seconds later, AVAIL light (green) illuminates.

   When pressed out:

   - FCU shut-off valve closes.
   - APU shuts down, and AVAIL light extinguishes.

External Service Panel

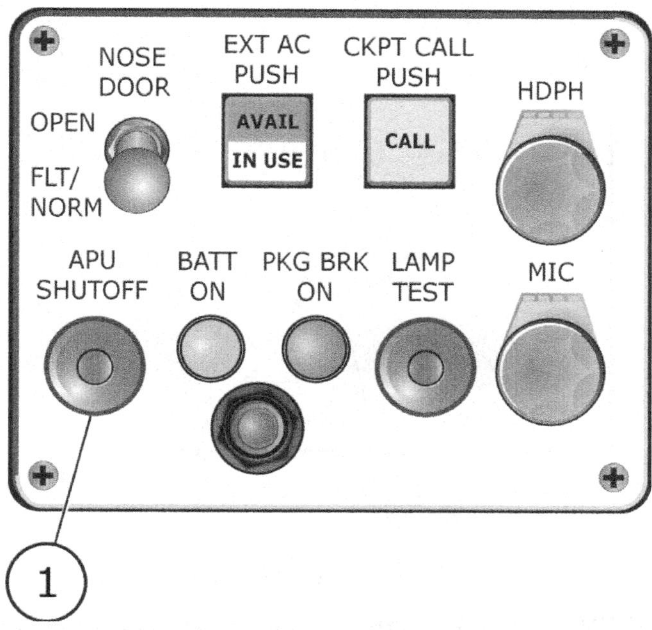

Figure 20 - 20: External Service Panel

1. **APU SHUT-OFF Switch –** Used by ground personnel to shut down the APU during ground servicing.

AFT Equipment Bay

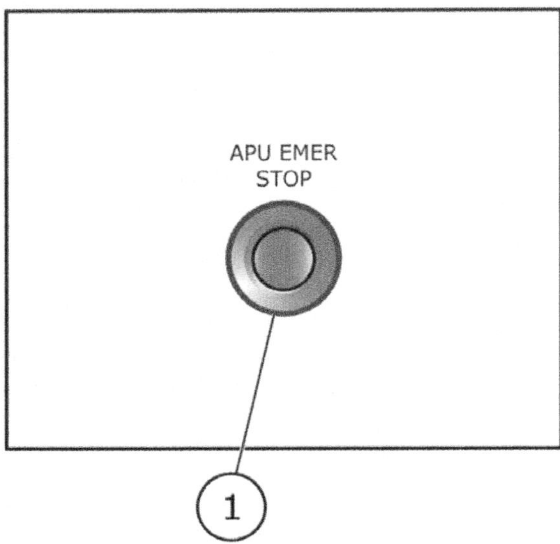

Figure 20 - 21: Aft Equipment Bay

1. **APU EMER STOP Switch** – Used by ground personnel to shut down the APU during ground servicing.

Status Page

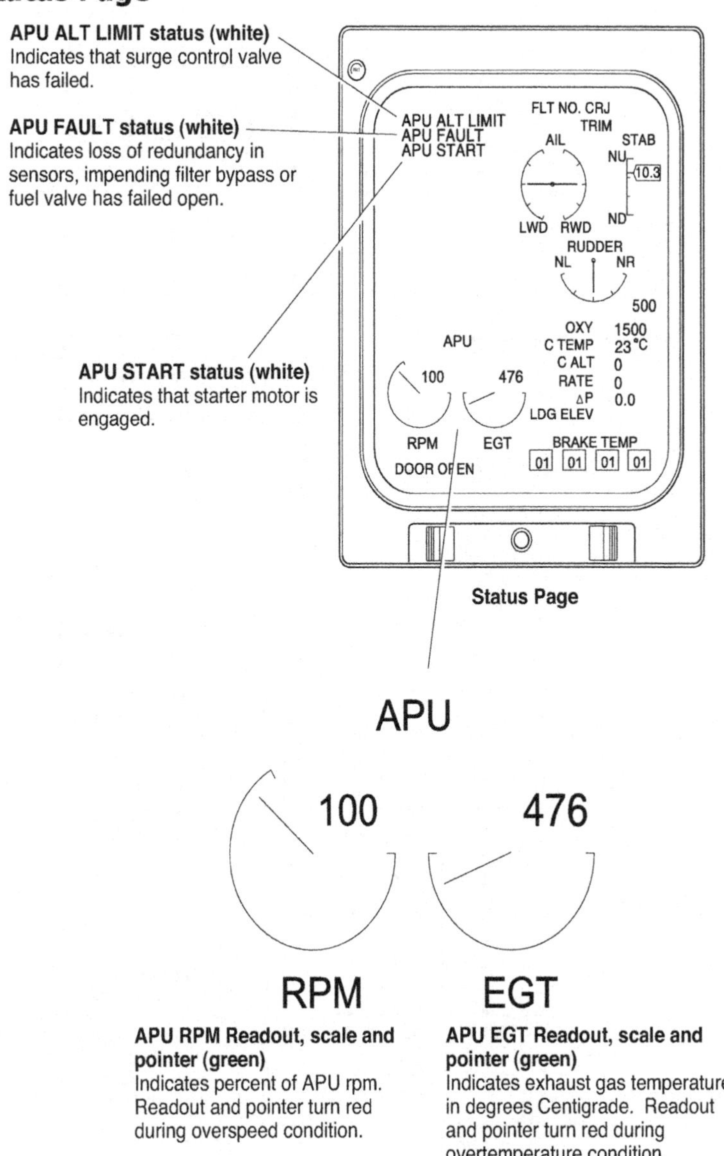

Figure 20 - 22: Status Page

Fuel Synoptic Page

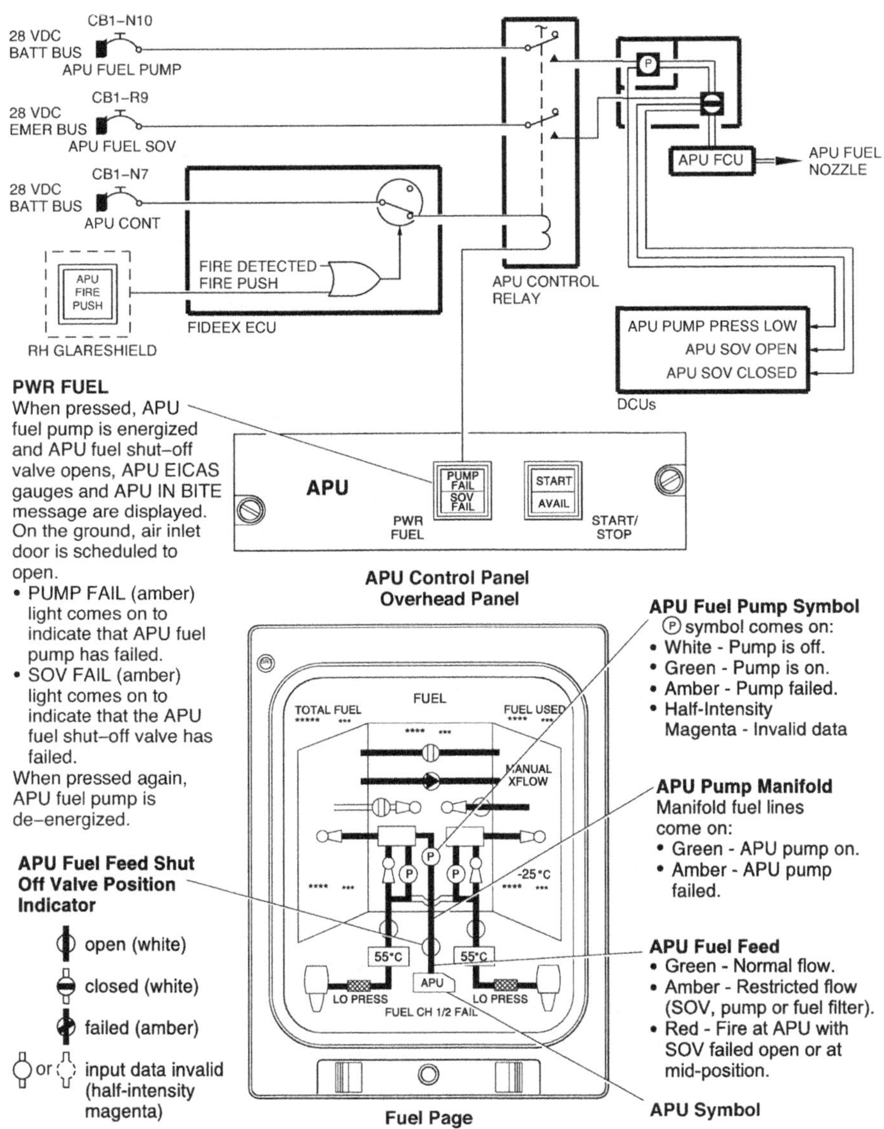

Figure 20 - 23: Fuel Synoptic Page

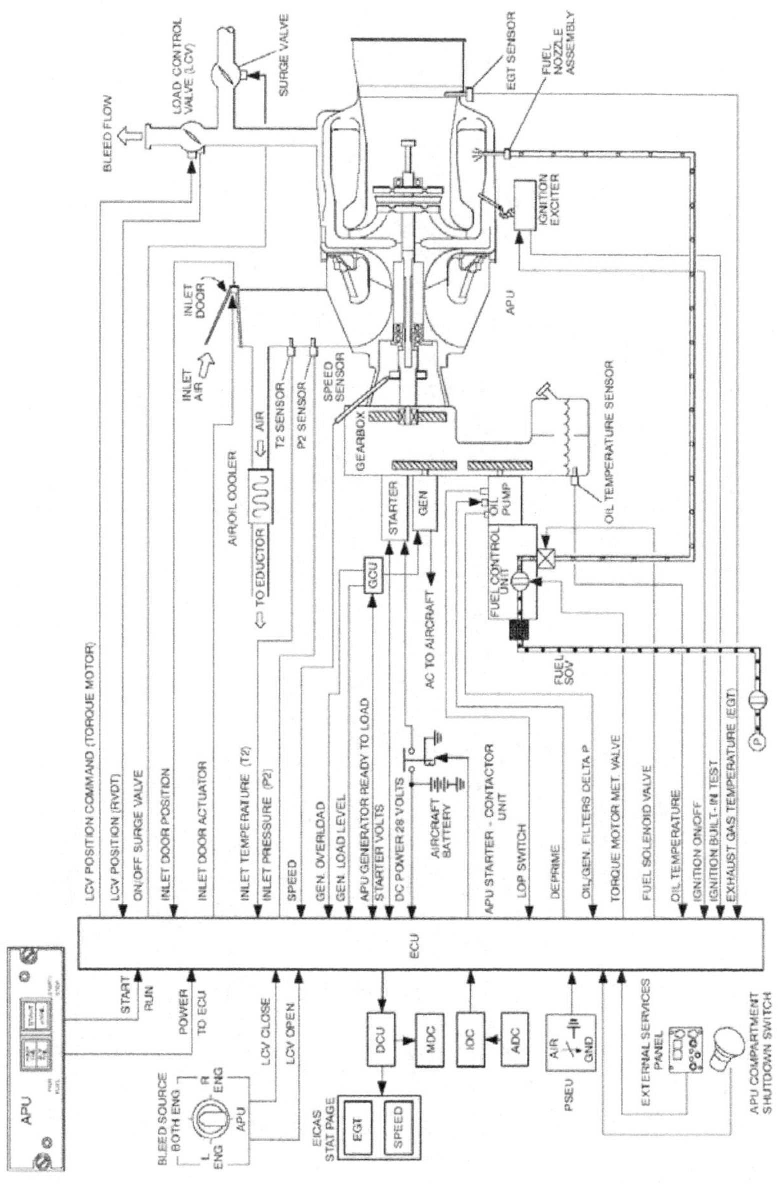

Figure 20 - 24: APU Controls and ECS Interface

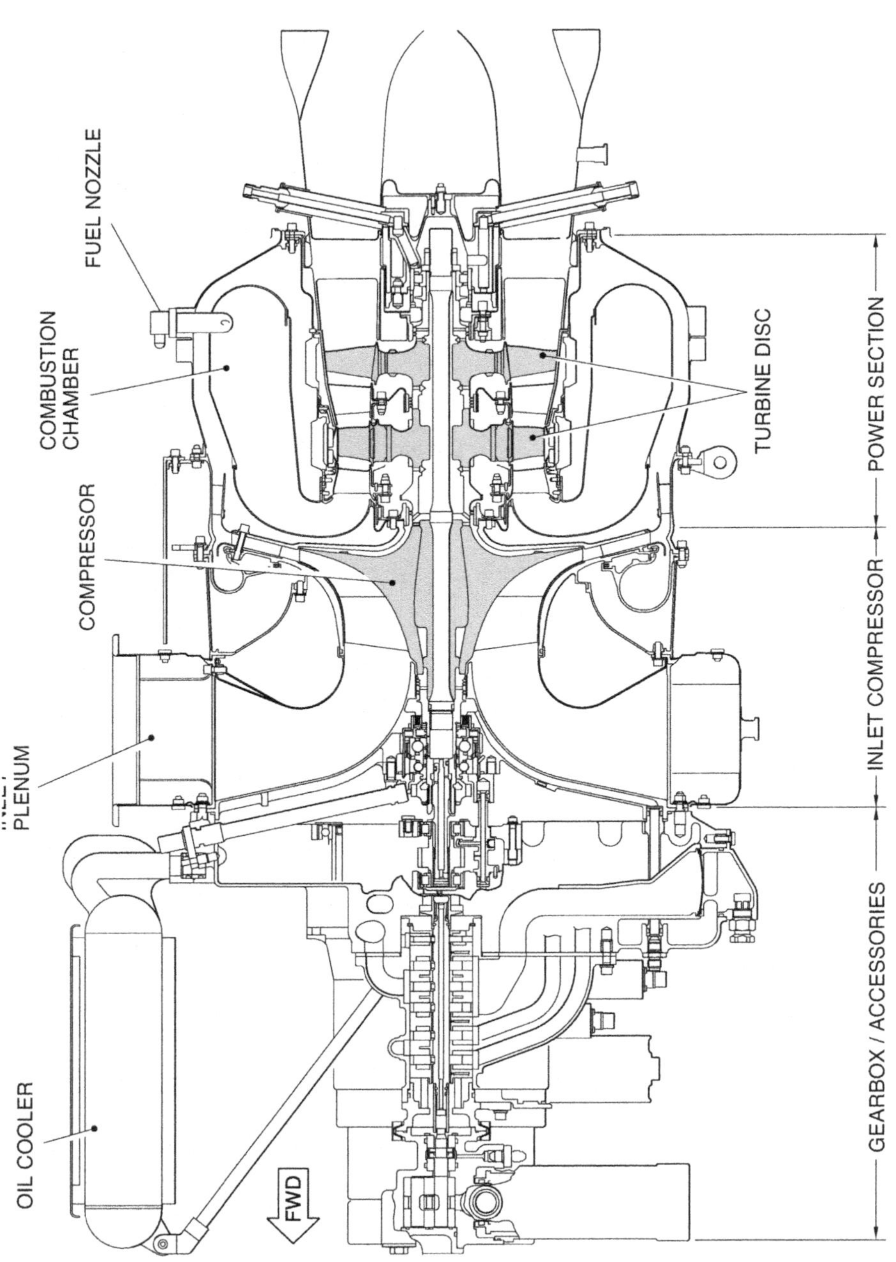

Figure 20 - 25: APU Cross Section

EICAS Messages

Master Warning	Aural	Cause
APU OVERSPEED	"APU"	APU is running greater than 107% rpm. APU shuts down automatically.
APU OVERTEMP	"APU"	EGT is greater than 743° C with APU RPM greater than 87% or greater than 974° C at all times. APU shuts down automatically (on ground).
Master Caution	**Cause**	
APU LCV FAIL	APU load control valve has failed (open or closed).	
X-FLOW/APU PUMP	Illuminates to indicate that X-flow/APU pump has failed (low pressure). Accompanied by XFER/APU PUMP Switch/light indicating FAIL.	
APU OIL PRESS	APU has low oil pressure. APU shuts down automatically. (On ground only).	
APU OIL TEMP	APU has high oil temperature (325° C). APU shuts down automatically. (On ground only).	
APU FAULT	Loss of APU control circuits (microprocessor, thermocouple or speed signal). APU shuts down automatically.	
APU SOV FAIL	APU Fuel Feed SOV not in a confirmed position (either open or closed).	
APU SOV OPEN	APU Fuel Feed SOV is open 10 seconds after an APU fire condition. APU FUEL SOV light, on APU panel, illuminates (amber).	
APU LCV FAIL	Load control valve has failed (either open or closed).	
APU BLEED ON	Load control valve is open and barometric altitude is greater than 15,000 feet.	

Status Message	Cause
APU IN BITE	Electronic control unit is running diagnostics (prior to start), for approximately 5 seconds.
APU START	APU start in progress.
APU ECU FAIL	Electronic control unit has failed. APU fuel feed SOV is confirmed open.
APU LCV OPEN	APU load control valve is open. 10th stage, APU LCV OPEN / Switch/light illuminates.
APU SOV OPEN	APU fuel feed SOV is open with the APU not ready and no APU fire detected.
Advisory Message	**Cause**
APU SOV CLSD	APU fuel feed SOV is closed during an APU fire condition.

System Schematic

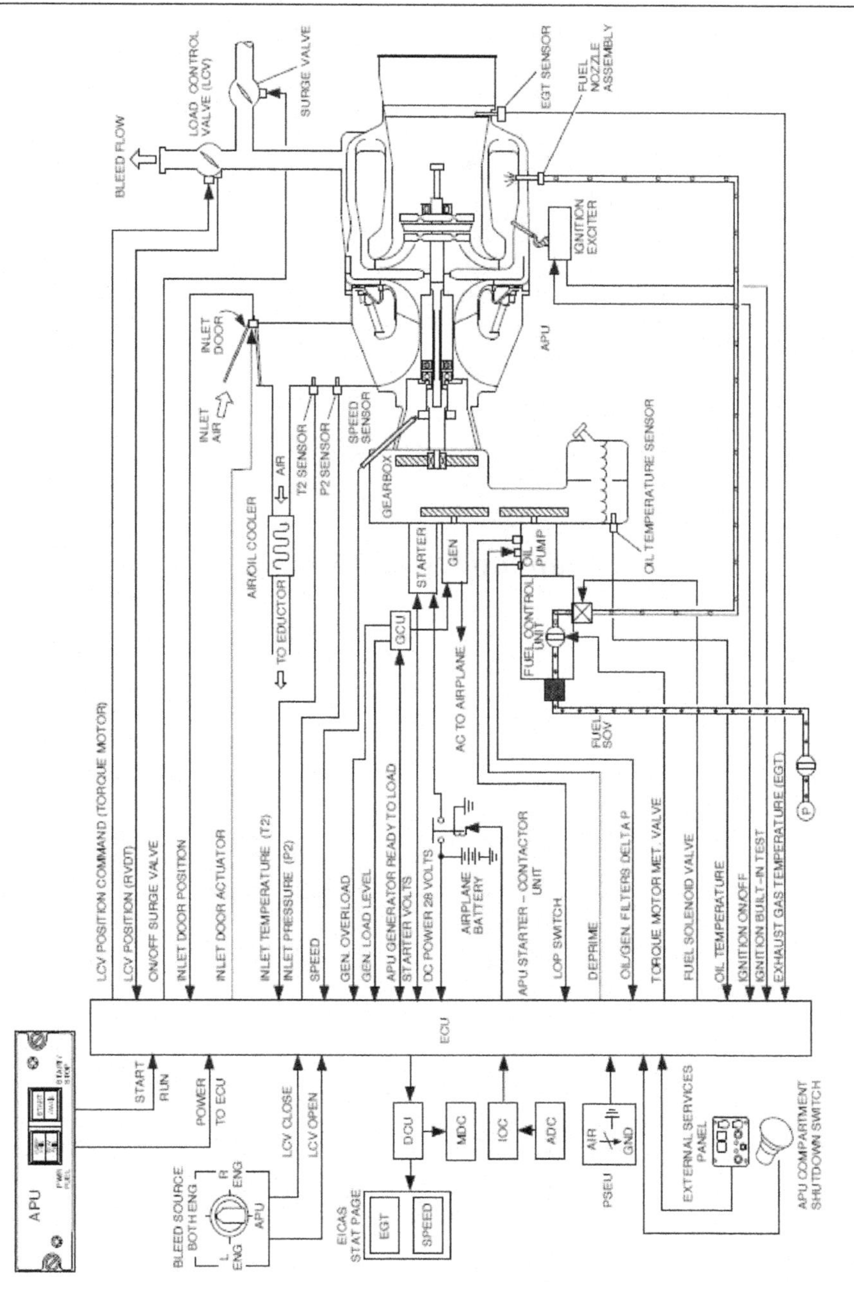

Figure 20 - 26: APU Controls & ECS Interface Schematic

Intentionally Left Blank

Chapter 21: Doors

Intentionally Left Blank

Chapter 21: Doors

General

The CRJ900 has 12 doors consisting of the passenger door, forward galley service door, four over-wing emergency exits, a flight deck overhead escape hatch, three cargo bay doors, an avionics bay door, and an aft equipment bay door.

The passenger door, galley service door, flight deck overhead escape hatch and the over-wing exits can be operated from inside or outside of the airplane. The passenger door and galley service door can also be used for emergency evacuation. The cargo bay doors, the aft equipment bay door and the avionics bay door can only be operated from outside the airplane All doors, except the aft equipment bay door and the flight deck overhead escape hatch are monitored by the proximity sensing electronic unit (PSEU).

Sensors and switches on the doors supply position signals to the PSEU. The PSEU processes the signals and then transmits the data to the EICAS system for display of warning and caution messages on the primary page and for door status indications on the DOORS synoptic page.

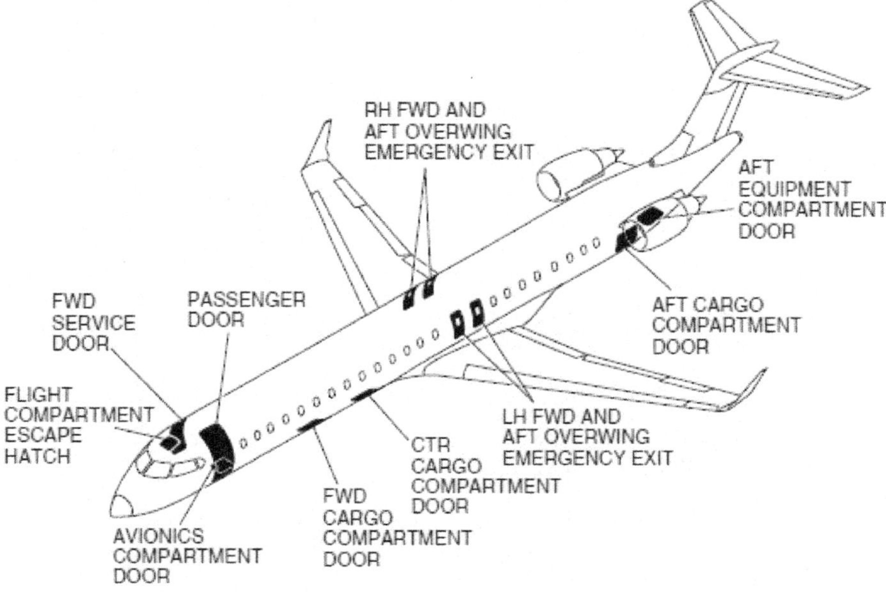

Figure 21 - 1: Door and Hatch Locations

Passenger Door

The passenger door, located at the forward left side of the fuselage, is the main entrance and exit to the cabin area.

The passenger door incorporates integral stairs with a retractable lower step and folding handrails. The door is hinged at the cabin floor level and opens outward. A counter balance mechanism with gas springs is used to take the weight of the door and to dampen the door movement. At the fully open position, the door rests on a support wheel.

Handrails are provided to assist passengers in boarding and disembarking. Mechanical linkages raise the handrails when the door is opened and collapse them when the door is closed. When a jetway is used, the handrails must be collapsed. Collapsing of the handrails is done by removing the forward and aft handrail quick-release pins.

Closing the passenger door from inside the airplane is normally accomplished using the power assist system which is controlled from a DOOR ASSIST switch light on the forward flight attendants panel.

The passenger door indications are displayed on the primary page and on the DOORS synoptic page.

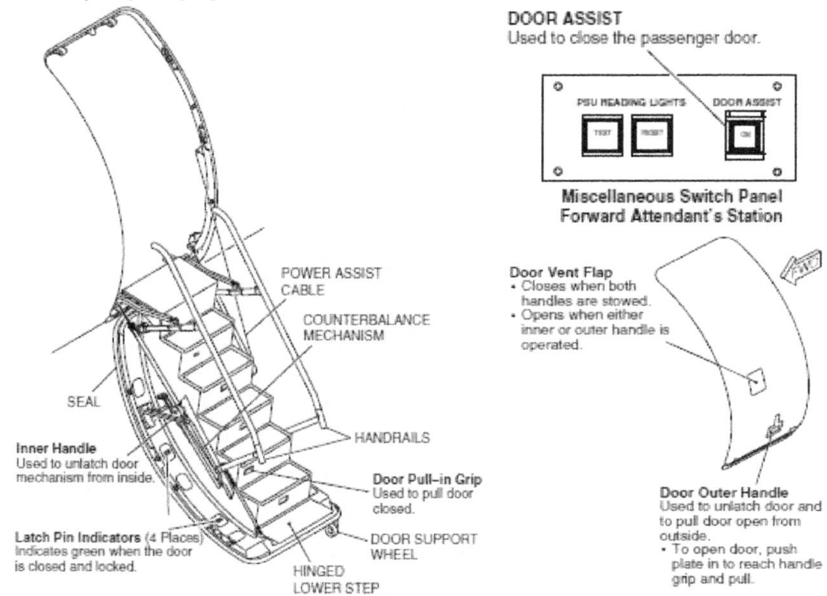

Figure 21 - 2: Passenger Door

Operation

1. To open the door from the inside:
   a. Lift the inner handle out of its cam recess
      (1) The outer handle ejects from its recess
      (2) The latch mechanism unlocks
      (3) The vent flap opens
   b. Continue the upward movement of the handle to the OPEN position
      (1) The latch cams and latch pins disengage from the door frame fittings
      (2) Fwd and aft pullout levers open the door to the near vertical (balanced) position
   c. Firmly push the door outward
      (1) The door descends in a gradual downward movement (dampened by the counterbalance mechanism gas springs).
      (2) The retractable lower step and folding handrails deploy.
      (3) The door support wheel extends and locks in place before reaching the ground

NOTE:
The maximum load capacity of the door is 1000 lb or a maximum of four passengers on the stairs at any time.

2. To close and latch the door from inside with power assist system:
   a. Press and hold the DOOR ASSIST switch on the forward flight attendant panel

NOTE:
Do not operate the door assist motor during electrical power system switching.

      (1) The electrical motor pulls the door up and stops automatically when the door reaches the near vertical position

NOTE:
The cam mechanism under the lower step includes a lock lever that prevents the inner handle from moving to the closed position until the door is fully pulled into the fuselage structure.

b. Take hold of the handle in the second step riser and pull the door fully closed

(1) The fwd and aft pull-in/push-out levers engage in respective cams to hold the door in this position

c. Push the inner handle down to the CLOSED position

(1) The latch cams and latch pins engage in the door frame fittings

(2) The inner handle, the outer handle and the door vent flap close simultaneously

d. Make sure that the visual indications of door latches are as follows

(1) The green marks on the two latch cams and the green mark on the lock pawl must align with the green marks on the door structure (located under the hinged lower step)

(2) The green marks on the four latch pins must align with green marks on the latch pin indicator windows

(3) The latch mechanism lower lock indicator flag changes from a red UNLOCKED to a green LOCKED indication

3. To open the door from outside:

a. Push-in the outer handle push plate

(1) The handle ejects

b. Take hold of the handle grip and pull outward then downward

(1) The door latch mechanism unlocks

(2) The vent flap opens

(3) The latch cams and latch pins disengage from the door frame fittings

(4) The fwd and aft pull-in levers open the door to near vertical position

(5) The door descends in a gradual downward movement (dampened by the counterbalance mechanism gas springs).

(6) The two folding handrails deploy

(7) The support wheel extends and locks in place before reaching the ground

CAUTION:
Door support cables must be installed when moving the airplane with the door open or when placing the airplane on jacks.

4. To close and latch the door from outside:.

   a. Manually raise the door up and push it fully closed

NOTE:
The gas springs will assist in retracting the door up to near vertical position.

(1) The fwd and aft pull-in levers engage in respective cams to hold the door in this position

   b. Push the outer handle down fully in its recess

(1) The latch cams and latch pins engage in the door frame fittings

(2) The inner handle, outer handle and vent flap close simultaneously

5. (5) Lowering the stair handrails:

   a. When the door is closed:

(1) Remove the two quick-release pins from the holes of the stair handrails

(2) Stow the quick-release pins in the storage holes of the brackets

(3) Open the passenger door

   b. When the door is open:

(1) Hold the stair handrails and remove the two quick release pins from the holes of the stair handrails

(2) Stow the quick-release pins in the storage holes of the brackets

(3) Lower the stair handrails and ensure that the clips on the stair handrails attach to the quick release pins.

CAUTION:
Do not use force to lift or lower the stair handrails. Make sure that the bottom step moves freely when the stair handrails are lifted or lowered.

---

6. Lifting the stair handrails:

   a. When the door is open:

CAUTION:
Do not close the door unless the two quick-release pins are installed in the handrails or inserted in the stowage brackets.

---

(1) a. Remove the quick-release pins from the storage holes of the brackets

(2) b. Lift the handrails into position

(3) c. Install the two quick-release pins in the holes of their handrails

The quick-release pins must be installed in the holes of the stair handrails before you move the airplane. This is necessary so that the stair handrails are in the upper position in the event of an emergency evacuation.

b. When the door is closed:

(1) Remove the two quick-release pins from the storage holes of the brackets

(2) Insert the quick-release pins into the holes of the stair hand-rails

---

CAUTION:
For all procedures concerning the handrails, the door must not be closed unless the quick release pins are either in the storage holes of the brackets or in the holes of the stair handrails.

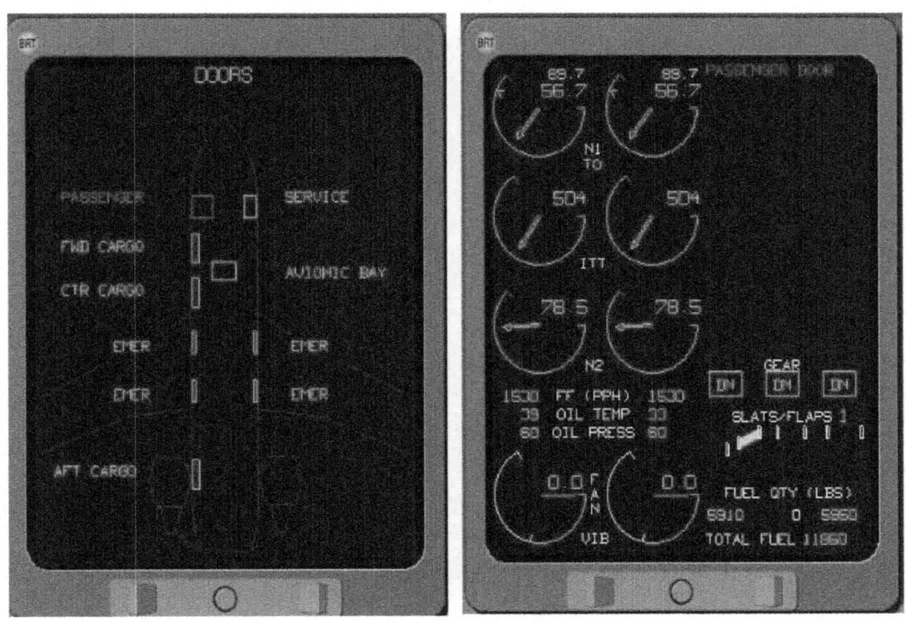

Figure 21 - 3: Passenger Door Warning

Emergency Exits

Emergency evacuation of the cabin area is done through the passenger door, the galley service door, the 4 over-wing emergency exits, and the flight deck overhead escape hatch. All emergency exits can be opened from the inside or outside of the airplane.

> ***NOTE:***
> Although the passenger door and galley service door can be used as emergency exits, their operation is not covered in this section. Refer to the respective sections of this chapter for the operation of the passenger and galley service doors.

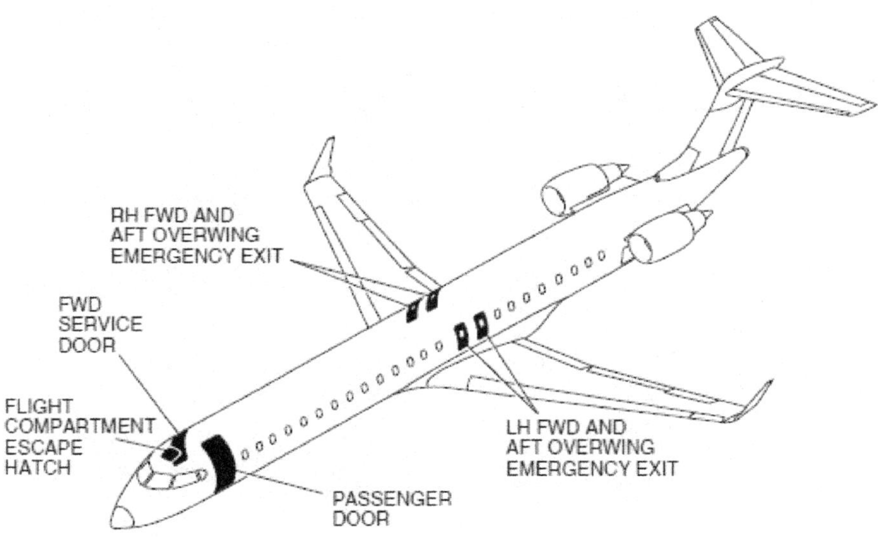

Figure 21 - 4: Emergency Exits

Flight Deck Escape Hatch

The flight deck overhead escape hatch provides an emergency exit for the pilots in case of emergency evacuation. The hatch is an inward opening plug type door and is removable at the hinge supports. The hatch can be opened from both the inside and outside of the airplane. The hatch is supported in the fuselage cut-out by two hinge hook fittings on the forward side and two lock pins on the aft side. Once the door is open, it can be lifted off the forward fittings and removed from the area.

The flight deck escape hatch is not monitored by the PSEU, therefore there is no EICAS indication of door position.

The hatch inner and outer handles are rotated to the OPEN position to unlatch, and rotated to the CLOSED position to latch.

1. To open the flight deck escape hatch from inside:
    a. Press the hatch inner handle release button
        (1) The inner handle ejects
    b. Rotate the inner handle to the OPEN position (left)
        (1) The hatch mechanism unlatches
    c. Manually move the hatch fully down
2. To close the flight deck escape hatch from inside:
    a. Manually lift the hatch into its opening
    b. Push the aft part of the hatch up to squeeze the seal, and rotate the inner handle to the CLOSED position (right)
        (1) The hatch mechanism latches as the inner handle reaches the end of its rotation
    c. Correctly align the inner handle with the door recess and push it fully in.
        (1) The inner handle locks in stowed position
3. To close the flight deck escape hatch from outside:
    a. Press the outer handle trigger plate
        (1) The outer handle ejects
    b. Rotate the outer handle to the OPEN position and carefully lower the hatch fully down

4. To close the flight deck escape hatch from outside:
   a. Using the door outer handle, manually lift the hatch into the fuselage opening
   b. Pull the outer handle up to squeeze the seal and rotate it to the CLOSED position
      (1) The hatch mechanism latches as the outer handle reaches the end of its rotation
   c. Correctly align the outer handle with the door recess and push it fully in.
      (1) The outer handle locks in stowed position

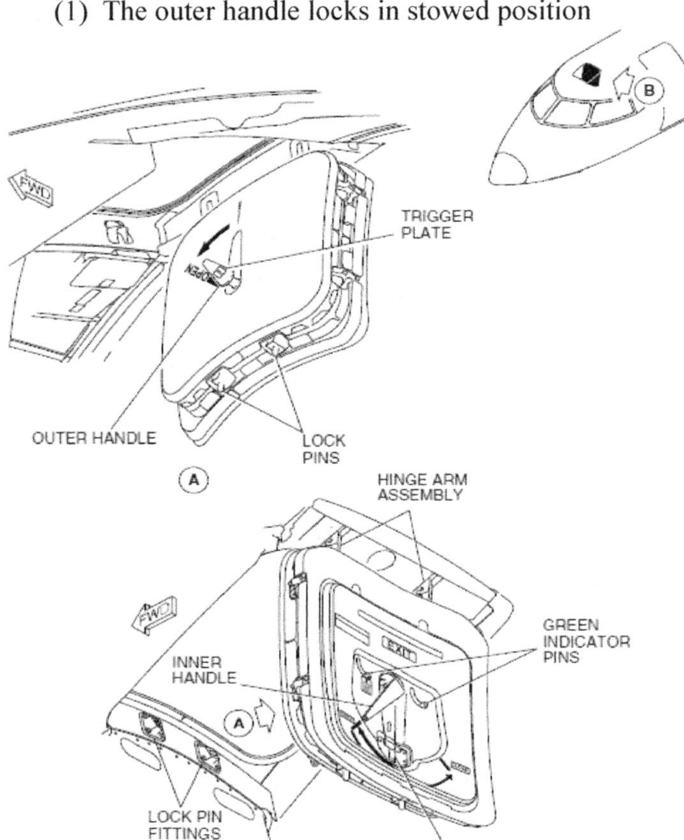

Figure 21 - 5: Crew Escape Hatch

Over-wing Emergency Exits

Over-wing emergency exits are provided for the evacuation of the cabin area, 4 for the CRJ900. The over-wing exits are located on each side of the passenger compartment, above the wings.

The over-wing emergency exits can be opened from inside or outside of the airplane. The door is supported in the fuselage cut-out by hinge and hook fittings on the bottom and by two shoot-bolts at the top.

The left and right over-wing emergency exits are opened by using the single action inner or outer handles located on the upper part of the door.

The exits open inward and are then lifted off the lower of hinge supports and moved out and free from the fuselage using the inner support handles. The over-wing emergency exit position indications are displayed on the primary page and on the DOORS synoptic page.

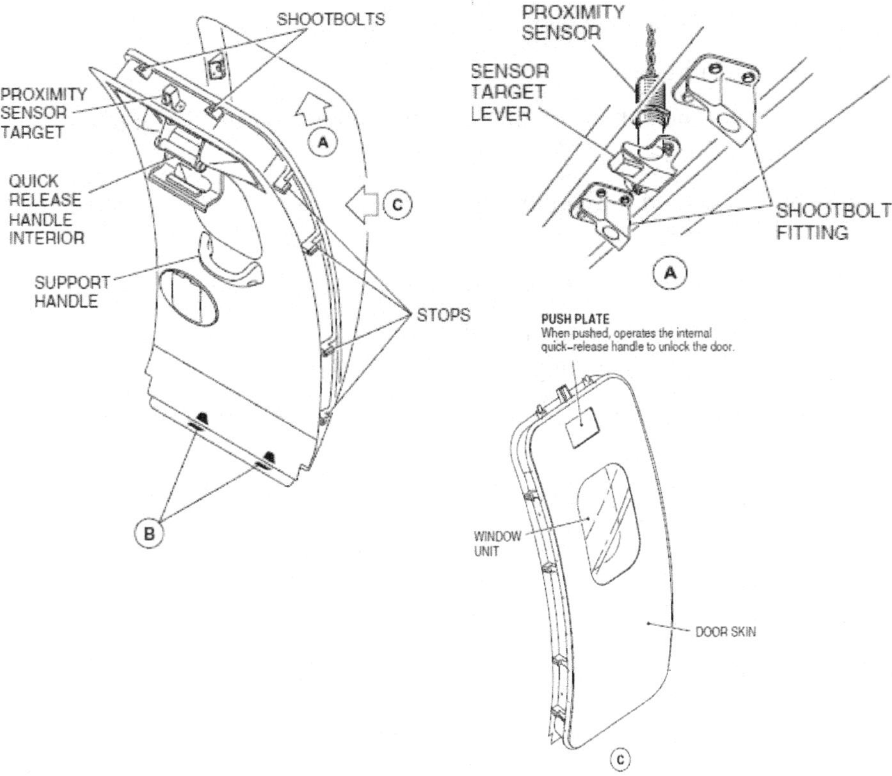

Figure 21 - 6: Over-wing Emergency Exit

1. To open the over-wing emergency exits from inside:
   a. Open the exit door emergency handle cover
   b. Take hold of the exit door inner handle and pull inward and down
      (1) The door shoot bolts retract

NOTE:
The door shoot bolts will be held in a retracted position by a latch lever under the inner handle.

      (2) The exit door opens inward and is then free to be moved away
   c. Take hold of the lower handle and move the exit door to a suitable location away from the emergency exit
2. To open the over-wing emergency exits from outside:
   a. Push in the red outer handle push plate
      (1) The exit door opens inward and is then free to be moved away
   b. Take hold of the inner handles and move the exit door away from the emergency exit
3. To close the over-wing emergency exits from inside:
   a. Manually lift and place the over-wing emergency exit door in front of its opening and set it on its hinge supports.
   b. Push the upper part of the emergency exit door fully outward to squeeze the seal
   c. Release the shoot bolt latch lever under the handle
   d. Push the inner handle up and outward to fully engage the shoot bolt
   e. Close the inner handle cover

Cargo Bay Doors

The CRJ900 airplanes have two cargo bays with three access doors.

Access to all cargo bays is from the left side of the forward and aft fuselage. The cargo bays are not accessible from the cabin or flight deck. The forward cargo bay is located on the left side of the forward fuselage. The aft cargo bay is located aft of the cabin area. The cargo door handles are operated from outside only. Position indications and messages for the cargo door are displayed on the primary page and on the DOORS synoptic page.

Aft Cargo Bay Door

The aft cargo bay door opens inward, and up inside the upper fuselage. The door movement is assisted by a balance spring and cable system. The aft cargo bay door handle is rotated to the OPEN position to unlatch, and to the CLOSED position to latch.

1. To open the aft cargo bay door.
   a. Press the control handle trigger plate
      (1) The control handle ejects
   b. Rotate the control handle to the OPEN position (counter-clockwise)
      (1) The door mechanism unlatches
      (2) The door moves inward, within guiding tracks
   c. Manually move the door fully up
      (1) The door moves up, guided by track rollers, and remains in full OPEN position
2. To close and latch the aft cargo bay door:
   a. Manually lower the door
      (1) The door moves down, guided by track rollers
   b. Pull the door outward against the stops. and rotate the control handle to the CLOSED position (clockwise)
      (1) The door mechanism latches as the control handle reaches the end of its rotation.

   c. Correctly align the control handle with the door recess and push it fully in

      (1) The control handle locks in position

Forward and Center Cargo Bay Doors

The forward and center cargo bay doors are identical in construction and operation. Each door outer structure incorporates a two part control handle and a vent flap. The doors will initially move inward to clear the door stops, then swing up to the locked open position, parallel to the lower fuselage. Two sets of balance springs are used to assist the door in up and down movements.

1. To open the forward or center cargo bay door:

   a. Press the secondary handle trigger plate

      (1) The vent flap opens and the secondary control handle ejects

      (2) The control handle and its access panel unlocks

   b. Pull the control handle

      (1) The door mechanism unlatches, and locks in the unlatched position

      (2) The door moves in and up to clear the door stops

   c. Using the control handle, manually move the door down around the lower fuselage until it latches in the full OPEN position

2. To close and latch the forward or center cargo bay door:

   a. Pull both door open latch levers simultaneously to unlatch the door from the full OPEN position

      (1) The door comes up to its balanced position

   b. Using the control handle, manually raise the door upwards until the latch shafts contact their respective roller fittings

      (1) The door mechanism unlocks (from the unlatched position)

   c. Position the door in front of the door stops, and push the control handle fully in

      (1) The door mechanism latches as the control handle reaches the end of it travel

(2) The control handle and its access panel are locked in position

d. Push the secondary handle fully in

(1) The control handle and its access panel are locked in position

(2) The vent flap closes

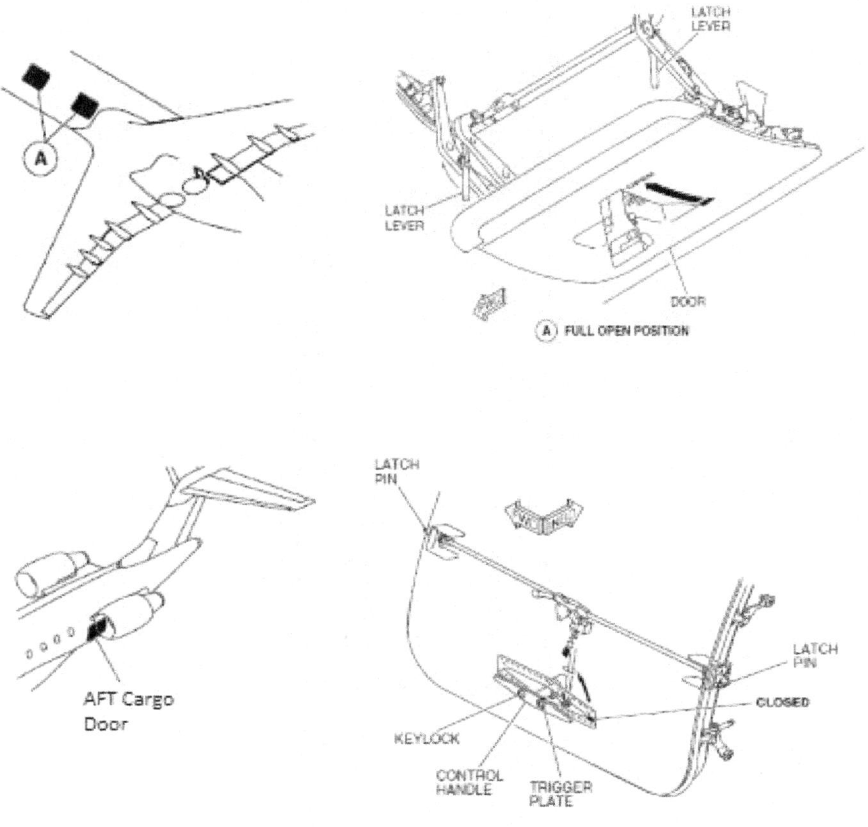

Figure 21 - 7: Forward and Center Cargo Bay Doors & Aft Cargo Bay Door

Galley Service Door

The galley service door is used for servicing the galley, and can also be used for emergency evacuation of the cabin area. It is located on the right forward fuselage adjacent to the galley. The outer structure of the door has a window, outer handle and a cabin pressure vent flap. The door initially moves upward to clear stops on the fuselage structure, then swings outward and fully forward to the lock open position, parallel to the fuselage. Position indications and messages for the galley service door are displayed on the primary page and on the DOORS synoptic page.

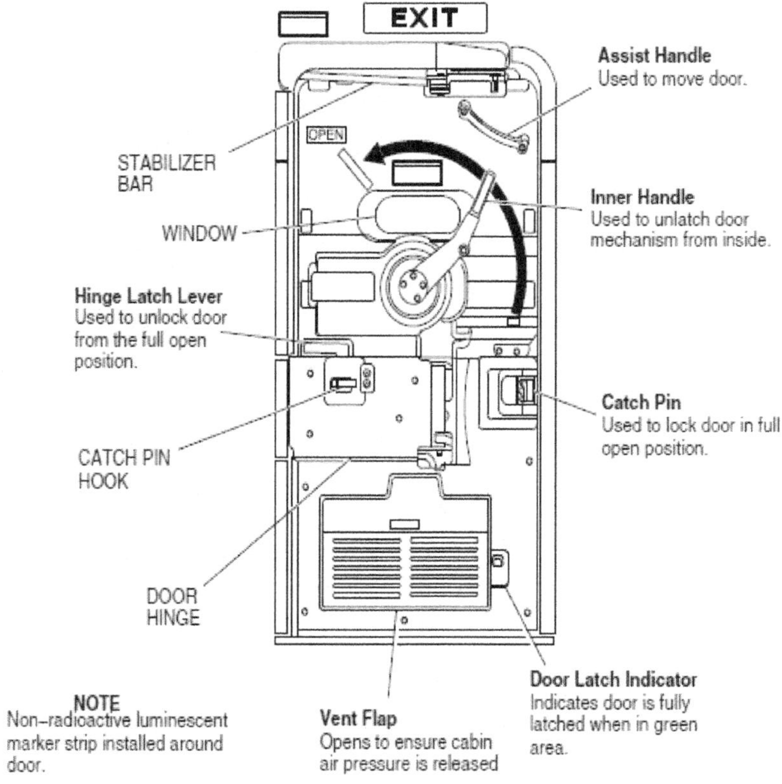

Figure 21 - 8: Galley Service Door

1. To open the galley service door from inside:
    a. Rotate the inner handle counter-clockwise to the OPEN position

(1) The door moves up to clear the door stop fittings (guided by door rollers within track fittings)

(2) The two lower latches disengage from the door lower frame latch fittings

(3) The vent flap opens

b. Push the door outward and forward until it locks in the OPEN position

2. To close and latch the galley service door from inside:

a. Pull the hinge latch lever to release the door from the locked OPEN position

(1) The door moves aft, in front of the door opening

b. Pull the door in to engage the rollers in the door track fittings, then rotate the inner handle to the CLOSED position (clockwise)

(1) The door slides down, behind the door stop fittings (guided by door rollers within track fittings)

(2) The two lower latch pins fully engage in the door frame latch fittings

(3) The vent flap closes

c. Verify the correct indication of door latch through the indicator window located at the lower aft corner of the door

3. To open the galley service door from the outside:

a. Release the outer handle from the door recess

(1) The door latch mechanism unlocks

(2) The vent flap opens

b. Rotate the outer handle fully clockwise to the OPEN position

(1) The door moves up to clear the door stop fittings (guided by door rollers within track fittings)

(2) The two lower latch pins fully disengage from the door frame latch fittings

c. Pull the door outward and move it forward until it locks into position

4. To close and latch the galley service door from outside:
   a. Pull the latch lever, to release the door from the locked OPEN position
      (1) The door moves aft into the door opening
   b. Push the door in to engage the door rollers in the track fittings, then rotate the outer handle counter-clockwise to the CLOSED position until it lines up with its recess
      (1) The door slides down. behind the door stop fittings (guided by door rollers within track fittings)
      (2) The two lower latch pins engage in the door frame latch fittings
   c. Release the handle
      (1) The outer handle springs into its recess
      (2) The vent flap closes as the outer handle nears the end of its travel

Avionics Bay Door

The avionics bay door is used to gain access to the equipment in the avionics bay. It is located on the centerline of the lower forward fuselage.

The door opens inward and moves up on four spring-loaded roller arms. The roller arms engage a set of tracks that allows the door to move forward or aft in the avionics bay. The door can not be opened from the inside.

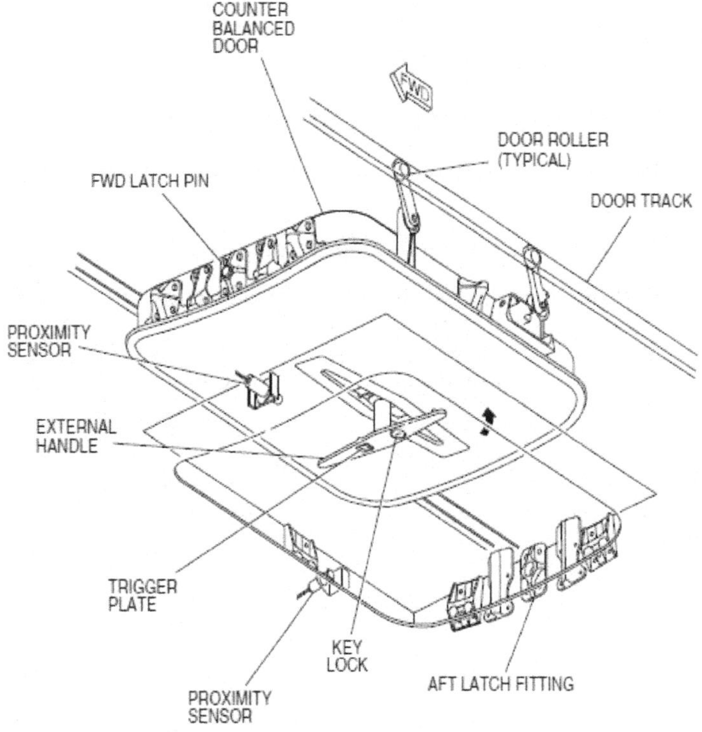

Figure 21 - 9: Avionics Bay Door

1. To open the avionics bay door:
   a. Press the outer handle trigger plate
      (1) The handle ejects from the door recess
   b. Rotate the handle 90° counter-clockwise to the OPEN position
      (1) The fwd and aft latch pins disengage from the door frame latch fittings

c. Push the door up

    (1) A latch on the roller locks the door in the up position

d. Rotate the outer handle to the CLOSED position and push the handle into the door recess

e. Slide the door fwd or aft as required

2. To close the avionics bay door:

    a. Slide the door above its opening

    b. Press the handle trigger plate

        (1) The handle ejects from the door recess

    c. Rotate the handle to the OPEN position to release the hold-open latch

        (1) The control handle locks in position

    d. Pull the door fully down to compress the door seal, and rotate the handle to the CLOSED position

        (1) The fwd and aft latch pins engage in the door frame latch fittings

    e. Push the handle into the door recess

        (1) The door handle locks in the stowed position

Aft Equipment Bay Door

The aft equipment bay door is located on the lower aft fuselage. It provides access to the equipment located in the unpressurized aft equipment bay. The aft equipment bay door is hinged at the front and opens downwards. The door is also removable through quick release hinge pins.

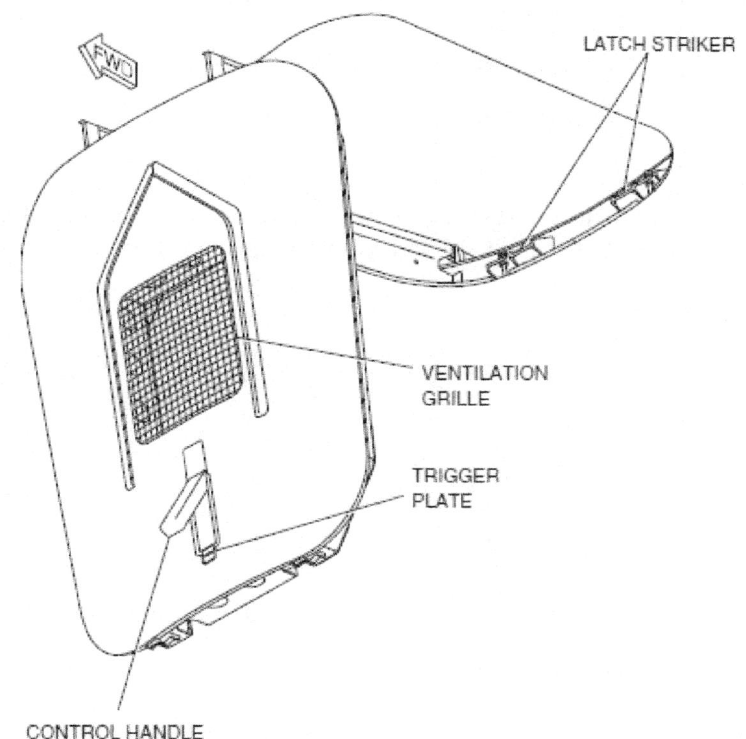

Figure 21 - 10: Aft Equipment Bay Door

(1) To open the aft equipment bay door:

a. Press the control handle trigger plate

The control handle ejects

b. Rotate the control handle

The door mechanism unlatches

c. Manually move the door fully down

(2) To close and latch the aft equipment bay door:

a. Manually move the door up in its opening

(1) The control handle ejects

b. Rotate the control handle

(1) The door mechanism latches as the control handle reaches the end of its travel

c. Push the control handle fully in

(1) The control handle locks in position

NOTE:
The aft equipment bay door is not monitored by the PSEU, therefore there is no flight deck EICAS indication of door position.

Reinforced Flight Deck Door

Information removed in this edition

Information removed in this edition

EICAS Messages

Master Warning	Aural	Cause
PASSENGER DOOR	"Door"	Indicates the passenger door is unsafe when two or more door sensors indicate the door is open.

Master Caution	Cause
AV BAY DOOR	Indicates that avionics bay door is unsafe.
CARGO DOOR	Indicates that cargo bay door is unsafe.
L (R) EMER DOOR	Indicates that applicable emergency exit is unsafe.
PAX DR LATCH	Indicates the passenger door is not latched (one or more of the pins/cams are unsafe).
PAX DR OUT HNDL	Indicates the passenger door outer handle is not stowed.
PAX DR STOW	Indicates the passenger door outer handle stow knob is not stowed.
SERVICE DOOR	Indicates that galley/service door is unlocked/ unsafe.

System Schematic

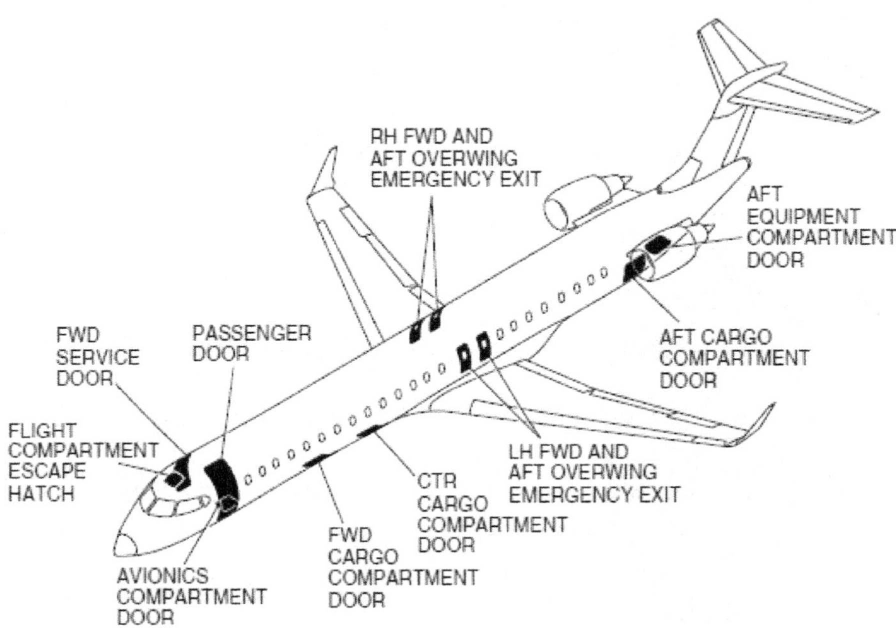

Figure 21 - 11: Door & Hatch Locations

Chapter 22: Power Plant

Intentionally Left Blank

Chapter 22: Power Plant

General

The aircraft are equipped with two high-bypass turbofan engines that supply thrust. electrical power, hydraulic power and bleed air for aircraft operation. Each power plant is controlled by a full authority digital engine control system (FADEC)

The dual-assembly engine consists of a fan rotor (N1) and a compressor rotor (N2). The N1 rotor is comprised of a single-stage fan connected through a shaft to the four-stage low pressure turbine. The N2 rotor is a 10 stage axial flow compressor connected through a shaft to the two-stage high-pressure turbine. The N2 compressor drives the accessory gearbox.

The CF34-8C5 is the baseline engine model for the CRJ900. An optional CF34-8C5A 1 engine is available. The CF34-8C5A 1 optional engine allows approximately 2% greater normal takeoff thrust.

Outside air temperature and pressure altitude are determining factors in achieving takeoff and APR thrust. Increases in ambient temperature or pressure altitude adversely affect the engine's ability to produce rated thrust. The engines are flat-rated to ISA+ 15°C (86°F) at sea level.

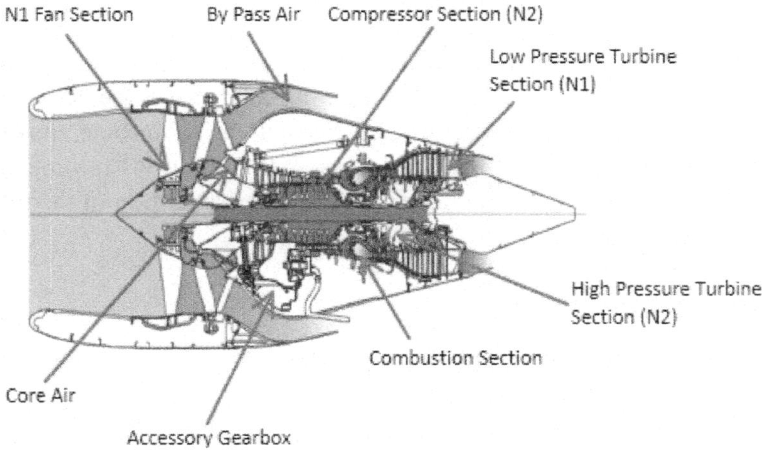

Figure 22 - 1: Power plant

Engine Construction

The power plant has two independent major assemblies. The N1 section consists of a fan rotor that is driven through a shaft by a four stage low-pressure turbine. The N2 section is comprised of a 10 stage axial flow compressor. combustor, accessory gearbox and a two stage high-pressure turbine. The high-pressure turbine drives the compressor.

Components and Operation

Airflow Paths

Engine airflow passes through the single-stage fan and is divided into two airflow paths: bypass and core air.

Bypass Air

Air is accelerated by the single-stage N1 fan and is ducted around the engine nacelle to produce most of the thrust generated at takeoff. Thrust reversers are used to divert the bypass air forward to assist in aircraft braking on the ground.

Core Air

Some of the air that is accelerated by the N1 fan enters the N2 core where it is compressed. mixed with fuel and ignited. The resulting combustion gases are then exhausted onto the high-pressure two-stage N2 turbine to drive the N2 assembly. After passing through the N2 turbine, the exhaust gases are discharged onto the low-pressure four-stage N1 turbine to drive the N1 fan. Jet-pipe thrust accounts for a smaller portion of the thrust generated at takeoff.

N1 Fan

The N1 fan is a single-stage fan that consists of 28 titanium alloy blades.

A Kevlar blanket is wrapped around the fan housing to contain possible damage from a failed N1 fan or blades. N1 fan rpm is indicated on the EICAS primary page.

N2 Compressor

The N2 compressor is a 10 stage variable geometry (VG) axial flow compressor that is mechanically driven by the high-pressure turbine assembly. The operability valve and variable geometry (VS) system are used to maintain the compressor's aerodynamic performance and stall margin throughout the operating range.

The VG system regulates airflow across the N2 compressor by changing the position of the inlet guide vanes and the stator vanes for the first four stages of the compressor. The VG system optimizes the angle of attack of the airflow at the compressor blades and provides compressor stall and surge protection.

The VG inlet guide vanes and stators are programmed by FADEC and positioned by actuators and mechanical linkage. High-pressure fuel from the engine's fuel metering unit (FMU) is used to hydraulically move the actuators.

The operability valve provides additional compressor airflow control. The valve extracts bleed air to offload the compressor during engine starts and during high aerodynamic loading of the compressor. The operability valve is programmed by the FADEC and hydraulically activated by high pressure fuel from the engine's FMU.

N2 rpm is displayed on the EICAS primary page. There are no EICAS indications for the variable geometry system or the operability valve.

Accessory Gearbox

The N2 compressor drives the engine's accessory gearbox (AGB). The AGB has a partition for an integral oil reservoir. Mounted on the AGB are

- Engine lubrication pumps
- FADEC alternator
- Hydraulic pump (EDP 1 or EDP 2)
- Engine fuel pump
- Integrated drive AC generator (IDG)
- Air turbine starter (ATS)

Engine Fuel System

Fuel is delivered to the fuel injectors at the pressures and flow rates necessary to maintain the desired engine thrust. The engine fuel system in addition to providing fuel for combustion is used to:

- Actuate the VG compressor linkage
- Operate the operability valve
- Cool the engine oil (fuel/oil heat exchanger)
- Actuate and lubricate fuel system components
- Provide motive flow for the scavenge and main ejector pumps

Combustion fuel can be interrupted by moving the thrust lever to SHUTOFF or by selecting the ENG FIRE PUSH switch-light. The SHUTOFF position shuts off the fuel at the fuel metering unit. The ENG FIRE PUSH switch-light closes the fuel shutoff valve and the fuel metering valve shutoff valve.

Components and Operation

Fuel Pump

The AGB mounted fuel pump is comprised of three separate pumps contained within a single housing. The fuel pump provides fuel under high pressure and at a flow rate that exceeds the requirements of the engine at any thrust setting. Excess fuel is bypassed back to the fuel filter.

Fuel pump pressure is also used to generate motive flow for the scavenge and main ejectors of the aircraft fuel system.

Fuel/Oil Heat Exchanger and Fuel Filter

A fuel/oil heat exchanger is used to warm engine fuel and cool engine oil.

The engine fuel temperature is indicated on the FUEL synoptic page. A micron fuel filter is used to remove solid contaminants from the fuel. The filter is depicted on the FUEL synoptic page. If the filter becomes clogged, the fuel bypasses the filter to ensure continued engine operation. When the filter becomes contaminated to a point that an impending fuel bypass condition exists, an EICAS caution message is presented and the fuel filter icon on the FUEL synoptic page changes from green to amber.

Fuel Metering Unit

The fuel metering unit (FMU) is an electro-hydraulic device that meters and distributes the fuel needed for combustion based upon control signals from the FADEC system. The FMU's primary components are the fuel metering valve and pressurizing and shutoff valve. The metering valve regulates fuel in response to commands from the FADEC to maintain combustion under all operating conditions. The pressurizing and shutoff valve controls the supply of fuel for combustion.

The FMU also supplies the high-pressure fuel used to actuate the operability valve, and VG inlet guide vanes and compressor stator vanes.

Fuel System Operation

The main fuel ejectors or DC-powered boost pumps deliver fuel from the collector tanks via fuel feed manifolds to the engines. Engine fuel shutoff valves (SOVs) are installed in the manifold to interrupt the supply of fuel to the engines during a fire. The ENG FIRE PUSH switch-lights control the SOVs.

At the engine, the fuel is pressurized, heated. filtered, metered and distributed to the combustion chamber. All fuel scheduling is controlled by the FADEC.

Engine Control System

The powerplants are controlled by the full authority digital engine control (FADEC) system. The FADEC provides a full range of engine control under all steady state and transient engine conditions. The FADEC controls the operation and performance of the engine through three subsystems: fuel control, compressor airflow management and engine starting/ignition control.

Components and Operation

FADEC Unit

Each powerplant has its own dual-channel FADEC computer. One FADEC channel operates as the in-control channel and processes information to provide engine control outputs. The other channel operates in standby.

The standby channel processes all the input information but does not provide control output but with one exception. Both the i n -control and standby channels will respond to an engine over-speed by commanding the shutoff valve in the fuel metering unit to close to return the engine to an on speed condition.

The in-control and standby channels continuously share command and status data through a crosstalk data bus. Should the designated in-control channel become unserviceable, the standby channel assumes the in control role.

During normal operations with two serviceable FADEC channels, FADEC software will cause the channels to alternate in-control and standby roles after every second engine start. During the engine start, switchover occurs at 7% N2 rpm.

Each FADEC unit has many levels of system redundancy. FADEC malfunctions are presented as either caution or status messages. Each message has a different impact upon the dispatchability of the aircraft.

The minimum equipment list (MEL) defines the dispatch implications for each of the following messages in detail:

- L or R FADEC FAULT 1 status message - 10 day dispatch maximum before fault rectification required

- L or R FADEC FAULT 2 status message - 30 day dispatch maximum before fault rectification required
- L or R FADEC caution message - no dispatch

There are no EICAS indications of active FADEC channel assignment.

Electrical Power

The FADEC system has its own source of electrical power. When N2 rpm is above 50%, a dedicated accessory gearbox-mounted FADEC alternator supplies power to both FADEC channels. Below 50% N2 rpm, the aircraft's electrical system powers the FADEC system.

Engine Idle

When the thrust lever is placed in IDLE, the minimum N2 idle rpm is programmed by FADEC. Idle rpm is dependent upon atmospheric information, bleed air loading and phase of flight.

The FADEC will always program the best idle speed for any phase of flight. When the aircraft transitions from one flight phase to the next, idle rpm is automatically adjusted. There are five different N2 idle settings: flight idle, approach idle, landing idle, reverse idle and ground idle.

Flight Idle

Flight idle refers to the idle setting used when the thrust lever is at idle, landing gear is up and the flaps are set to 20 or less. N2 flight idle rpm increases with altitude.

Approach Idle

Approach idle is a corrected N2 core speed that permits the engine to accelerate to go-around thrust in the minimum amount of time. N2 approach idle is activated when the thrust lever is at idle and the approach bit is set (landing gear down. or flaps greater than 20).

Landing Idle

Landing idle is an intermediate N2 speed between reverse idle and ground idle that occurs on touchdown with weight-on-wheels or wheel rotation.

The purpose of landing idle is to keep the N2 speed elevated until reverser deployment is. commanded to allow the engine to achieve reverse idle in an acceptable amount of time.

If the thrust reversers are not activated after weight-on-wheels or wheel spin-up is detected, the FADEC maintains landing idle for only 5 seconds. At the end of the 5 second period, the FADEC sets ground idle. Because landing idle is set to a lower rpm than reverse idle, the landing distance penalty associated with not using the thrust reversers is minimized.

Reverse Idle

The purpose of reverse idle is to raise the N2 core speed to permit the engine to achieve full reverse thrust in an acceptable amount of time.

Reverse idle is selected when:

- Weight is on wheels or wheels rotation
- Applicable thrust reverser lever is raised
- Thrust reverser cowl not stowed (moving to the open position)

Ground Idle

N2 ground idle is the minimum thrust setting. It is selected when weight is on wheels and the thrust lever is in idle. Ground idle varies with altitude and temperature. Increases in altitude or ambient temperature result in an increase in N2 idle rpm. Decreases in altitude or ambient temperature result in a decrease in N2 idle rpm.

One-Engine-Inoperative (OEI) Fuel Scheduling

The FADEC computers continuously crosstalk to share data on the health of each engine. The FADEC will automatically increase the thrust on the operating engine when the other engine fails. The value of the increased thrust is dependent on the position of the thrust lever at the time of the engine loss.

The FADEC system monitors the N1 rpm of both engines. If the mismatch of N1 fan speeds exceeds 15%, the FADEC automatically increases the N1 speed of the operable engine.

If engine failure is detected due to the fan speed mismatch, the engine failure signal is cleared when fan speed mismatch becomes less than 13%.

This automatic increase in thrust is related to thrust lever position

- APR thrust when the applicable thrust lever is in the TOGA detent
- Automatic increase to maximum continuous thrust (MCT) when the applicable thrust lever is in the CLIMB detent
- Proportional increase in thrust when operating with the applicable thrust lever in the cruise range

Automatic Performance Reserve (APR)

When an engine fails, the automatic performance reserve (APR) feature of the FADEC will automatically increase the thrust for the operating engine to the emergency thrust setting of APR.

The automatic increase in thrust occurs when:

- Thrust levers are in the TOGA detent
- Either normal takeoff, flex takeoff thrust or go-around is programmed

The automatic thrust feature is armed during takeoff when the N1 rpm of both engines are within 8% of scheduled takeoff thrust. On the approach, APR is armed for the go-around when the approach bit is set (either engine is available, flaps greater than 20 or the landing gear is down).

The green APR icon appears in the center of the N1 gauge of the operating engine when:

- APR system is activated by an engine failure, or

- Thrust lever for the operable engine is set in the MAX POWER detent

APR thrust for both engines can also be selected by advancing both thrust levers to MAX POWER. A green APR icon appears in the center of each N1 gauge.

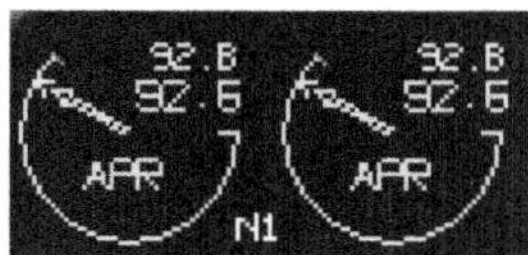

Figure 22 - 2: APR N1 Gauge Indication

Climb Thrust/Maximum Continuous Thrust

When climb thrust is set (CLIMB detent) and an engine fails, the operating engine N1 rpm is increased to the emergency thrust rating of maximum continuous thrust (MCT). MCT is indicated on the primary page by the thrust mode annunciator located below the N1 gauges and by the increase in N1 thrust on the operating engine.

Cruise Range

When an engine fails with the thrust levers in the cruise range, the N1 rpm of the operable engine is increased. The fuel scheduling is changed to provide a smooth transition to MCT as the thrust lever for the operable engine is advanced to the CLIMB detent The actual percentage of increase in N1 rpm is a function of the position of the thrust lever for the operating engine.

High Power Schedule Switch-light

If MCT is required with both engines operating and lever is in climb detent, the pilot must select the HIGH PWR SCHEDULE switch-light on the ENGINES/miscellaneous test panel. Both FADECs then schedule the MCT thrust level and an advisory message ENGS HI PWR SCHED is displayed on the status page. When selected, the HIGH PWR SCHEDULE switch-light displays a green 'ON' legend.

NOTE:

There is no procedure that permits the use of this switch-light

Thrust Calculations

FADEC automatically computes a target thrust rating for selected thrust lever position and presents the target or maximum value on the N1 gauge.

On the ground, normal rated takeoff thrust is automatically calculated when electrical power is first applied to the aircraft. Normal rated N1 takeoff rpm is continuously updated for changes in the Mach number, delta ambient temperature and pressure altitude. Although the FADEC is capable of calculating normal takeoff N1 with the engine off, the calculation is more accurate with the engine at idle speed.

During the takeoff roll, the N1 normal rated takeoff thrust value is continuously updated as the aircraft accelerates down the runway. At 65 knots the thrust calculation is locked in. The calculation is unlocked when the aircraft's altitude exceeds 400 feet AGL or the thrust lever is moved from the TOGA detent.

The cyan color target N1 appears on the N1 gauge as:

- N1 caret (dough-nut for cruise thrust)
- N1 digital reference
- Thrust mode annunciation

Thrust Mode Annunciation

The thrust mode annunciation identifies the position of the thrust lever during most phases of operation or a specific armed condition. With both engines operating, the following thrust mode annunciations are presented:

- CRZ -- thrust levers in the cruise range
- CLB -- thrust levers in the climb detent
- TO -- ground operations or thrust levers in TOGA detent for takeoff
- GA -- thrust levers in the TOGA detent for in-flight go-around

During ground operations or during the approach phase, the thrust mode annunciation may not indicate the position of the thrust lever but an armed condition.

Prior to lift-off, TO is the default mode. It appears automatically and is independent of engine operation or thrust lever position.

On approach, when the landing gear is down or flaps are greater than 20, the GA mode is armed and presented as the thrust mode annunciation. With both engines operating, the thrust levers are normally moved together and a single thrust mode annunciation is presented.

When the thrust levers are moved separately, the thrust mode annunciation changes to reflect the position of each thrust lever. When the thrust lever is moved to SHUTOFF, the primary page N1 thrust mode annunciation for the applicable engine is blank.

Prior to takeoff, only the default TO setting or programmable FLX thrust mode annunciations are presented on the N1 gauges

Normal-Rated Takeoff N1 Thrust (TO)

Normal-rated takeoff thrust is automatically calculated by FADEC when electrical power is applied to the aircraft on the ground. Takeoff thrust is set on the takeoff roll by placing the thrust levers in the TOGA detents.

The programmed normal-rated takeoff thrust value is displayed on the primary page by the following indications:

- Cyan caret
- Cyan N1 digital reference
- Cyan TO thrust mode annunciation for takeoff thrust

Go-Around Thrust (GA)

During flight, go-around thrust is automatically calculated by FADEC. Go-around thrust indications appear on the primary page when the approach discrete is set (landing gear down or flaps greater that 20) or when the thrust lever is placed in the TOGA detent.

The programmed go-around thrust is displayed on the primary page by the following indications:

- Cyan caret
- Cyan N1 digital reference
- Cyan GA thrust mode annunciation

Climb Thrust (CLB)

When the thrust lever is placed in the CLIMB detent, the FADEC electronically programs the engine to accelerate or decelerate to the climb thrust setting.

The programmed climb thrust value is displayed on the primary page by the following indications:

- Cyan caret
- Cyan N1 digital reference
- Cyan CLB thrust mode annunciation

Max Power (APR)

When either or both thrust levers are placed in the MAX POWER detent, the FADEC increases the engine thrust to automatic performance reserve (APR) thrust. APR is an emergency thrust setting.

To protect against inadvertent APR activation, the pilot must apply increased force to move the thrust lever forward of the TOGA detent into the MAX POWER detent.

APR thrust is indicated on the associated N1 gauge by:

- Green APR icon
- Cyan caret
- Cyan N 1 digital reference

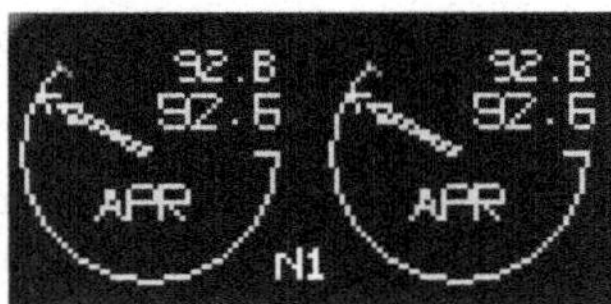

Figure 22 - 3: APR Indication N1 Gauge

Cruise Range (CRZ)

Cruise thrust is unlike other thrust settings in that it is not associated with a thrust lever detent.

Cruise thrust is set by positioning the thrust lever in the quadrant range that exists between the IDLE and CLIMB detents. In this range, the pilot manually sets the thrust levers to maintain cruise speed, or sets thrust for the descent and approach.

In cruise, FADEC does not set the cruise thrust to match the FADEC generated maximum cruise thrust target value. To distinguish maximum cruise (CRZ) thrust from other detent-related thrust settings, a cyan N1 dough-nut is used instead of a caret.

Maximum cruise (CRZ) thrust is indicated by the following:

- Cyan dough-nut
- Cyan N1 digital reference
- Cyan CRZ thrust mode annunciation

Bleed Air Extraction

When the thrust levers are in any of the detents and new air conditioning or anti-icing demands are placed on the engines, the N1 target rpm will change to compensate for the new bleed air demands.

When the thrust levers are in the cruise range and the bleed air or anti icing configuration is changed, the N1 digital reference and N1 caret are changed to reflect the new maximum cruise (CRZ) thrust limit.

If operating at or near maximum cruise thrust, the pilot may be required to adjust the thrust levers to prevent exceeding the new maximum cruise thrust limit

Flex Thrust (FLX)

Flex thrust (FLX) is routinely used for takeoff when weather and runway conditions are favorable. Flex thrust takeoffs significantly reduce fuel consumption and extend the usable life of the engine.

Flex thrust is selected by entering an assumed temperature on the PERF MENU page of the FMS CDU. By entering an assumed temperature (ISA +35°C max) the pilot provides the FADEC with the data needed to complete the FLX calculation.

Flex thrust can only be entered on the ground when the following conditions are met:

- Thrust levers at the IDLE or SHUTOFF detents
- WOW for at least 1 minute
- Assumed temperature is greater than actual OAT
- Airspeed less than 65 KIAS

The assumed temperature can be cleared by sequentially selecting the DEL key, then the FLX TEMP line key on the FMS CDU or entering an assumed temperature that is less than actual OAT.

Flex thrust is removed when any of the following conditions occur.

- Any thrust lever is retarded to CLIMB
- Any thrust lever is advanced to APR
- Cowl anti-ice is selected ON
- Wing anti-ice is selected ON

To clearly differentiate flex thrust from other thrust settings, the FLX indications are displayed in magenta on the N1 gauges.

Flex thrust is identified by

- Magenta caret
- Magenta N1 digital reference
- Magenta FLX thrust mode annunciation

When programmed, FLX is set during the takeoff roll with the thrust levers in the TOGA detents. Unlike normal-rated thrust, flex thrust is not updated by FADEC prior to 65 KIAS during the takeoff roll.

N1 and N2 Synchronization

The FADEC can synchronize the fan or core speed of the two engines when operating the engines with the thrust levers in the cruise range. The left engine is designated the master engine and the right engine is the slave.

Engine synchronization is selected at the ENGINES control panel.

Components and Operation

N1 Fan Synchronization

When the selector switch is at N1 and the thrust levers in the cruise range, the FADEC matches the fan speed of the right engine to the speed of the left engine. Synchronization is disabled if the right engine's N1 reference is not within 1.5% of the left engine for at least one FADEC update.

N2 Synchronization

When the selector switch is at N2 and the thrust levers in the cruise range, the FADEC matches the core speed of the right engine to the speed of the left engine. Synchronization is disabled if the right engine's N2 reference is not within 2.5% of the left engine for at least one FADEC update.

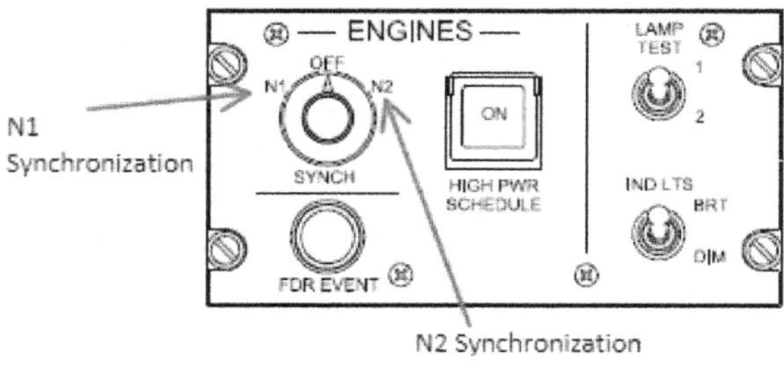

Figure 22 - 4: ENGINE SYNCH Selector

Engine Oil System

Each engine has an independent lubrication supply system. Each system consists of an oil pump and an oil reservoir which is integral to the accessory gearbox. The pressure pump draws oil from the reservoir and supplies it to the various engine components for cooling and lubrication.

Components and Operation

Oil temperature and pressure indications are displayed on the EICAS primary page. Oil filter impending bypass and chip detector indications are provided on the engine fault panel in the aft equipment bay.

The AGB driven main lubrication pump pressurizes the lubrication system. The oil flows from the pump, passes through an oil filter and the oil/fuel heat exchanger. The oil then continues through the engine, for cooling and lubricating, then to the engine sumps. Scavenge pumps return the oil to the reservoir after passing through a chip detector and a deaerator.

Sensors for the oil pressure indication and EICAS messages are located in the oil filter module mounted on the forward side of the oil reservoir. The chip detector is also mounted on the AGB, in the scavenge oil return line.

During engine start, the oil pressure indications on the EICAS primary page are displayed with an analog gauge and a digital readout. When both engines are started and oil pressure is normal, the oil pressure gauges revert to N1 fan vibration gauges. The digital oil pressure indication remains.

The left and right engine oil reservoir quantities are displayed on the EICAS MENU page.

Engine Oil Level and Replenishment System

The engine oil reservoir is manually replenished through an oil fill cap on the top of each engine. Access to the fill cap is gained by opening the engine cowlings. There is an oil level indicator adjacent to the fill cap.

An optional engine oil replenishment system, located in the aft equipment bay, enables the engine oil reservoir to be filled remotely. The system includes an oil storage tank with sight glass level indicator, an electric pump, a control panel and an engine selector valve.

When engines are operating and oil quantity is below 57% or with engines not operating and oil quantity below 80%, the L or R OIL LEVEL LO status message is displayed.

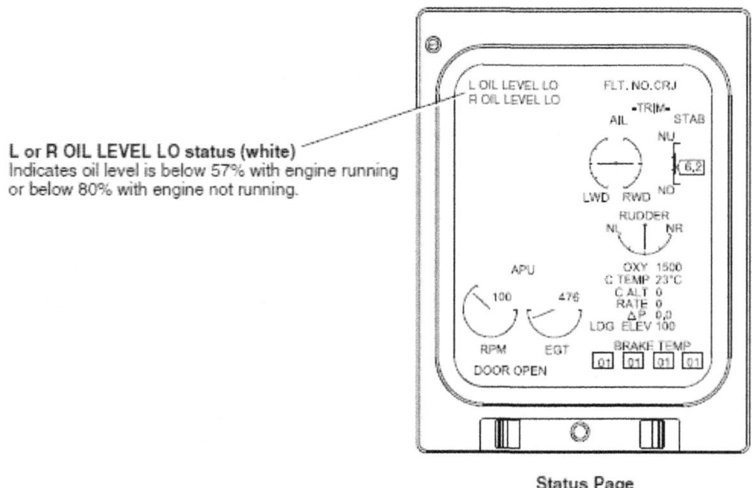

Figure 22 - 5: L or R OIL LEVEL LO Status

Engine Bleed Air System

Bleed air is extracted from the 6th or 1 0th stage of the engine compressor to provide pneumatic services for:

- Wing and cowl anti-icing
- Air-conditioning and pressurization
- Crossbleed engine starts

Two air-conditioning system controllers (ACSC 1 and ACSC 2) control the bleed air extraction from each engine through a pressure regulating

shutoff valve (PRSOV). The left engine PRSOV is controlled by ACSC 1. The right engine PRSOV by ACSC 2. Bleed air extraction is monitored by the FADEC system.

Bleed air is also extracted from the compressor by the operability bleed valve to regulate airflow through the engine core. Core air regulation is controlled by FADEC.

Components and Operation

Engine bleed air components consist of a bleed air duct, high-pressure valve (HPV) and intermediate check valve.

Engine bleed air duct assembly directs 6th or 10th stage compressor bleed air to the common bleed manifold that is located in the aft equipment bay. Normally, the 6th stage of the engine compressor supplies the bleed air to operate the aircraft's pneumatic system. When system demands exceed the capability of the compressor's 6th stage, the HPV is commanded opened by the associated ACSC to provide 10th stage bleed air to the manifold.

An intermediate check valve is installed downstream of the 6th stage compressor port. When the HPV is open and 10th stage bleed air is supplying the manifold, the intermediate check valve in the 6th stage duct is closed. Closing the intermediate check valve prevents high pressure 10th stage bleed air from back-flowing to the lower pressure 6th stage port.

The nacelle bleed air ducts are monitored for bleed air leakage by the engine FIDEEX system.

Engine Starting System

The starting system is used to accelerate the N2 high pressure section of the engine to a speed at which combustion lightoff and self sustaining rpm occurs. The start sequence is initiated by pressing the START switch-light on the control panel but is controlled by the FADEC system.

DC electrical power and air from the bleed air manifold are required to open the start valve and engage the air turbine starter. The bleed air manifold can be pressurized by the:

- APU
- External air cart
- Bleed air from the opposite engine (crossbleed start)

The bleed air manifold pressure from any of the three pneumatic sources is displayed digitally on the ECS synoptic page.

Components and Operation

Start Valve

The start valve is mounted next to the air turbine starter (ATS) inside the engine nacelle. The valve normally requires 28 vdc power and pneumatic pressure to open. When the START switch-light is selected, the FADEC supplies the 28 vdc power to energize the solenoid to allow pneumatic pressure to open the valve and engage the ATS.

When N2 rpm reaches approximately 50%, the FADEC removes power from the solenoid, the start valve closes and the ATS disengages. The valve can be manually actuated should the start valve solenoid fail. On the ground, an override socket drive allows the start valve to be opened mechanically. There is no requirement to open the engine cowls for this mechanical procedure.

Air Turbine Starter

The air turbine starter (ATS) is mounted on the accessory gearbox. The ATS converts pneumatic energy into mechanical motion. The ATS, through a sprag clutch, mechanically engages the accessory gearbox to accelerate the engine to idle speed. During the start sequence, the start valve automatically closes and the ATS disengages when the N2 rpm reaches approximately 50% or the STOP switch-light is selected.

For subsequent starts. the starter sprag clutch does not require that the engine rotation be completely stopped before reengaging the starter. The ATS may be reengaged at any rpm up to 45% N2 rpm.

Start Sequence

Each engine has a set of ST ART and STOP switch lights on the engine IGNITION/START panel. When the engine START switch-light is pressed the:

- Isolation valve (ISOL) opens
- Start valve on the associated engine opens to allow pressure from the bleed air manifold to engage the air turbine starter (ATS)
- White light illuminates in the associated START switch-light
- L or R ENGINE START status message is displayed

When the engine's N2 rpm reaches approximately 50%, the:

- Start valve closes and the ATS disengages
- White START switch-light extinguishes
- Associated L or R ENGINE START status message is removed

If the start valve fails to close after achieving approximately 50% N2 rpm, a NO STRTR CUTOUT caution message is presented.

On all new production CRJ900 airplanes (SIN 15260), the NO STRTR CUTOUT caution message has been replaced by an applicable L STRT VLV OPEN or R STRT VLV OPEN caution message for a starter does not cut-out (start valve fails open) condition.

The starter is capable of dry-motoring the engine up to approximately 30% N2 rpm.

Ignition System

Two independently-controlled alternating current (AC) ignition systems are provided for each engine. Each system consists of one ignition exciter and one igniter plug. The ignition systems are controlled by FADEC but when required can be manually activated by the pilot.

The AC essential bus powers ignition A and ignition B is powered by the battery bus through a static inverter.

Components and Operation

Engine Starting

When the FADEC 'in-control' channel is determined during N2 spool-up, the in control channel closes the associated ignition relay. When the thrust lever is moved to the IDLE position, the FADEC selected, AC powered ignition exciter is energized. The exciter converts AC voltage to DC and sends high-energy pulses through the ignition lead to the igniter. Ignition is de-energized by the FADEC at starter cutout.

The L or R AUTO IGNITION advisory message is displayed during the engine start sequence. On the ground the message advises the pilots that the FADEC has energized either ignition A or B. When starting the engine in-flight, the advisory message indicates that the FADEC has energized both igniters.

Aerodynamic Stall Protection

At an excessively high angle of attack (AOA), there is a possibility that turbulent airflow from the wing root could disrupt the flow of air into the engine's intake. Disruption of airflow at the intake could lead to an engine compressor stall. The FADEC unit and the stall protection computer provide two levels of compressor stall protection.

As the aircraft's angle of attack reaches the stick shaker firing angle, the stall protection computer signals the FADEC to energize both ignition channels. The CONT IGNITION status message is presented. If airflow through the engine is disrupted, the FADEC provides a second signal to energize both the ignition channels and opens the operability bleed valve to offload the compressor. The advisory messages L AUTO IGNITION and R AUTO IGNITION are presented.

Continuous Ignition

Continuous ignition is selected by pressing the CONT switch-light on the start/ignition panel. Both ignitions A and B are energized. Continuous ignition must be used during the following flight conditions:

- Takeoff and landings on contaminated runways
- Flight through moderate or heavier-intensity rain
- Flight through moderate or heavier-intensity turbulence
- Flight in the vicinity of thunderstorms

Flame-out Protection

If the FADEC detects an engine flame-out, it will automatically initiate a re-light. Both ignition systems are energized and the L AUTO IGNITION and R AUTO IGNITION advisory messages are presented.

If the N2 rpm falls below the relight envelope, the auto relight is canceled and the FADEC shuts off the fuel and ignition to the engine. The L or R ENG FLAMEOUT caution message is displayed.

When the affected engine's thrust lever is moved to the SHUTOFF detent, the L or R ENG FLAMEOUT caution message is replaced by the L or R ENG SHUTDOWN status message.

Thrust Levers

The thrust lever quadrant contains the thrust levers, thrust reverser levers, micro-switches, internal locks and stops necessary to control the engines in forward and reverse thrust.

Components and Operation

Thrust Levers

Thrust lever quadrant settings are SHUTOFF, IDLE, CLIMB, TOGA and MAX POWER. In addition, the range of movement of the thrust lever between IDLE and CLIMB is defined as the cruise range.

The thrust levers incorporate both mechanical stops and soft detents. Retractable mechanical stops are used at the SHUTOFF and IDLE positions. When the thrust lever is at SHUTOFF, the thrust lever is mechanically locked in position. When the thrust lever is at IDLE, the mechanical stop prevents the thrust lever from being accidentally moved to SHUTOFF. The mechanical stops are retracted by the idle/shutoff release latch on the thrust lever.

Self-centering soft detents prevent the thrust levers from moving inadvertently. Soft detents are used for CLIMB, TOGA and MAX POWER.

Increased force is required to move the thrust lever from the TOGA to the MAX POWER detent, which is a mechanical stop.

A friction knob allows the pilot to change the friction setting for the thrust levers.

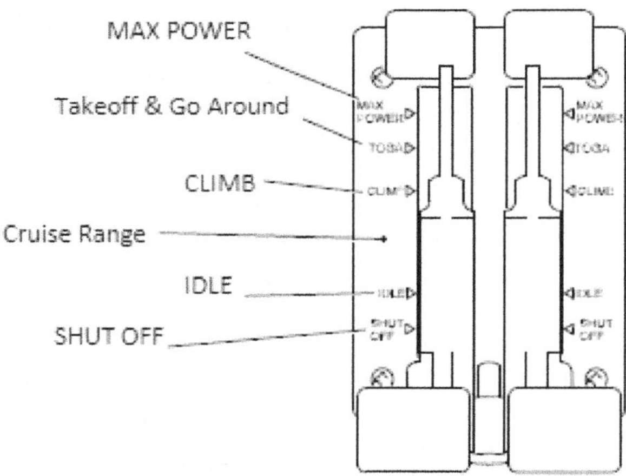

Figure 22 - 6: Thrust Lever Quadrant

Thrust Lever Position Measurement

Actual thrust lever positions are electrically measured by rotary variable differential transformers (RVDTs) or sensed by micro-switches that are housed within the thrust quadrant. The information is provided to the FADEC, the flight control computers (FCCs) and to the data concentrator units (DCUs). Other aircraft systems are influenced by thrust lever position. They include:

- Landing gear warning system
- Takeoff configuration warning system
- Cabin pressurization
- Ground lift device system SSCU

Takeoff/Go-Around (TOGA) Switches

Takeoff and go-around (TOGA) switches are included with each thrust lever. When pressed, the TOGA switch signals the flight control computers to modify flight director commands and the FMS position is updated through the flight management computer(s).

Thrust Reverser Levers

The thrust reverser levers control the operation of the thrust reverser system. See the thrust reverser description in this chapter for further detail.

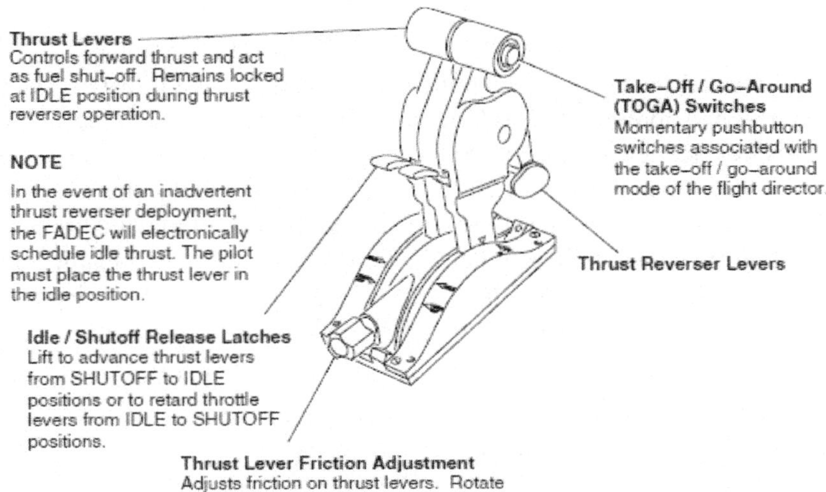

Figure 22 - 7: Thrust Levers

Thrust Reverser System

The thrust reversers assist in stopping the aircraft during landing roll-out or during a rejected takeoff (RTO). The system is operable on the ground only; the reversers are locked out in-flight.

Components and Operation

The thrust reverser system responds to electrical signals from FADEC and the thrust reverser levers. The system is armed when the appropriate switches on the THRUST REVERSER panel are selected to ARMED and the respective hydraulic systems are operative.

Reverse thrust is accomplished by moving the translating cowl assemblies aft to block the rearward discharge of fan bypass air. As the translating cowl moves rearward, the fan discharge airflow is redirected forward through the uncovered cascade vanes.

Hydraulic actuators operate the engine translating cowls. The left engine reverser is powered by hydraulic system 1, the right engine by hydraulic system 2. The trans-cowl assemblies when stowed are hydraulically and mechanically locked to prevent inadvertent thrust reverser deployment.

Reverser deployment is accomplished by squeezing the thrust reverser triggers and applying upward pressure on the thrust reverse levers. The thrust reverse lever movement is restricted to approximately 15 degrees by a solenoid stop that acts as a reverse thrust lever lock. This action simultaneously locks the thrust lever in the IDLE position and prevents thrust from being applied. When the translating cowl is in transition and reaches full reverse travel, the solenoid stop is released, unlocking the thrust reverser lever and allowing reverse thrust to be applied.

A reverse idle soft detent is incorporated in each thrust reverser lever. The detent provides the pilot with a tactile reference to assist in establishing the reverse idle thrust setting.

Vibration Monitoring System

The powerplant consists of two major rotating assemblies, the N1 fan and N2 core sections. Each assembly is continuously monitored for vibration.

Indications are displayed on the primary EICAS page.

Components and Operation

N1 Fan

After both engines are started on the ground and oil pressure is normal, the analog oil pressure gauges are replaced by the N1 fan vibration (FAN VIB) gauges. When the N1 vibration level is above a target value (within the FAN VIB gauge amber arc), the green digital readout and pointer changes to amber.

There are no associated EICAS messages.

N2 Fan

N2 core vibration levels are continuously monitored but are presented only when vibration levels exceed a target value. An amber VIS icon appears in the middle of the N2 gauge when the vibration target value is exceeded.

There are no associated EICAS messages.

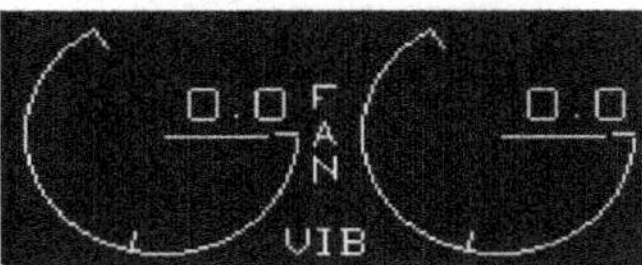

Figure 22 - 8: FAN VIB Gauge

System Schematic

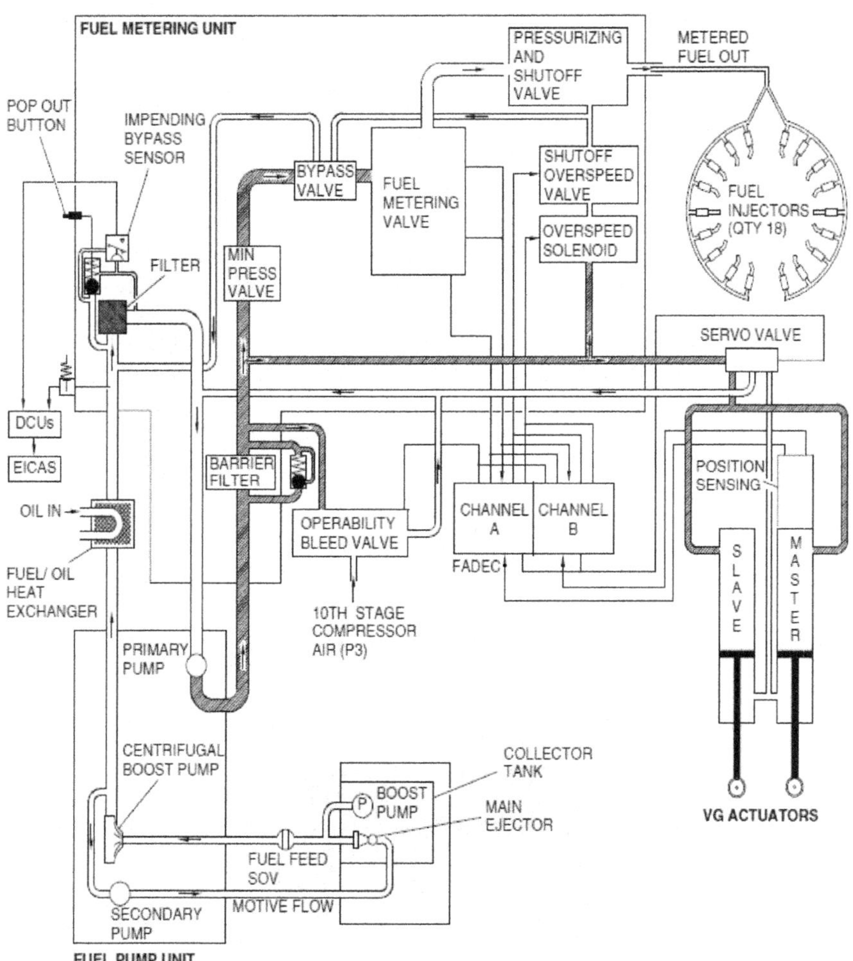

Figure 22 - 9: Fuel Distribution System Schematic

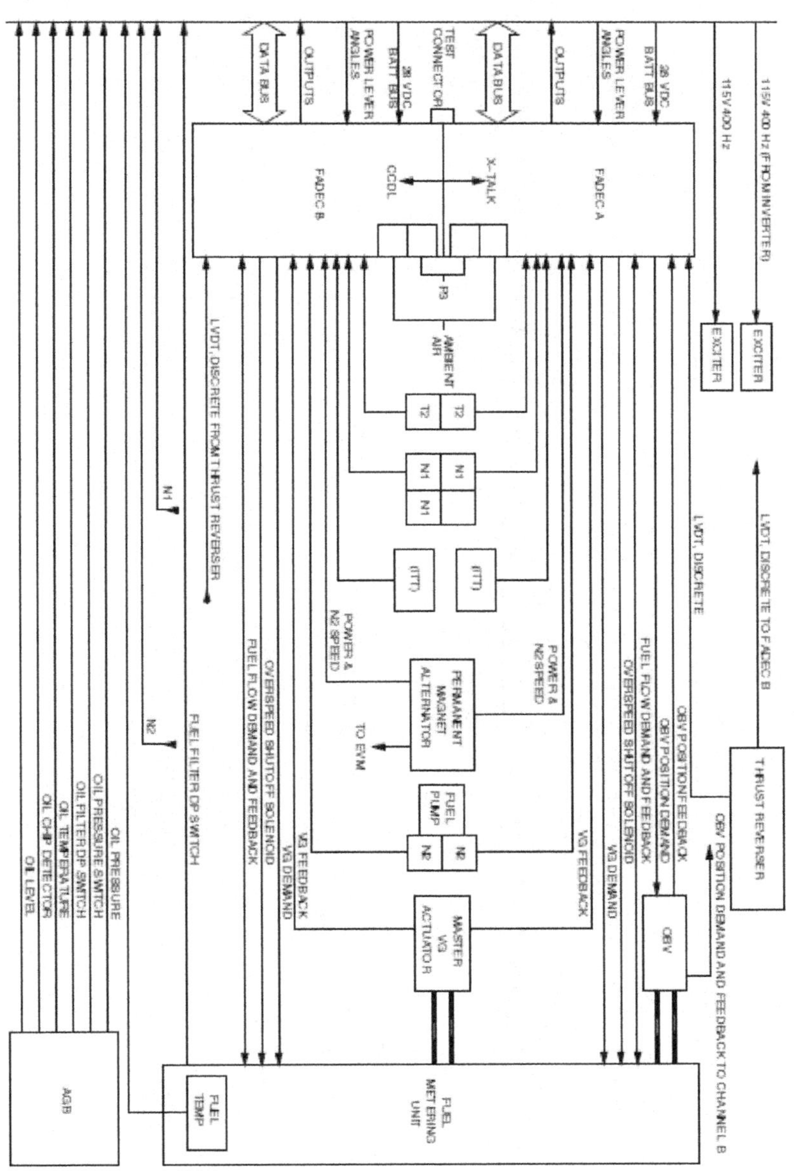

Figure 22 - 10: FADEC / FMU Interface

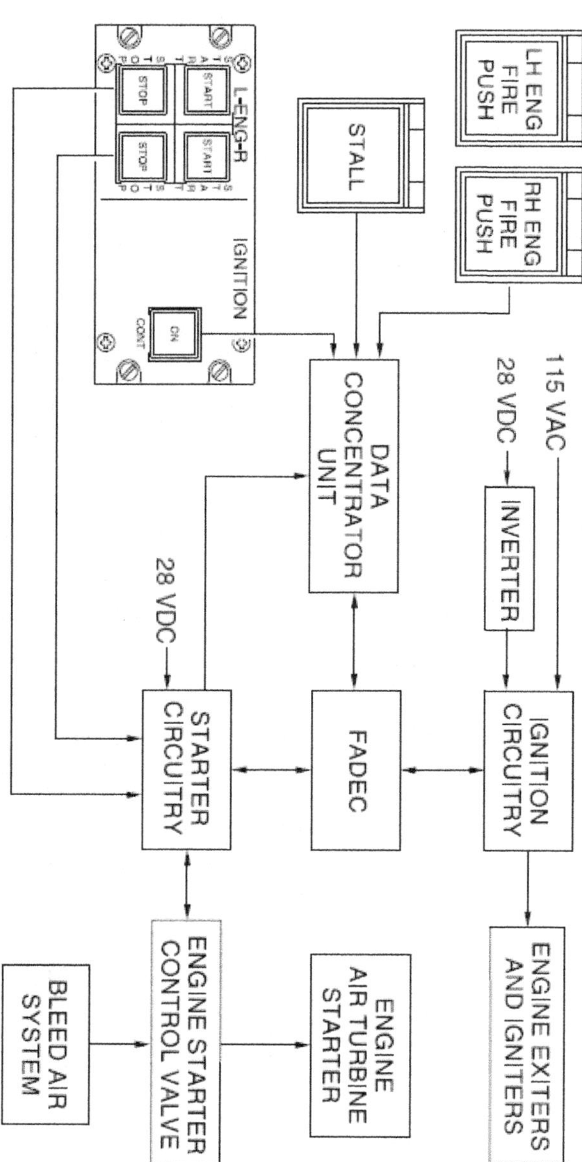

Figure 22 - 11: Ignition System Schematic

Chapter 23: Flight Instruments

Intentionally Left Blank

Chapter 23: Flight Instruments

General

Flight instruments include the electronic flight instrument systems, standby instruments and clocks. Data for the flight instruments is provided by an air data system. radio altimeter and either an attitude and heading reference system (AHRS) or an inertial reference system (IRS). Flight instruments provide the following basic information to the flight crew:

- Altitude (barometric/radio)
- True airspeed
- Airspeed (Mach/KIAS)
- Temperature data
- Airspeed trend
- Airplane attitude
- Vertical speed
- Heading information
- Overspeed warning
- Navigation information

Electronic flight instruments consists of a primary flight display (PFD) and a multi-functional display (MFD) for each pilot. An integrated standby instrument (ISI) provides standby attitude, altitude and airspeed information to the flight crew. An independent standby compass provides aircraft heading in relation to magnetic north. An electronic clock provides the time source for the aircraft avionics equipment.

Air data provided by a pilot-static system and a temperature probe provide the flight instruments with speed, altitude and temperature data. The radio altimeter provides an accurate measurement of height above terrain at low altitudes. The AHRS or IRS systems provide attitude, heading, position, angular rate and linear acceleration information.

Electronic Flight Instrument System

All basic flight information is presented to the flight crew on electronic flight instrument system (EFIS) displays. Each pilot instrument panel contains a primary flight display (PFD) and a multi-functional display (MFD). All four displays are electronically identical to permit transfer of display data.

Each PFD has the primary function of pictorially showing aircraft attitude, altitude, airspeed, flight director commands and flight mode annunciations.

Each MFD acts as a navigation system display and has a primary function of showing current heading (compass) and course information. The MFDs can also display moving map navigation pictorials, navigation sensor data, weather radar targets, and TCAS traffic. Cross-side compass information and backup navigation information can be superimposed on either display.

EICAS information can also be displayed on either MFD.

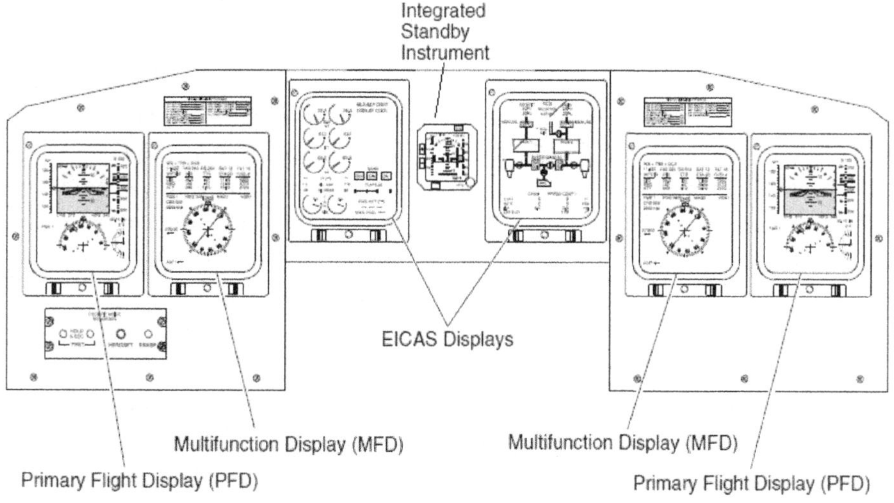

Figure 23 - 1: Electronic Flight Instrument System

Primary Flight Display

Located on the main instrument panels, two primary flight displays (PFDs) provide the operator with the information necessary for the safe operation of the aircraft. Cooling is provided for the screens and is discussed in detail in Environmental Control System chapter, of this manual.

The following data is displayed on the PFD:

- Airspeed {Mach/KIAS)
- Vertical speed
- Altitude
- Attitude (ADI)
- Heading (HSI)
- Flight director (FD) command bars and status
- Navigation information
- Flight mode annunciation (FMA)
- Autopilot information (AP)

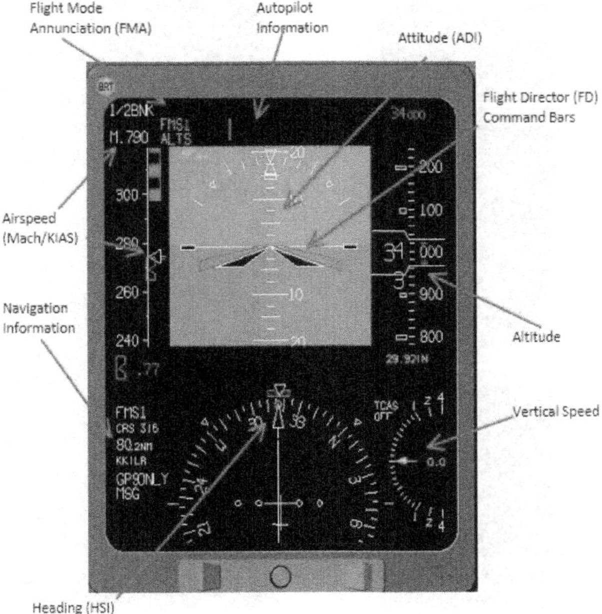

Figure 23 - 2: Primary Flight Display

Multifunction Display

The multi-functional displays (MFDs) combine HSI display and map display data on selectable navigation formats. The MFD can also be used as a reversionary display should the PFD or engine indication and crew alerting system (EICAS) display fail.

The following MFD navigation formats may be selected:

- HSI
- NAV SECTOR
- FMS MAP
- TCAS
- FMS PLAN MAP
- RADAR

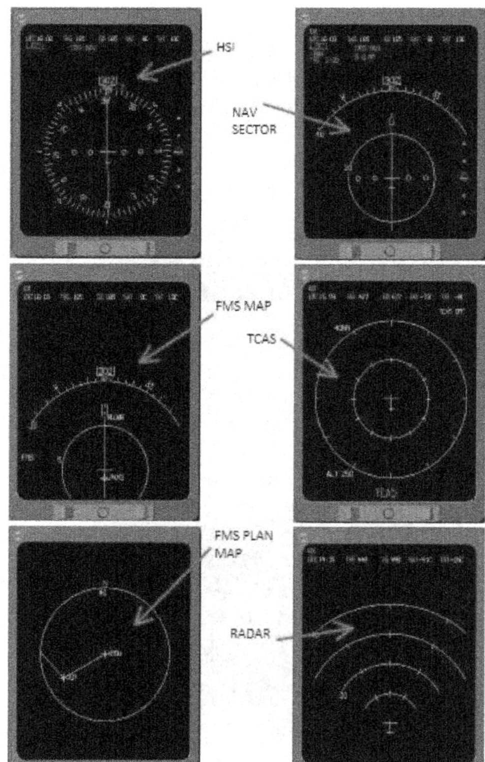

Figure 23 - 3: Multifunction Display Formats

EFIS Control Panels

Information related to speed, barometric setting, navaid source and MFD format is inserted through controls on the following control panels:

- Air data reference panel (ARP)
- Display control panel (DCP)
- Flight control panel (FCP)

Components and Operation

Air Data Reference Panel

The air data reference panel (ARP) is used to present the following pilot selectable information on the associated PFD:

- Reference speeds
- MDA reference altitude
- Decision height
- Barometric altitude setting

The RA TEST pushbutton on the ARP is used to conduct the radio altitude test.

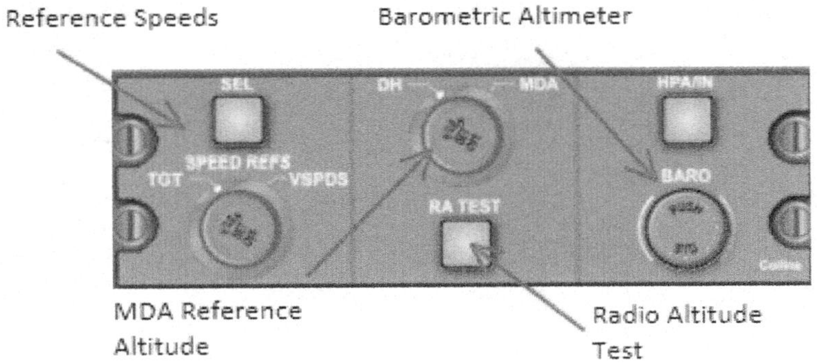

Figure 23 - 4: Air Data Reference Panel

Display Control Panel

Each display control panel (DCP) provides control for the following:

- MFD format selection
- NAV source selection
- Bearing pointer selection
- Cross-side NAV data and course display
- Weather radar or EGPWS presentation
- TCAS presentation

If one DCP should fail, pilot selection of the DSPL CONT knob on the source selector panel permits the operable DCP to control both EFIS displays.

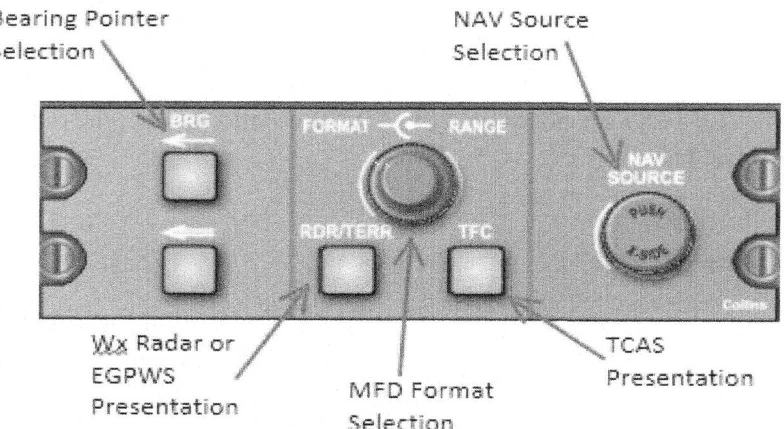

Figure 23 - 5: Display Control Panel

Pitot Static and Air Data System

The pitot static system supplies variable air pressure inputs to the air data computers (ADC 1 and ADC 2), stall protection system, the cabin pressurization system and the integrated standby instrument (ISI).

Components and Operation

There are three independent pitot and static systems:

- P1 is the normal pitot input to ADC 1
- S1 is the normal static input to ADC 1
- P2 is the normal pitot input to ADC 2
- S2 is the normal static input to ADC 2
- P3 is the pitot input to the ISI
- S3 is the static input to the ISI

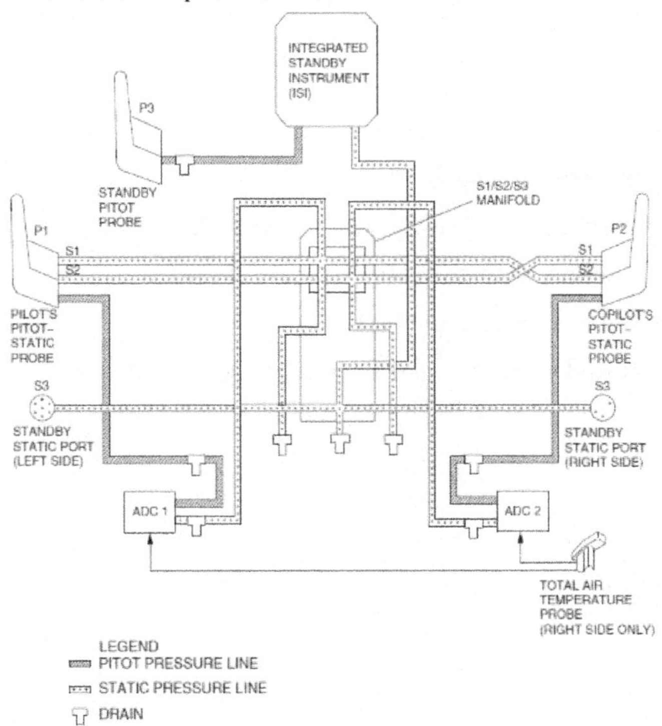

Figure 23 - 6: Pitot Static System

Air Data Computer

Each air data computer (ADC) receives pitot static pressure and temperature from the pitot, static and TAT probes and pilot-selected inputs from the air data reference panel (ARP). The ADC also receives inputs from the integrated avionics processing system (IAPS).

The ADC calculates the following air data system parameters:

- Pressure altitude and barometric-corrected altitude
- Static air temperature (SAT)
- Total air temperature (TAT)
- Temperature variations from standard (ISA)
- Vertical speed
- Indicated airspeed
- Mach number
- True airspeed
- IAS reference
- Vertical speed reference
- Airspeed trend vector
- Barometric setting alerts (preselected to FL 180)
- Static source error correction
- Maximum speed (V_{MO} and M_{MO})
- Overspeed warning and the cabin pressurization system

Total Air Temperature Probe

An electrically-heated total air temperature (TAT) probe is mounted on the fuselage beneath the copilot side window. The probe measures the TAT and supplies this information to the air data computers (ADC 1 and 2). The ADCs use the TAT probe data to calculate the static air temperature (SAT), true airspeed (TAS) and the total air temperature (TAT). The information is then sent to the EFIS for display.

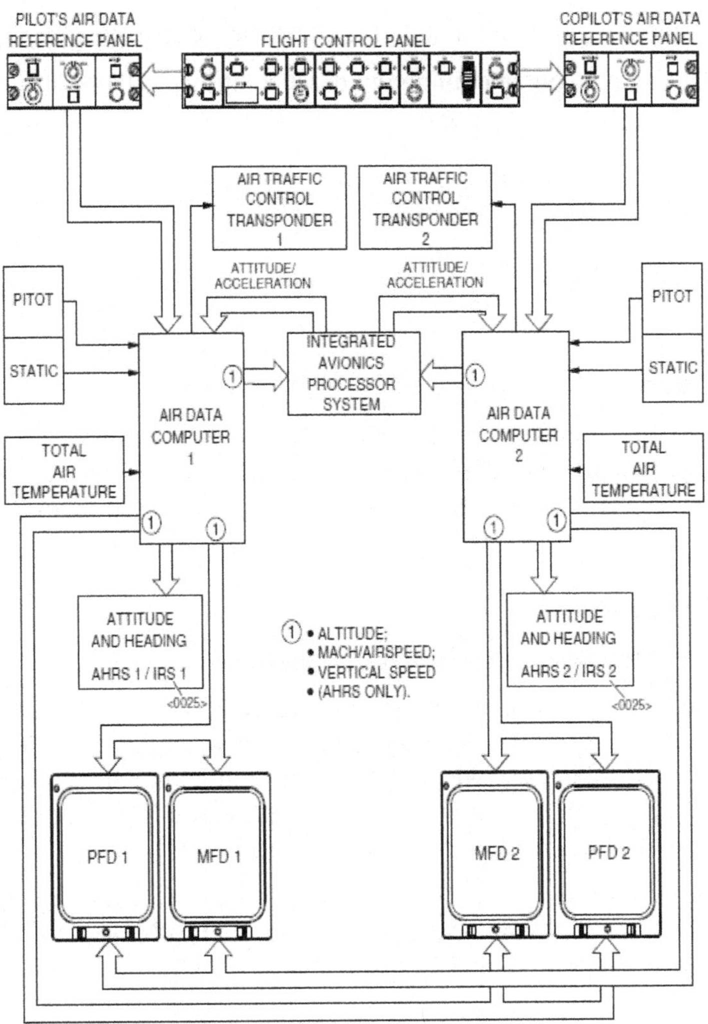

Figure 23 - 7: Air Data System (ADS) Block Diagram

Air Data Reference Panel

The air data reference panels (ARPs) are microprocessors that permit the pilots to select the information presented on the PFDs. The left ARP controls the selectable information displayed on the left PFD. The right ARP controls the selectable information displayed on the right PFD.

The ARP is used to select:

- Barometric altitude setting (BARO)
- Altimeter presentation in hectapascels or inches of mercury (HP/IN)
- Decision height and minimum descent altitude (DH/MDA)
- VFTO and V-speeds (TGT/SPDS)
- Radio altimeter test (RA TEST)

HPA/IN Pushbutton

The HPA/IN pushbutton selects either hPa or inches Hg for display on the PFD.

BARO Knob

The BARO knob is rotated to change the barometric pressure setting displayed on the PFD. Pressing the center of the BARO knob changes the altimeter setting to the standard barometric pressure setting of 29.92 inches of mercury, or 1013 hPa.

DH/MDA Knob

The DH/MDA knob consist of three separate controls.

The outer collar is used to initiate the change process. Turning the outer collar counter-clockwise displays DH (decision height) on the PFD. Turning the collar clockwise displays MDA (minimum descent altitude). The inner control knob is used to change the numerical value of the DH or MDA readout (feet) as displayed on the PFD.

If DH or MDA are not displayed on the PFD, pressing the center of the inner knob will present the previously selected DH/MDA value. A second press removes the DH/MDA readout from the display.

TGT/SPDS

PFD reference speeds are selected at the ARP. Pilot selection at either panel changes the reference speeds presented on both PFDs. The outer collar of the SPEED REF mode knob is used to initiate the change process. Turning the outer collar counterclockwise displays Vr in the change window of the PFD speed reference field. Turning the outer collar clockwise presents the V_1, V_R, and V_2 reference speeds.

To modify the displayed V_{FTO} speed, turn the outer collar to TGT. Rotating the inner knob changes the numerical value of the V_R speed in the change window.

To change the V_1, V_R, or V_2 reference speeds, turn the outer collar to VSPDS. The SEL pushbutton is used to cycle the PFD change window through the 3 selectable (V_1, V_R, V_2) values. When the correct V-speed is presented in the change window, rotating the inner knob changes the numerical readout. Pushing the SEL a second time cycles the change window to the next selectable V-speed and the inner knob is used to modify the presented value. The process is repeated a third time until all V-speeds are set.

Pressing the center of the inner knob replaces the previously set value in the change window with a dashed line. A second press returns the readout to the previously set value.

The change window is only presented on the PFD when modifying the reference speeds. When the changes are complete, the change window is removed from view after the required time delay has been observed.

RA Test

A radio altimeter test (RA TEST) pushbutton when held, initiates the radio altimeter self-test. A valid RA TEST is indicated on the PFDs when 50 feet AGL is displayed on both the digital and analog readouts. Releasing the pushbutton terminates the test and returns the PFD indications to 0.When the optional second radio altimeter is installed, RAD ALT #1 is tested at the pilot's ARP and is displayed on the pilot's PFD. RAD ALT #2 is tested and displayed at the copilot's station.

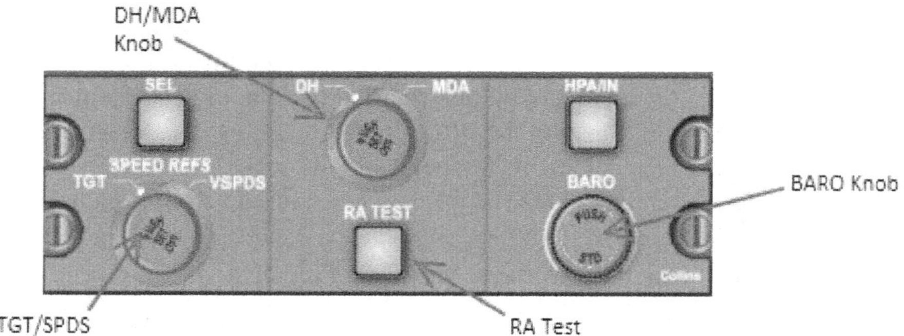

Figure 23 - 8: Air Data Reference Panel

Air Data System Indications

The PFD display presents airspeed, altitude and vertical speed indications. Airspeed information is displayed on the left side of the PFD. Altitude and vertical speed information is displayed on the right.

Components and Operation

Airspeed information displayed on the PFD includes:

- Airspeed.and Mach number display
- Reference speeds
- Low/high-speed cues
- Flap over-speed cue
- Low-speed awareness cue (green line <FAA>)
- Airspeed trend vector

The ADC computes the indicated and true airspeeds that are displayed on the PFD. The V_1, V_R, and V_2 reference speeds are set through the ARP. The SPEED knob on the FCP is used to set the magenta speed bug on the indicated airspeed tape.

Airspeed Display

The airspeed display is shown on the top left side of the PFD as a vertical moving tape. The airspeed pointer is a fixed arrow aligned with the aircraft symbol.

The indicated airspeed tape has an airspeed range from 40 to 400 knots. The tape has a scale that is marked every 10 knots and numerically marked every 20 knots. A 5 knot mark is added between 40 and 200 knots.

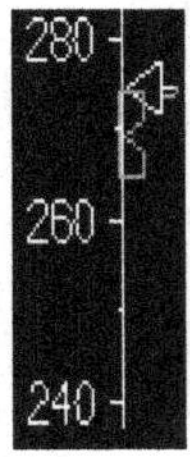

Figure 23 - 9: Display Airspeed

Reference Speeds

The takeoff reference speeds are presented on the speed scale as cyan 1, R, 2, and T characters. Each character is accompanied by a thin solid cyan line that is overlaid on the speed tape. Takeoff reference speeds are selected at the ARPs.

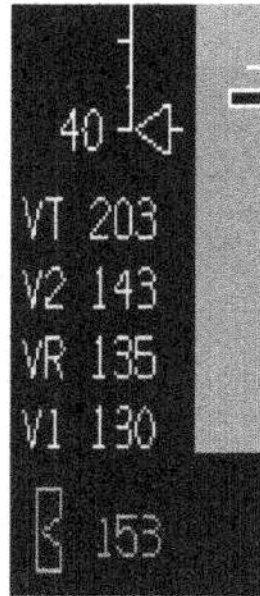

Figure 23 - 10: Reference Speeds

Speed Bug

The SPEED knob o n the flight control panel (FCP) is used to set the airspeed reference bug to the desired speed on the airspeed tape. When the center of the SPEED knob is pressed, the speed reference toggles between Mach and speed. When the SPEED pushbutton on the FCP is pressed, the speed bug automatically aligns itself with the airspeed pointer.

The speed bug is sometimes referred to as the speed bucket because of its unique design. The bug is 10 knots wide. Measuring from the center, there is a ±5 knot speed deviation to the outer edges of the indicator.

Figure 23 - 11: Speed Bug

High-Speed/Flap Overspeed Cue

An alternating red and black checkerboard cue is used to visually represent both the maximum operating speed and when the flaps are extended, VFE for the selected slats/flaps setting.

When the aircraft is operated near the top of its speed envelope, the cue indicates the airspeed at which at VMO/MMO will be exceeded. The over-speed clacker sounds if VMO/MMO is exceeded.

When the slats/flaps are extended, the checkerboard cue is overlaid on the airspeed indicator. The bottom of the cue represents the VFE speed for the existing slats/flaps selection. The over-speed clacker sounds if VFE is exceeded.

Low-Speed Cue

An alternating red and black checkerboard cue is used to visually alert the pilot of an impending stall shaker. The cue descends from the stick shaker speed to the bottom edge of the tape window and appears on the PFD 3 seconds after lift-off. In the landing configuration (gear down, flaps 45), the top of the cue represents a calculated airspeed of 1.06 V5.

When AOA data fails or is missing (which may or may not be due to a stall system failure), the low-speed cue changes to its default setting. The alternating red and black checkerboard cue stops at 100 knots and a yellow line extends upwards from the top of the checkerboard to 120 knots.

The intent of this defaulted failure indication is to provide some remaining indication of low speed awareness while indicating that the low-speed cue is in a degraded and defaulted mode.

The defaulted low speed cue should not be relied upon to provide an accurate indication of proximity to stall shaker.

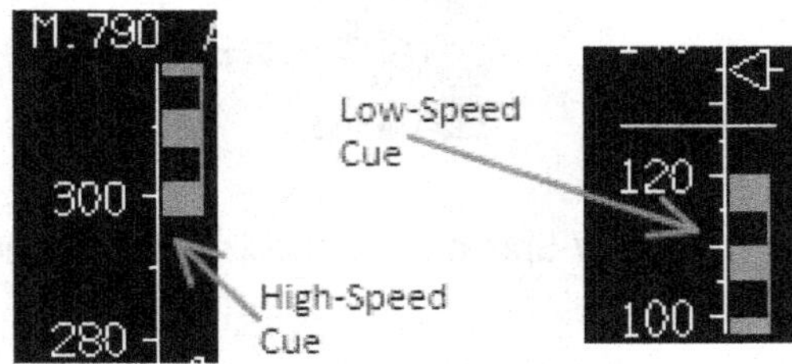

Figure 23 - 12: High-Speed/Flap Overspeed & Low-Speed Cue

Low-Speed Awareness Cue (Green Line)

The green line is provided as a reference only. The pilots must always use the appropriate QRH or AFM V-speeds to ensure adequate safety and performance factors.

The purpose of the low-speed awareness cue or green line (25% margin above stick-shaker speed) is to promote a visual awareness of the approach speed in relationship to stick-shaker activation. The green line is also helpful in determining the maneuver margin available when the landing gear is extended and/or for any flap setting during approach and departure.

During cruise, the green line should not be consulted. At higher altitudes or airspeed, slight differences in the airflow over the left and right angle-of-attack (AOA) vanes caused by side-slippage or turbulence may cause splits in the left and right PFD green line positions. This does not constitute a stall protection system failure.

Trend Vector

The airspeed trend vector originates from the tail of the indicated airspeed arrow. The magenta trend vector moves vertically in response to aircraft acceleration or deceleration. The trend vector continuously predicts the speed of the aircraft that is likely to occur 1 0 seconds into the future if conditions remain unchanged.

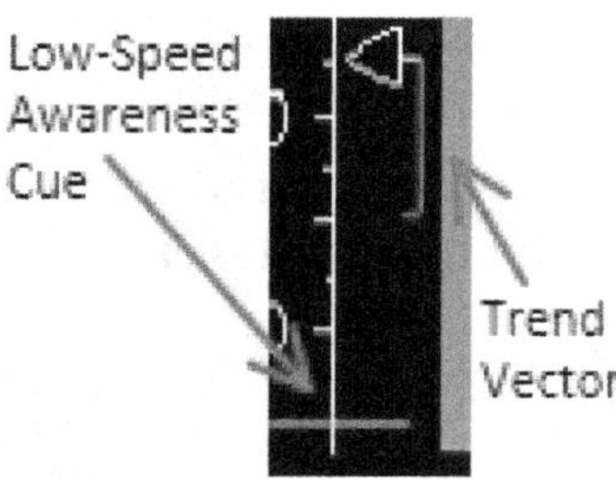

Figure 23 - 13: Low-Speed Awareness Cue & Trend Vector

Mach Number

When the aircraft speed exceeds Mach .45, a Mach window is displayed in the top left corner of the PFD. When aircraft Mach decreases below .40, the window is removed from view.

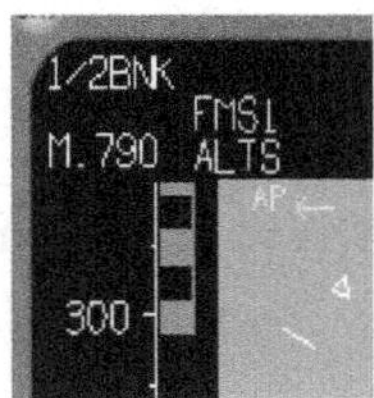

Figure 23 - 14: Mach Number (PDF)

Air Data Flags

When ADC airspeed, altitude or vertical speed data is invalid, red air data flags are presented to replace the affected system's information.

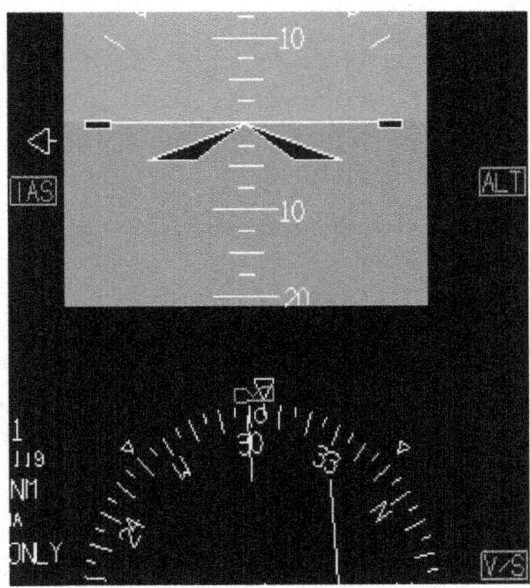

Figure 23 - 15: Air Data Flags

Negative Altitude Warning

A yellow NEG label is displayed when the altitude is less than 0 feet.

Altitude

Barometric altitude information displayed on the PFD consists of:

- Altitude display and readout
- Preselect altitude
- Altitude alert
- Minimum descent altitude (MDA) or decision height (DH)
- Barometric pressure setting

The ADC calculates the pressure altitude and barometric-corrected altitude. The preselected altitude is set via the FCP. Barometric pressure readout and DH/MDA is set through controls on the ARP.

Components and Operation

Altitude Display

The altitude display is shown on the right side of the PFD as a moving tape with a fixed digital readout window. The barometric altitude readout ranges from -1000 to 50,000 feet. The digital readout in the fixed altitude window indicates the actual altitude.

The fine altitude display graduations on the tape are marked every 20 feet and numbered every 100 feet.

The coarse altitude display graduations on the tape provide gross altitude awareness and assist the pilot in capturing altitudes. The coarse graduations are not numbered but are marked with large rectangles for increments of 1000 feet and small rectangles for increments of 500 feet.

Preselect Altitude

The preselect altitude bug and digital readout are selected at the ALT knob on the FCP. The preselect altitude is a reference for the altitude alert and flight director systems. The magenta preselect altitude readout is displayed above the altitude tape and has a selection range of -1000 to 50,000 feet in increments of 100 feet.

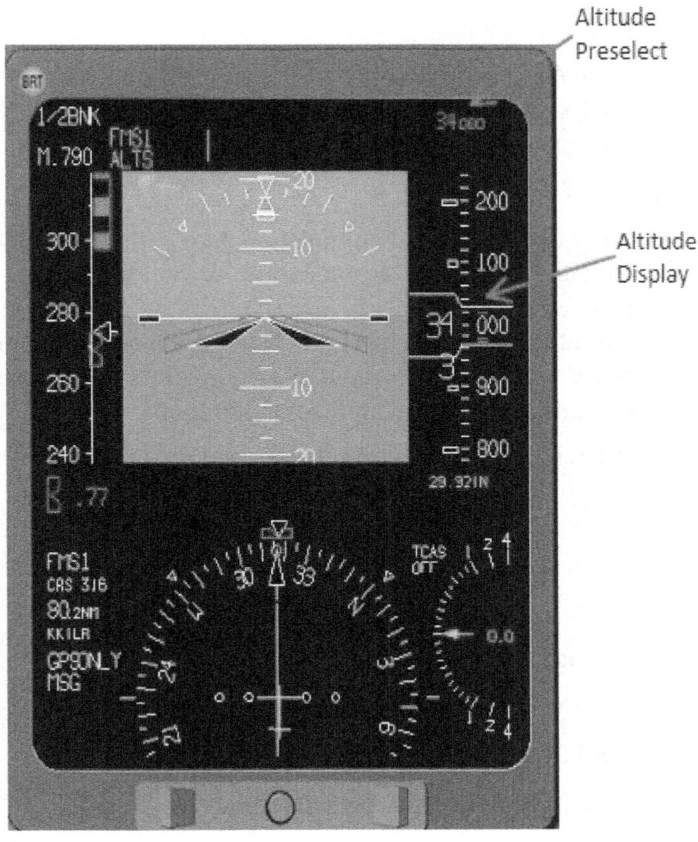

Figure 23 - 16: Altitude Preselect & Altitude Display (PFD)

Altitude Alert Modes

1. Acquisition Mode
   a. During the climb or descent, both the pilot and copilot's preselected altitude bugs flash magenta and a "C chord" tone sounds when the aircraft is within ±1000 feet of the preselected altitude.
2. Cross-side Tracking
   a. Each ADC compares the preselected altitude value from both ADCs. If the values are not equal the display digits change from magenta to cyan.
3. Deviation Mode
   a. After the preselected altitude is captured, if the altitude deviates from the preselected altitude by more than ±200 feet, a deviation alert warning (aural "C chord") will be set and the preselected altitude readout and bug will change from magenta to amber and begin to flash. The readout and bug will return to normal once the altitude is back within deviation limits.
   b. A deviation alert will also be made if the airplane has gone within the acquisition limits on an altitude capture but then deviates by more than 100 feet from the preselected altitude.
4. Advisory Vertical Navigation
   a. When advisory vertical navigation is selected (VNAV), the following are displayed:
      (1) FMS reference altitude
      (2) Vertical deviation (snowflake) and scale
      (3) Suggested vertical speed (dough-nut

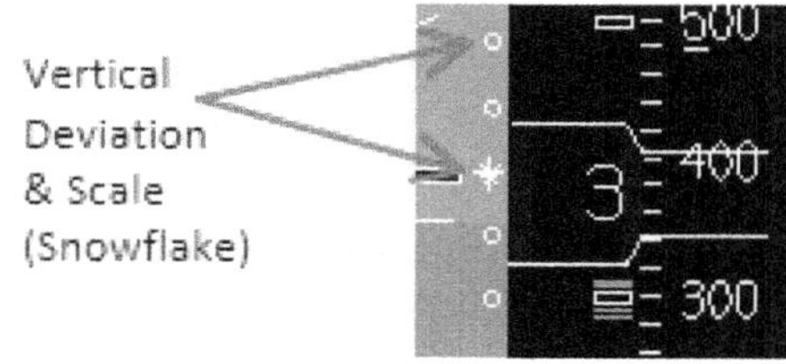

Figure 23 - 17: Advisory Vertical Navigation VNAV (Snowflake)

Barometric Pressure Setting Knob

The HPA/I N pushbutton permits the pilots to display the barometric pressure in either inches of mercury or hectapascels. The barometric altimeter readout is displayed below the altitude tape.

Minimum Descent Altitude

For Cat I approaches, the pilots set the required MDA by using the associated ARP. The selected MDA readout is cyan and appears on the top of the PFD presentation.

As the altitude tape changes with decreasing altitude, a cyan MDA tape marker comes into view. When the aircraft reaches the MDA, a flashing yellow MDA label is presented slightly above the horizon line to the right of center.

Decision Height

Decision height (DH) is displayed as a digital readout and a reference line on the radio altimeter and is used during Cat II approaches.

The digital range can be selected from 0 to 999 feet and is set on the air data reference panel (ARP) using the DH/MDA knob. The cyan reference line is displayed on the radio altimeter tape at the selected altitude.

When the aircraft descends to the decision height, the amber DH label comes into view.

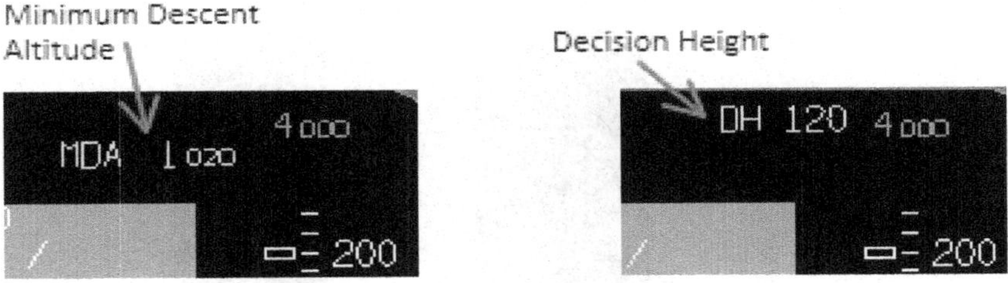

Figure 23 - 18: Minimum Descent Altitude (MDA) & Decision Height (DH)

Radio Altitude Display

The PFD radio altitude display is presented as both a digital and moving tape readout.

The digital readout is displayed from 0 to 2500 feet above ground. The digital display color is green for altitudes from 2500 feet to decision height (DH) and amber for altitudes equal to or lower than DH.

The tape is an altitude scale that indicates altitudes from 0 to 1,100 feet. The scale numbers on the tape change from green to amber at DH and turn green at touchdown. When the radio altitude is between 0 and 200 feet, the crosshatch amber tape is displayed.

Radio Altitude

If the radio altitude (RA) data is invalid, a red boxed RA label is displayed and the radio altitude tape and digit readouts are removed.

Vertical Speed

Vertical speed is displayed on the lower right corner of the PFD as a half-circle scale with a speed pointer and digital readout.

The vertical speed scale has a range of ±4000 feet-per-minute. Located in the middle of the speed scale is the digital vertical speed display that ranges from 0 to 15,000 feet.

The vertical speed scale is also used to display TCAS advisory deviations and the advisory VNAV dough-nut

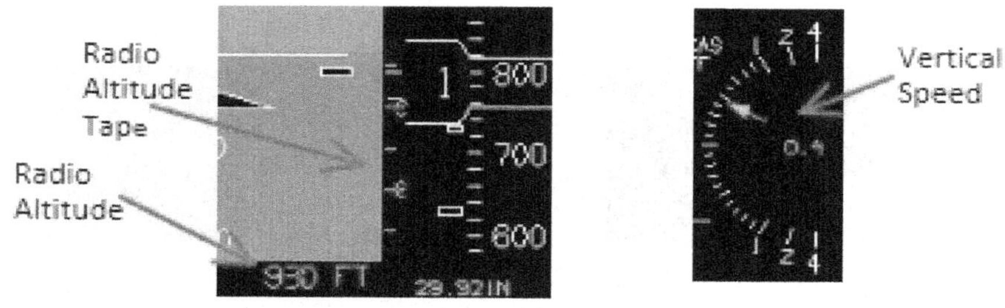

Figure 23 - 19: Radio Altitude Display & Vertical Speed

Attitude and Heading Reference System

The aircraft attitude and heading reference systems (AHRS) AHRS-85E, AHRS-3000 and AHRS-4000 are strapped gyro reference systems that generate angular rate and linear acceleration information about the aircraft body axes. The AH RS interface with the following systems:.

- EFIS
- Weather radar
- Automatic flight control system (AFCS)
- Flight data recorder (FOR)
- Flight control computer/lAPS (FCC)
- Enhanced ground proximity warning system (EGPWS)
- Traffic alert and collision avoidance system (TCAS)
- Stall protection system
- Aircraft fuel system

Components and Operation

AHRS-85E

The AHRS-85E system combines acceleration gyro information with heading data from the flux detector to provide highly accurate three-axis AHRS information.

The AHRS-85E consists of two computers, two flux detector units and two remote compensator units.

AHRS mode selections are made at the COMPASS control panel. The AHRS normally operates in magnetic mode (MAG). Directional gyro mode (DG) may be selected in areas where magnetic field disruptions exist.

AHRS Computers

Each AHRS computer has a dual-sensor assembly. Each sensor assembly has four pairs of piezoelectric accelerometer sensors mounted on a rotating wheel. These sensors provide the rate and acceleration data. The sensor data combined with ADC and heading information is processed by the AHRS computers and transmitted for use by the integrated avionics processor system (IAPS).

Flux Detectors

Each of the two flux detectors senses the earth's magnetic field and provides corrected heading information to the AHRS computers.

Compensator units carry out corrections for any north/south, east/west or misalignment of the flux detectors.

Compass Control Panel

There are two compass control panels located on the lower pedestal. Each control panel contains two switches: A MAG/DG switch and a SLEW switch.

In the MAG mode, heading computations are slaved to the AHRS computer and the flux detectors. In DG mode, flux detector data is removed. DG mode is not intended for long-term heading reference. The momentary action SLEW switch is operational in slaved (MAG) and DG mode.

In DG mode, the selection of the SLEW switch causes the heading computations and the displayed heading to slew in the selected direction. In the MAG mode, the selection of the SLEW switch causes the compass rose to momentarily rotate to a new value. As soon as the SLEW switch is released the compass rose returns to its original position.

Ground Alignment

AHRS initialization commences automatically when electrical power is established on the aircraft. Initialization takes approximately 70 seconds in MAG and 10 minutes in DG.

During the ground alignment process, the PFD displays the red ATT and MAG flags along with the message ATT/HDG ALIGNING DO NOT TAXI. When the alignment is complete, the messages are removed and the sky/ ground raster is displayed. The aircraft can be moved at this time.

NOTE:
If excessive airplane motion is detected during a ground alignment, the AHRS automatically re-initializes.

In-Flight Alignment

If a power interruption or a transient system fault occurs in flight, the system will take between 10 to 35 seconds to realign. During the in-flight alignment process, the aircraft must maintain straight and level flight.

AHRS AHRS-3000

The AHRS-3000 is a solid-state strapped gyro reference system that generates angular rate and linear acceleration information about the aircraft body axes.

The AHRS is equipped with quartz-based inertial sensors that provide acceleration gyro information with heading data to produce highly accurate three-axis AHRS information.

The AHRS-3000 is designed as a replacement for the AHRS-85E and the system operation is identical to the AHRS-85E.

The AHRS-3000 consists of two computers, two flux detector units and two external compensation units.

AHRS mode selections are made at the COMPASS control panel. The AHRS normally operates in magnetic mode (MAG). Directional gyro mode (DG) may be selected in areas where magnetic field disruptions exist.

AHRS Computers

Each computer has an inertial measurement unit using quartz rate sensors and micro mechanical quartz accelerometers which are strapped down to the principle aircraft axes. These sensors provide rate and acceleration data.

The sensor data combined with ADC and heading information is processed by the AHRS computers and transmitted for use by the integrated avionics processor system (IAPS).

Flux Detectors

Each of the two flux detectors senses the earth's magnetic field and provides corrected heading information to the AHRS computers. External compensation units store electronic correction data for any north, south, east/west "hard iron" errors or misalignment of the flux detectors.

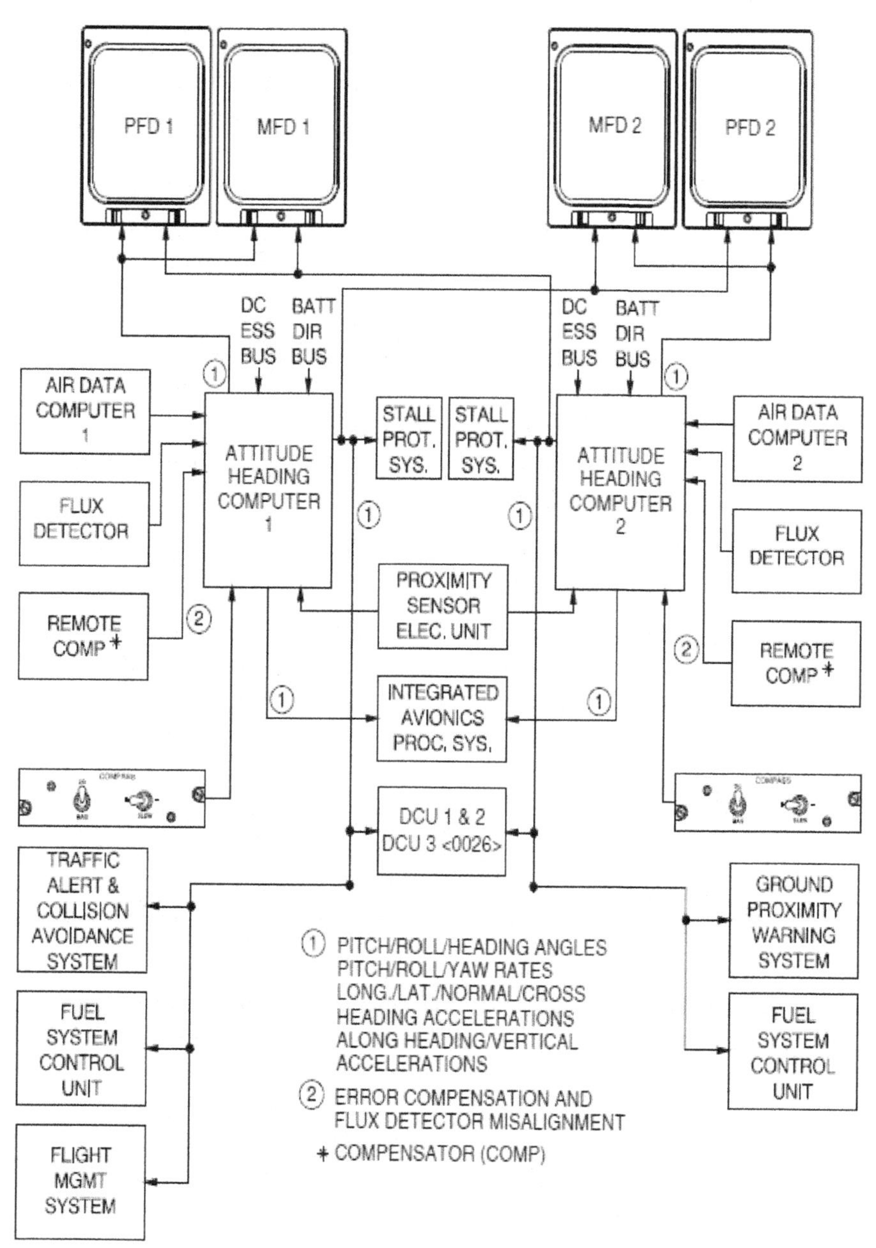

Figure 23 - 20: Altitude Heading Reference System (AHRS) - Interface Diagram

Compass Control Panel

There are two compass control panels located on the lower pedestal. Each control panel contains two switches: A MAG/DG switch and a SLEW switch.

In the MAG mode, heading computations are slaved to the AHRS computer and flux detectors. In DG mode, flux detector data is removed. DG mode is not intended for long- term heading reference.

The momentary action SLEW switch is operational in slaved (MAG) and DG mode. In DG mode, the selection of the SLEW switch causes the heading computations and the displayed heading to slew in the selected direction.

In the MAG mode, the selection of the SLEW switch causes the compass rose to momentarily rotate to a new value. As soon as the SLEW switch is released the compass rose returns to its original position.

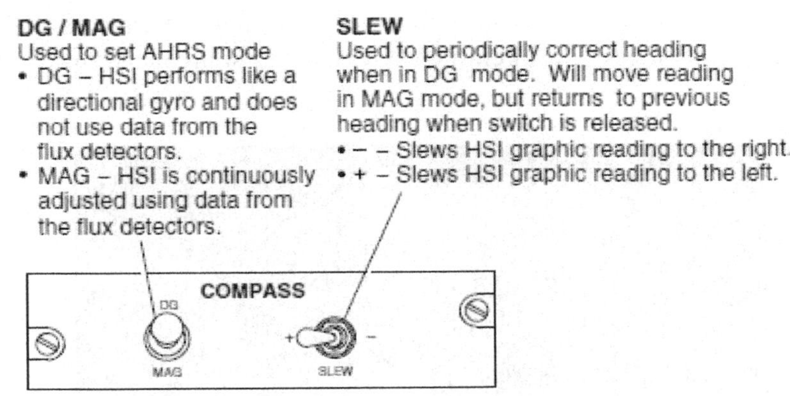

Figure 23 - 21: Compass Control Panel

Ground Alignment

AHRS initialization commences automatically when electrical power is established on the aircraft. Initialization takes approximately 50 seconds in MAG and 5 minutes in DG.

During the ground alignment process, the PFD displays the red A TI and MAG flags along with the message ATT/HDG ALIGNING DO NOT TAXI.

When the alignment is complete, the messages are removed and the sky/ground raster is displayed. The aircraft can be moved at this time.

NOTE:
If excessive airplane motion is detected during a ground alignment, the AHRS automatically re-initializes.

In-Flight Alignment

If a power interruption or a transient system fault occurs in flight, the system will take between 30 to 50 seconds to realign. During the in-flight alignment process, the aircraft must remain in a straight and steady level flight.

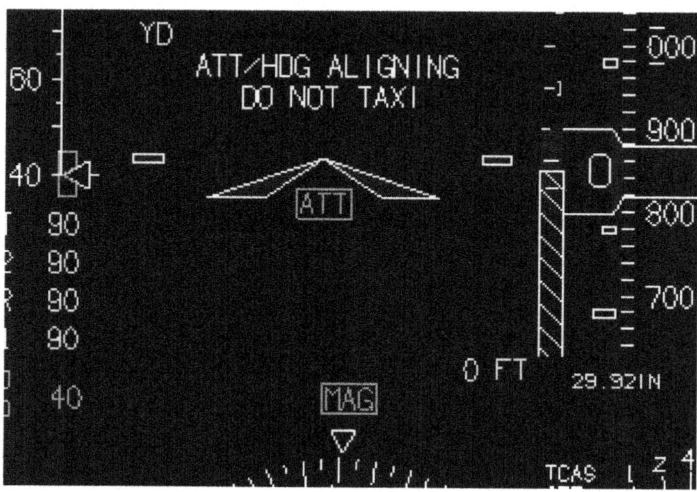

Figure 23 - 22: AHRS Alignment PFD Message

Inertial Reference System

The inertial reference system (IRS) uses ring laser gyros to compute aircraft attitude. The ring laser gyros have a single degree of freedom, which generates the attitude, heading and three-axis rate and acceleration information.

The IRS interfaces with the following aircraft systems:

- EFIS
- Weather radar
- Automatic flight control system (AFCS)
- Flight data recorder (FDR)
- Flight control computer/IAPS (FCC)
- Enhanced ground proximity warning system (EGPWS)
- Traffic alert and collision avoidance system (TCAS)
- Stall protection system
- Aircraft fuel system
- Optional heads up guidance system (HGS)

The IRS is composed of two laser inertial reference units (IRUs) with internal sensors and modules. IRS 1 and IRS 2 mode selection occurs through a single-mode selector unit. The IRS mode selection control panel is mounted on the pedestal.

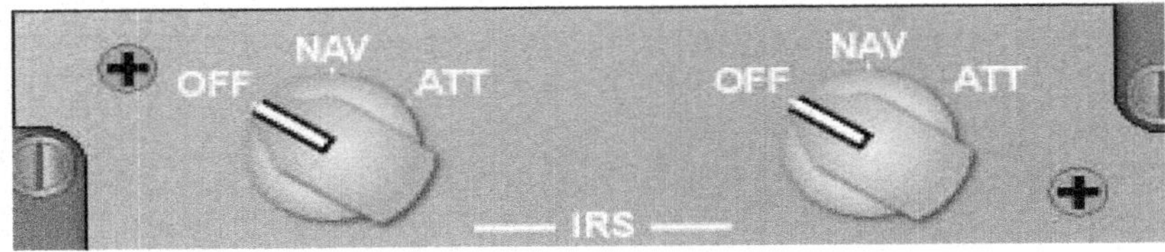

Figure 23 - 23: IRS Mode Select Unit (Center Pedestal)

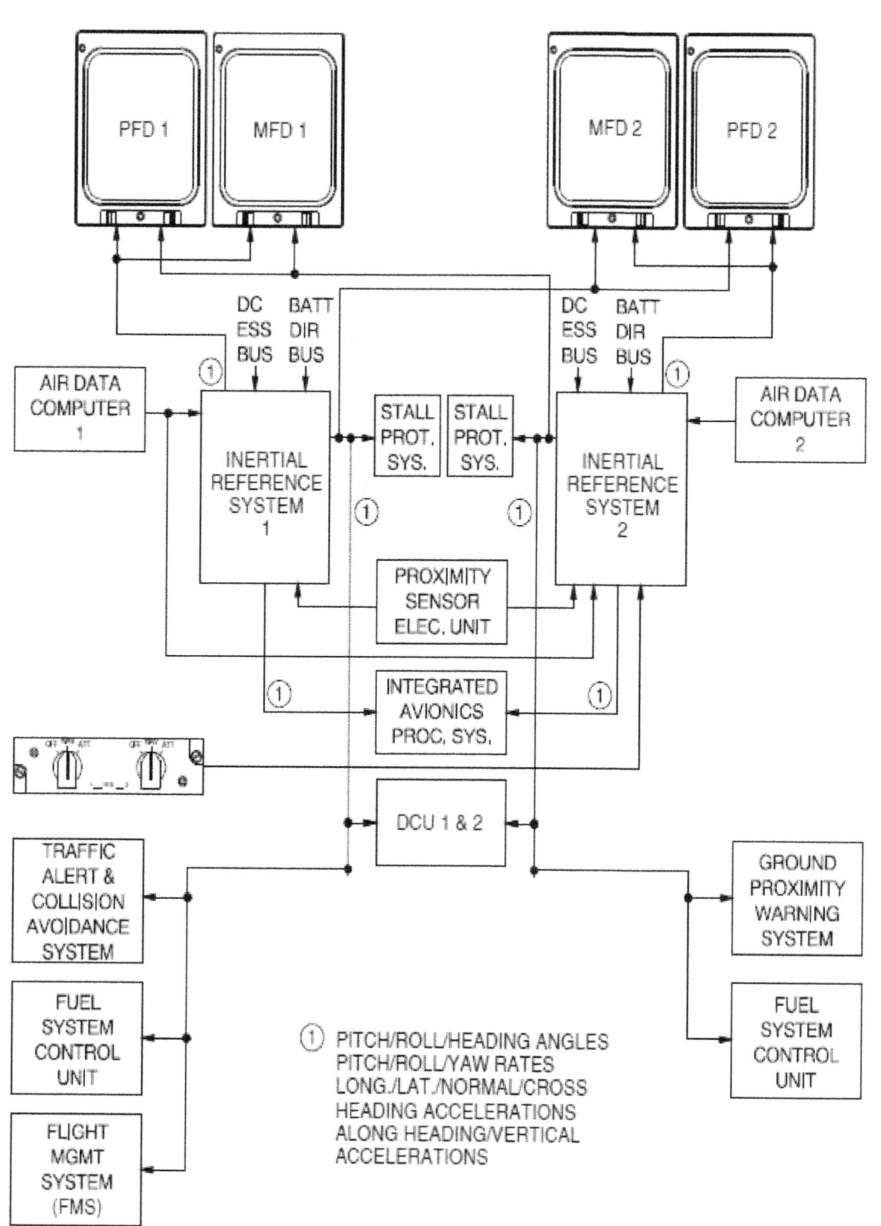

Figure 23 - 24: IRS Interface

Attitude Mode

In attitude (ATT) mode the IRS measures the aircraft's three-axis rate and acceleration and provides attitude and heading data to a number of aircraft systems.ATT mode may be used when navigation (NAV) mode is not available and it does not provide positional information.

Navigation Mode

In navigation (NAV) mode the IRS is capable of generating all the same data as ATT mode but can also provide positional information for the FMS.In NAV mode the IRS requires that the stationary aircraft's ground position be entered through the FMS control display unit (CDU) via the POS INIT page.

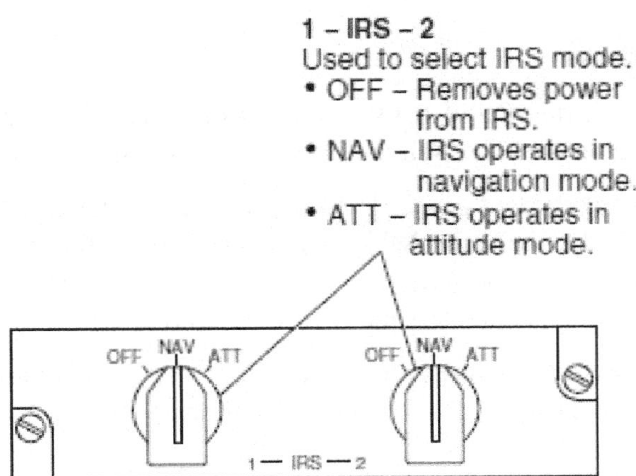

Figure 23 - 25: IRS Mode Select Unit (Center Pedestal)

Components and Operation

Inertial Reference Unit

Each inertial reference unit (IRU) is equipped with a three-ring laser gyro that calculates rate/acceleration information about the three axes. This laser gyro information along with ADC data is transmitted to the integrated avionics processor system (IAPS) and the flight management system (FMS).

IRS Control Panel

The IRS control panel contains a three-position rotary switch for each IRU. Both IRS 1 and IRS 2 normally operate in NAV mode. ATT mode is selected when NAV mode is not available.

Alignment

NAV Selection

Alignment of the IRS in NAV mode requires that the aircraft initial position be entered through the POS INIT page of the FMS.

During the alignment process on the ground, the message 'ATT/HDG ALIGNING DO NOT TAXI' is displayed on the primary flight displays. The aircraft should be stationary. Excessive aircraft movement may extend the time needed to successfully align the IRS.

After successful alignment, the message on the PFD is removed and the IRS automatically sequences into navigation mode.

Rapid Realignment

On the ground, rapid realignment is accomplished by selecting the NAV switch to OFF then back to NAV within 5 seconds. It is necessary to enter the present position in the FMS again to complete the realignment process. In-flight realignment is not possible when NAV is selected.

ATT Selection

Attitude alignment takes 1 minute from power off to ATT mode or 34 seconds from NAV to ATT mode provided the aircraft is stationary on the ground or straight and level in flight If excessive motion is detected, the attitude alignment time will be extended.

During this process, the message 'ATT/HDG ALIGNING DO NOT TAXI is displayed on the primary flight displays.

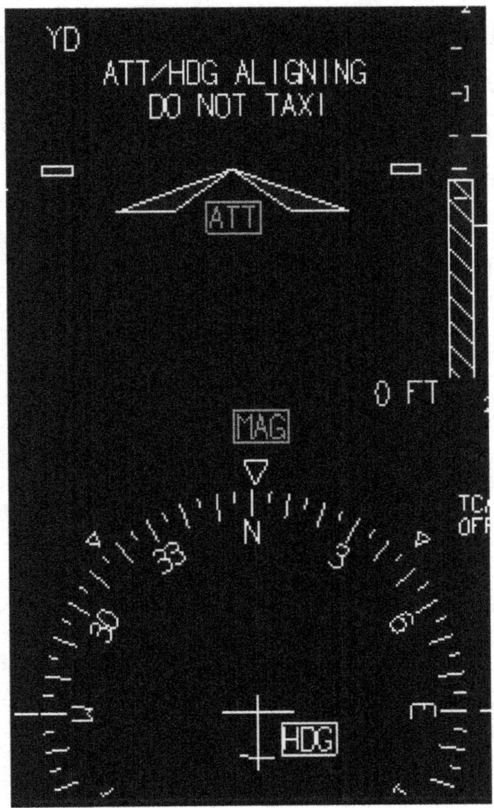

Figure 23 - 26: ATT/HDG ALIGNING Message (PFD)

Attitude

The PFD's pitch, roll and yaw indications are generated by the AHRS or IRS computers.

Components and Operation

Attitude Information Field

Attitude information displayed o n the PFD includes:

- Aircraft and wing symbol
- Horizon line
- Sky/ground raster display
- Pitch tape and chevrons
- Roll scale and pointer
- Slip/skid indicator

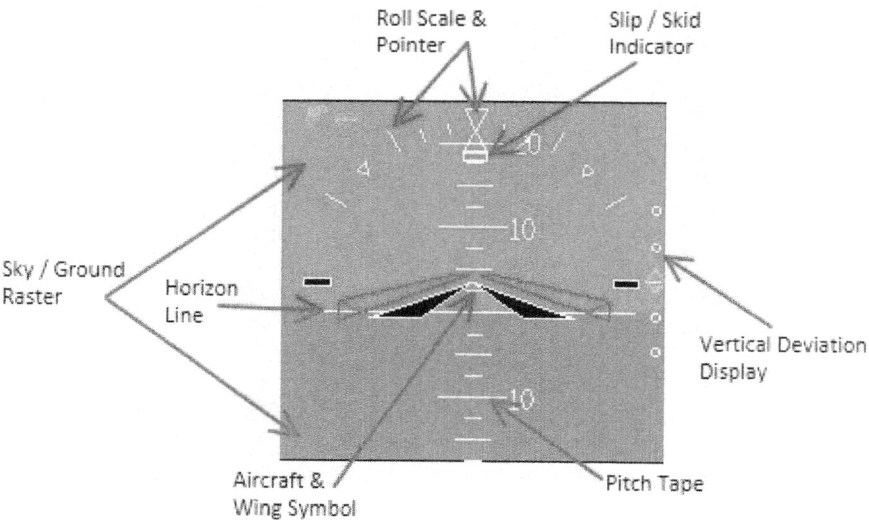

Figure 23 - 27: PFD Attitude Information Field

Fixed Aircraft and Wing Symbol

The white aircraft symbol is a stationary representation of the aircraft. Two small white rectangular wing symbols are shown to the left and right of the aircraft symbol.

Horizon Line

The horizon line is the midpoint indication of aircraft attitude. With 0 degrees of pitch and roll, the white horizon line touches the apex of the aircraft symbol and passes through the center of the aircraft wing symbol.

Sky/Ground Raster

This blue-and-brown image provides a distinct division between sky and ground.

Pitch Tape

The pitch tape is fixed on the raster and moves with the horizon line. The tape has wide line markings every 10 degrees, medium line markings every 5 degrees and small line markings every 2.5 degrees.

Roll Scale and Pointer

At the top of the blue raster is a roll scale and pointer. The scale range is ±60 degrees from a wings level attitude.

The scale is marked as follows:

- Large tick marks at ±30 and ±60 degrees
- Small tick marks at ±10 and ±20 degrees
- Small triangle at ±45 degrees

The roll pointer moves in reference to the center of the aircraft symbol and indicates the aircraft roll angle.

Slip/Skid Indicator

The slip/skid indicator is a small rectangle located below the roll pointer.

The slip/skid indicator is driven by lateral acceleration. It turns with the roll pointer and moves laterally with respect to the base of the roll pointer.

One brick-width deflection of the rectangle is equivalent to a displacement of two ball-widths on a conventional slip/skid indicator.

Vertical Deviation Display

The glide-slope vertical deviation display is shown on the right of the attitude raster. It is a white fixed scale employing a moving diamond deviation icon. The color of the deviation icon is green when displaying the onside data and is amber when displaying cross-side data.

The scale has two dots above and below a reference line with each dot representing a 0.25 degrees of deviation above or below the glide-slope beam. The icon moves vertically on the deviation scale and if an excessive vertical deviation condition exists, the icon fl ashes amber until the condition is corrected.

Glide-slope Flag

During an ILS with invalid glide-slope information, a red boxed GS flag is displayed to replace the pointer and scale.

Windshear

The alpha margin indicator (AMI) for the windshear indication is supplied by the enhanced ground proximity warning computer and is displayed during a windshear alert or warning.

The position of the AMI represents the maximum pitch attitude allowed before the stick shaker is activated. An amber AMI is presented during windshear alerts and warnings. The AMI is displayed for a minimum of 60 seconds.

The windshear alert is displayed when an increased performance windshear is encountered. An amber WINDSHEAR message is presented on the PFD.

The windshear warning is displayed when a decreased performance windshear is encountered. A red WINDSHEAR warning message is presented on the PFD and a "WINDSHEAR" aural alert sounds.

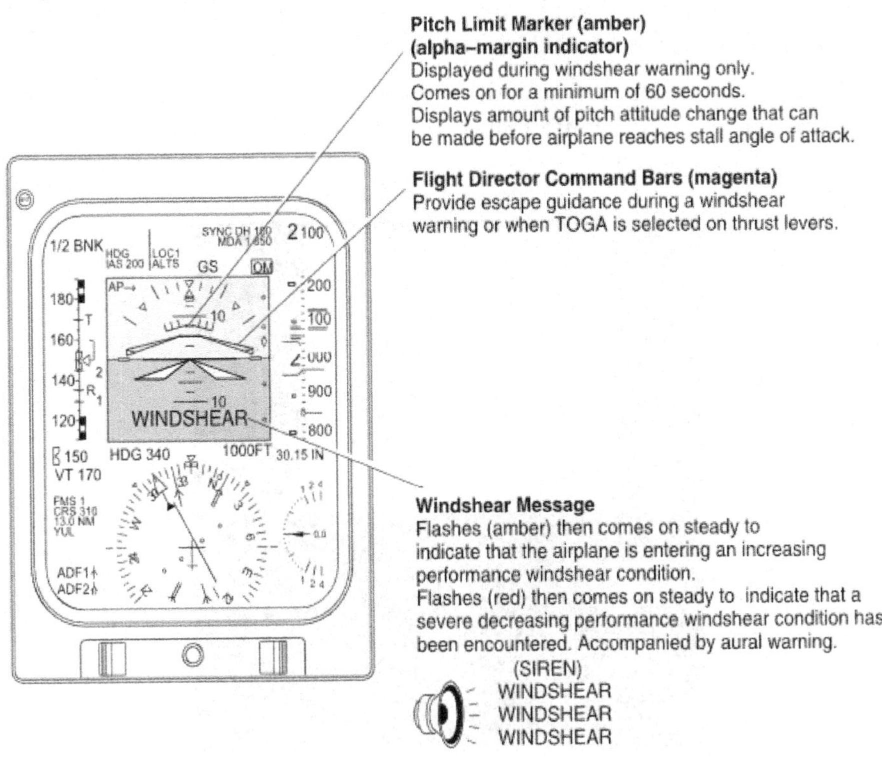

Figure 23 - 28: Windshear PFD Indications

Declutter

Declutter is used to clear all nonessential information from the PFD during aircraft upset or unusual attitudes.

When the pitch angle exceeds +30 or -20 degrees, or roll exceeds 65 degrees left or right, only the following information is displayed on the PFD:

- Altitude
- Airspeed
- Attitude
- Vertical speed
- Compass
- FCC transfer indicator

Red chevrons appear during maneuvers of extreme pitch attitude. The chevrons always point in the direction of level flight.

When the pitch and roll attitudes are less than the prescribed exceedances, all PFD information previously removed during the declutter is restored.

Figure 23 - 29: PFD Declutter

Heading

The aircraft heading is displayed on a partial compass card located at the bottom of the PFD. Data required to produce the display is supplied by the AHRS system (IRS system). With an AHR S installation, magnetic or directional gyro may be displayed.

With an IRS installation. magnetic or true heading may be displayed.

Components and Operation

Heading Information

Heading information displayed on the PFD includes:

- Compass card
- Lubber line
- Selected heading bug and vector
- Selected heading display
- Aircraft symbol
- Drift angle pointer

Heading Indicators

The heading indications are represented on the compass card, which display ±120 degrees from the current heading. The compass card has 10 degree major indices's, 5 degree minor indices's, cardinal point characters at N, S, E, W and numeric labels at 30 degree intervals.

At the top center position of the compass card is the fixed lubber line. The lubber line indicates the aircraft heading reference.

To assist heading visualization, two fixed index marks are presented 45 degrees on either side of the lubber line.

An aircraft symbol is centered on the compass card and is aligned with the lubber line.

Selected Heading Reference

The selected heading reference is presented on the heading display as a permanently displayed heading bug and temporarily as a bug vector line and digital readout.

The heading bug indicates the heading selected at the HDG knob on the FCP. When a new heading is selected, the heading vector line and digital heading readout is displayed for a few seconds and then removed from view.

Attitude and Heading Flags

When attitude or heading data from the AHRS/IRS becomes invalid, warning flags are presented.

A red boxed attitude flag 'ATT' is displayed when the attitude information is invalid. All other attitude information disappears from the PFD.

A red boxed heading flag (MAG, DG or TRUE) is displayed when the heading source is invalid. When the flag is in view, the heading display is removed from view.

Drift Angle Pointer

The drift angle pointer (green dough-nut) indicates the aircraft drift angle in relationship to a programmed FMS route. The amount of aircraft drift is indicated by the angular displacement of the green dough-nut from the aircraft heading indicator.

The drift angle pointer is green when the onside FMS is the nav source and amber for cross-side information.

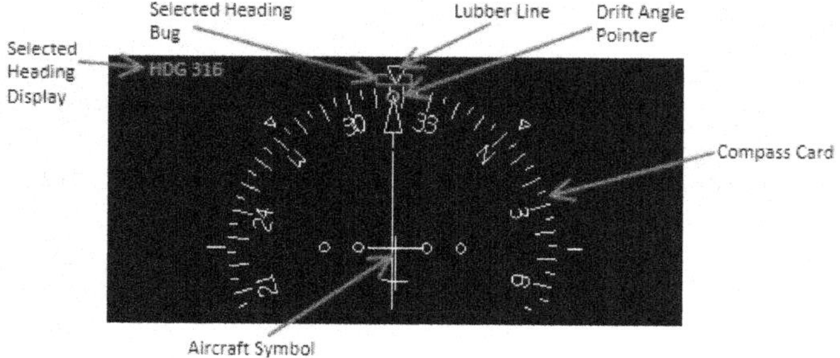

Figure 23 - 30: Heading Information PFD

Navigation Display Indications

The navigation indications on the PFD provide course, course deviation, distance and bearing information from the selected navaid. The data is presented in a conventional HSI format

Components and Operation

Navigation Data

Navigation data is displayed on the lower left side of the PFD. The display includes:

- Navigation source indicator
- Course display
- Distance display
- Station identifier

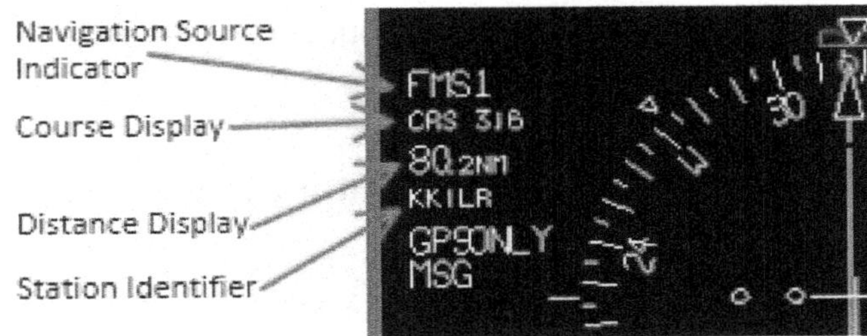

Figure 23 - 31: Navigation Data PFD

Navigation Source Indicator

The navigation source indicator is presented in digital format. Navigation source is selected at the NAV SOURCE knob on the DCP. One of the following sources may be selected:

- VOR 1/LOC 1
- VOR 2/LOC 2
- FMS 1/FMS 2
- OFF (no display)

When a cross-side source is selected, the presentation turns amber in color.

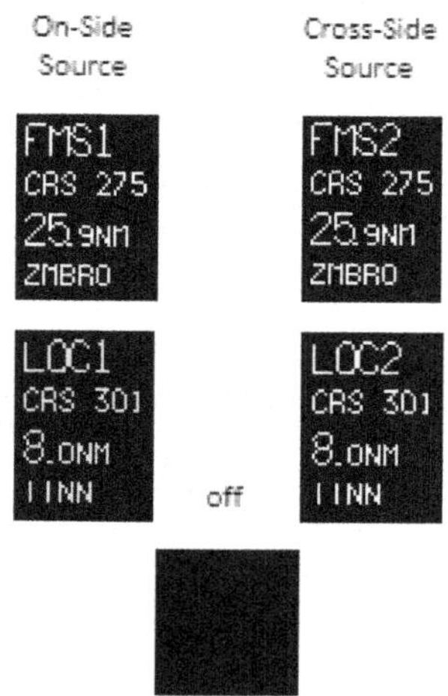

Figure 23 - 32: Navigation Source Indications

Course Display

The selected course display is a three-digit display preceded by a CRS label.

The course range is from one to 360 degrees and is selected from the CRS knob on the FCP.

Distance Display

The distance display shows the DME distance to the navaid station or to the next waypoint when FMS is the NAV source.

When the DME transceiver is in hold mode, an amber 'H' replaces the green 'NM'.

Station Identifier

The navaid identifier is shown in the navigation data field below the distance display. Data for the identifier is supplied by the DME.

In DME hold, the identifier is removed from view.

Bearing Indications

Bearing indications consist of bearing source indicators and bearing pointers.

The bearing source indicators are displayed in the lower left portion of the PFD. The pointers are displayed on the compass card.

The source indicators and pointers are selectable through the BRG switches of the DCP.

Like the navigation source indicator, the bearing source indicators and pointers show which bearing source is in use:

- VOR 1/ VOR 2
- ADF 1/ADF 2
- FMS 1/FMS 2
- OFF (no display)

The bearing pointers display bearing information from a selected navaid. The bearing from the RTU 1 navaids are displayed on a single-lined magenta pointer while the RTU 2 navaids are depicted as a double-lined cyan pointer.

The bearing pointer is removed from view when the BRG switch is selected off or when the bearing source is invalid.

Navigation Source Warning Flag

A red boxed navigation source flag (VOR, LOC or FMS) is displayed during a failure of the selected NAV source. The navigation display is removed from view when the flag is displayed.

Bearing Flag

Invalid data from the selected bearing source replaces the bearing source annunciator with a red boxed warning flag. The bearing pointer is removed from view when the flag is displayed.

Horizontal Situation Indicator

The horizontal situation indicator (HSI) represents the compass card with overlaid selectable navaids, bearing and course pointers.

HSI indications include:

- Course pointers
- Lateral deviation bar
- Lateral deviation scale
- TO/FROM Indicator

Course Pointer

The course pointer is an arrow that points to the pilot-selected course. The selected course is set via the CRS selector knob on the FCP.

The onside color is green and cross-side is amber.

Lateral Deviation Bar

The lateral deviation bar indicates aircraft deviation from the selected course. The lateral deviation bar moves from side to side and its position is read against the deviation scale.

Lateral Deviation Scale

The lateral deviation scale has two dots on each side of the course arrow. When in VOR each dot represents 5 degrees of deviation. In LOC, each dot represents 1 degree of deviation.

In FMS the lateral deviation sensitivity is:

Enroute:

- ±5 nm

In terminal area within 30 nm from ARP:

- ±1.0 nm for RNAV approach
- ±1.0 nm for GPS approach (±0.3 nm at the FAF)

TO/FROM Indicator

The TO/FROM indicator is a small triangle indicating the relative direction of the selected VOR station or FMS waypoint.

The color of the TO/FROM indicator is green for VOR onside information and amber for cross-side information.

In FMS mode the NAV data is white.

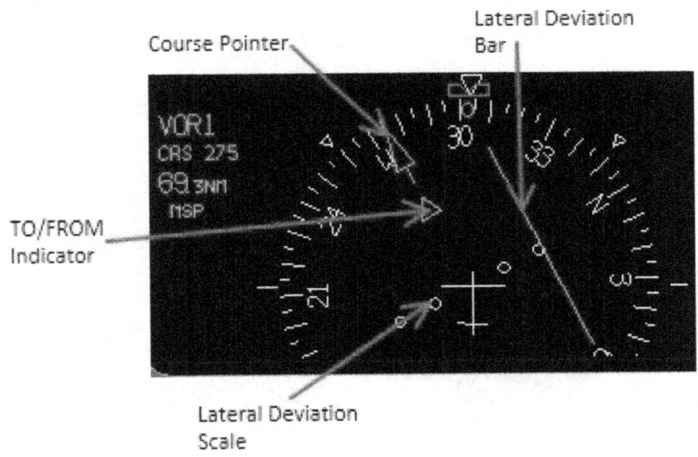

Figure 23 - 33: Horizontal Situation Indicator

Automatic Flight Control System Display

Automatic flight control system (AFCS) indications are presented on the PFD. In the FMA field of the PFD, the flight director (FD) lateral and vertical modes are presented. Other AFCS related PFD indications include the FD command bars and autopilot operating modes

Flight Mode Annunciator

The flight mode annunciator (FMA) is located above the attitude raster. The FMA displays the current and impending FD status separated by a cyan vertical line. The left side of the line represents the active or captured field while the right side of the line represents the armed field.

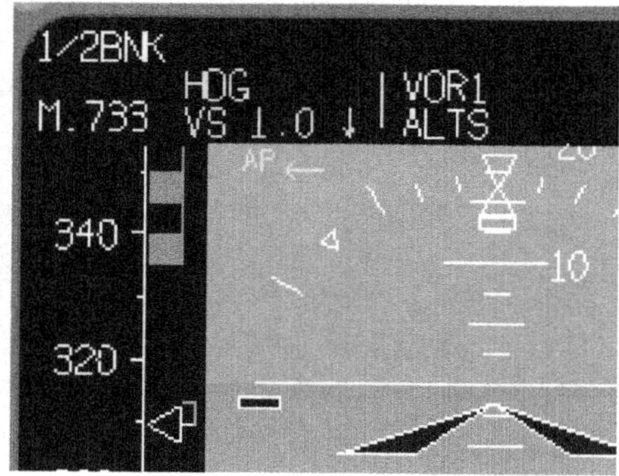

Figure 23 - 34: Flight Mode Annunciator

Modes

Lateral mode annunciation is displayed above the vertical mode annunciation. Armed modes are displayed in white and captured/active modes in green. When an armed mode transitions to a captured or active mode, the new mode flashes green for 5 seconds.

Flight Director Command Bars

The flight director command bars provide the pilots with instructions from the active flight control computer. The magenta flight director command bars are an inverted V-shaped cue symbol that is directed to move horizontally and vertically about the apex of the white stationary aircraft symbol.

1/2 BNK Indicator

When the aircraft climbs above 31,600 feet, the half bank flight director commands are automatically generated. The flight director 1/2 BANK mode can be manually selected on the FCP.

When 1/2 BANK mode is active, a white 1/2 BNK label is displayed to the left of the FMA. Selecting the 1/2 BANK pushbutton a second time disengages the mode.

Flight Director Label

A red boxed FD label appears when either the pitch or the roll attitude input to the flight director becomes invalid. The flight director command bar is removed when the FD label is displayed.

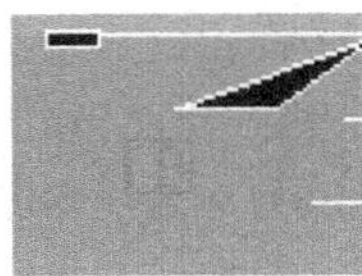

Figure 23 - 35: Flight Director Label

Autopilot Indication

Autopilot engagement is identified on the PFD by a green AP symbol. Active flight director is identified with a horizontal arrow that points in the direction of the active flight director.

A white → symbol indicates that the autopilot is not engaged and control of the FD will be transferred to the indicated side when engaged.

A green AP→ symbol indicates the autopilot has been engaged and the horizontal arrow points to the FD coupled side.

A flashing red AP← symbol indicates that the autopilot has disengaged. If the autopilot is disengaged by the pilot, the AP← symbol flashes and the aural warning sounds for 2 seconds.

If the autopilot disconnects due to a system fault, the AP→ symbol flashes and the aural warning is heard continuously. To cancel the flashing symbol and aural warning, press the AP/SP DISC switch on either control wheel or press either TOGA switch.

Autopilot Trim Indications

Autopilot indications related to yaw damper status, elevator and aileron mistrim conditions are presented on the sky portion of the raster.

Elevator Trim

An amber boxed 'E' indicates that an elevator out-of-trim condition has been detected by the flight control computer with the autopilot engaged.

Aileron Trim

An amber boxed 'A' indicates that an aileron out-of-trim condition has been sensed by the flight control computer with the autopilot engaged.

YAW Damper

An amber YD symbol indicates that both yaw dampers have been disengaged. The YD symbol is displayed on the left side of the raster.

Figure 23 - 36: YAW Damper Symbol

DCP Warning Flag

A display control panel (DCP) failure results in the display of a red boxed DCP flag. This flag appears on the PFD and MFD displays.

Figure 23 - 37: DCP Warning Flag

Over-temperature Warning

An over-temperature sensor is installed inside each of the EFIS displays. When the display overheats, the display information is removed from view and the red DISPLAY TEMP message is presented. When the display cools to an acceptable temperature, normal display presentation is returned.

Multifunction Displays

The multifunction display (MFD) can display several navigation formats.

The following formats may be selected:

- HSI
- NAV SECTOR
- FMS MAP
- FMS PLAN MAP
- RADAR
- EGPWS terrain information
- TCAS

The display reversionary control panel permits the MFD to present PFD or EICAS information.

Maintenance diagnostics computer (MDC) data can be displayed on the MFD. MDC presentation is selected at a switch located on CBP-1 and page scrolling is accomplished by using the EICAS control panel (ECP). FMS text-based data can be presented on the MFD. Presentation of the material is controlled through the FMS CDU.

Navigation sources for HSI, NAV and FMS PLAN MAP formats are selected at the NAV SOURCE knob.

Components and Operation

Display Operation

While each format display has a unique purpose, there are indications that are common to most of the displays. They include:

- Information lines
- Radar/terrain information line
- Range rings
- Compass card
- Aircraft symbol

Information Lines

Across the top of the display are two horizontal data fields that present the following information:

- Radar or terrain status
- Universal coordinated time
- True airspeed
- Ground speed
- Static air temperature
- Total air temperature

Radar/Terrain Status

The radar/terrain status line is presented on the top of the page and identifies which of the two is selected for presentation. This information line is present in all formats.

Universal Coordinated Time (UTC)

UTC time is presented on the second line. Time is taken from the pilot's clock. Both clocks provide time data to the DC Us. If the pilot's clock is unserviceable, the copilot's clock will supply the information for display.

True Airspeed

The true airspeed (TAS) display receives data from the onside ADC. The TAS data is normally green but changes to amber if the cross-side ADC is selected.

Ground Speed

The ground speed (GS) display is a white numerical readout followed by a green GS title. The FMS computer provides the ground speed data. The display is blank when the input data is invalid.

Static Air Temperature

The static air temperature (SAT) presentation is green when the onside ADC is supplying the data. The data is amber for cross-side ADC information. If the data is invalid, the display and title are blanked.

Total Air Temperature Display

The total air temperature (TAT) is a numerical readout preceded by a TAT title. When the onside ADC is used, the TAT label is green. Cross-side ADC data is amber. If the data is invalid, the display and title are blanked.

Display Temp Warning

The red DISPLAY TEMP warning is displayed on the radar line when the display temperature is excessive. When the message is displayed, all information is removed.

Range Rings and Compass Card

Range rings appear on all MFD formats with the exception of HSI. The HSI format displays a full compass card in place of the range rings.

Range rings are a visual representation of the distance from the fixed aircraft symbol.

The range scales are selectable on the DCP from 5, 10, 20, 40, 80. 160, and 320 nm. When the radar is overlaid on the MFD, the range scale can be extended to 640 nm.

Aircraft Symbol

The fixed aircraft symbol is common to all formats and represents the visual position of the aircraft

Weather Radar

The weather radar can be overlaid on most formats except HSI and FMS plan.

The radar format page is selected by the DCP format knob or by pressing the RDR/TERR pushbutton.

EGPWS Terrain

Terrain data can be overlaid on all formats except HSI and FMS plan. TERRAIN is the default selection for the information line at the top of the MFD page but can be selected off manually by selecting the RDR/ TERR pushbutton on the DCP.

Bearing Source and Pointers

The bearing source and pointers are common to all formats except the FMS plan format. Sources 1 and 2 are selected by the BRG pushbuttons on the DCP.

FMS Wind Display

When FMS -sensed winds are more than 5 knots, a white wind vector arrow and digital readout are displayed on the MFD pages.

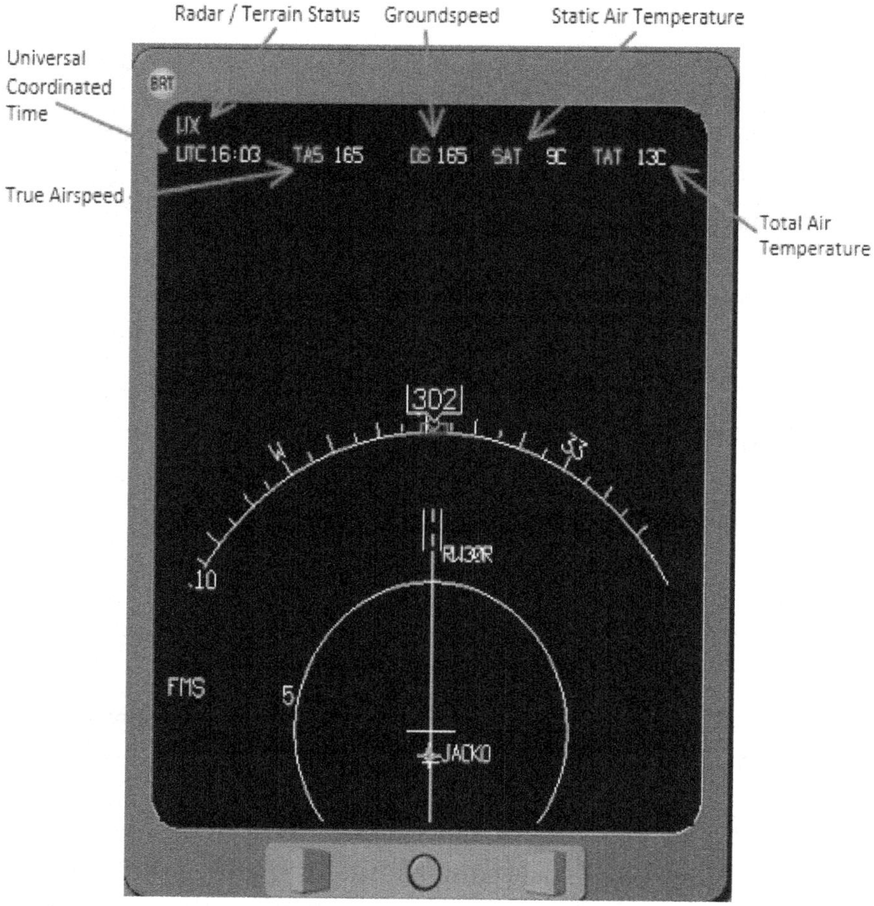

Figure 23 - 38: Multifunction Display

Display Control Panel

The DCP controls the information displayed on the MFD. It contains the FORMAT and NAV SOURCE rotary selector switches as well as the BRG, RDR/TERR and TFC pushbuttons.

The FORMAT selection is discussed in the PFD section of this chapter.

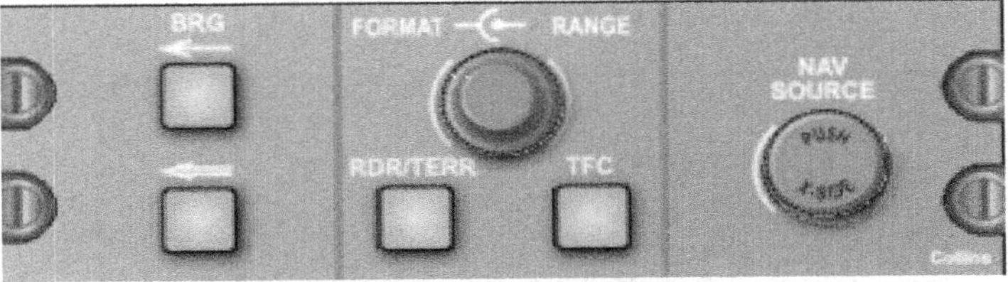

Figure 23 - 39: Display Control Panel

Components and Operation

NAV SOURCE Selector

The NAV SOURCE selector switch is a rotary switch with no mechanical stop in either the clockwise or counterclockwise direction. The page selections available are:

- HSI
- NAV SECTOR
- FMS MAP
- FMS PLAN MAP
- RADAR

HSI

The horizontal situation indicator (HSI) is a traditional display electronically presented on the MFD. The HSI repeats the information displayed on the NAV portion of the PFD with a larger display and a 360 degree compass card.

Below the information line on the HSI is a field consisting of the navigation source, station identifier, course display, and distance display.

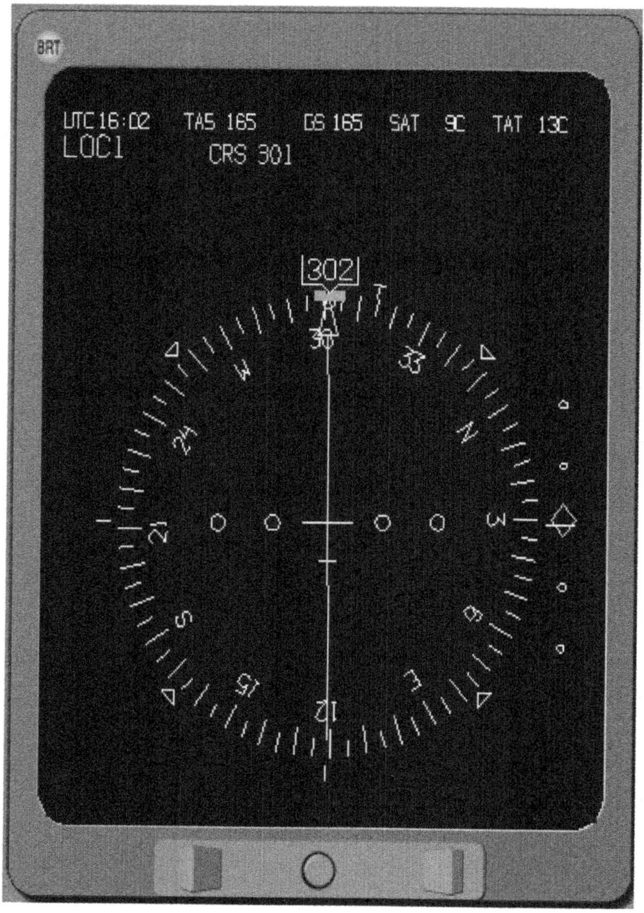

Figure 23 - 40: Horizontal Situation Indicator

NAV SECTOR

The NAV SECTOR page is a navigation display with a partial compass card and a background map. The navigation display is an expanded HSI with a course pointer and deviation bar. This format is a heading oriented map for displaying VOR bearing and DME distance information.

A green navaid symbol is displayed to indicate the position of the tuned navaid. For correct operation, the navigation system must be tuned to a valid VOR with a valid DME. If DME-H (DME hold) is selected on the radio tuning unit (RTU), the navaid symbol is removed from the display.

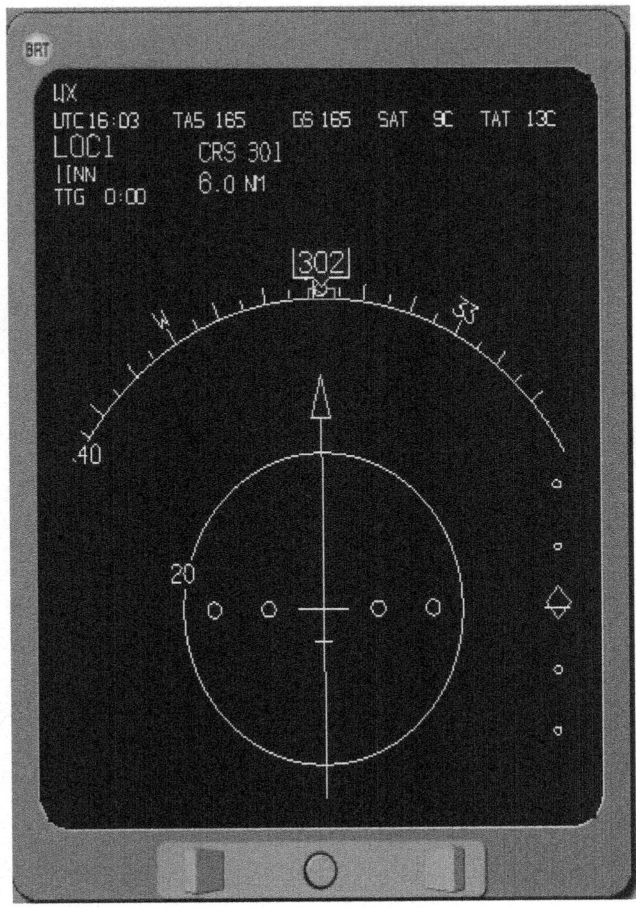

Figure 23 - 41: NAV SECTOR

PUSH X-SIDE

Common to the HSI and the NAV SECTOR pages is the cross-side course information For easier identification, cross-side information is colored cyan.

When the PUSH X-SIDE pushbutton on the DCP is pressed, the following cross-side navigation information is displayed:

- Navigation source indicator
- Course pointer
- Course display
- Lateral deviation bar
- Station identifier
- Distance display

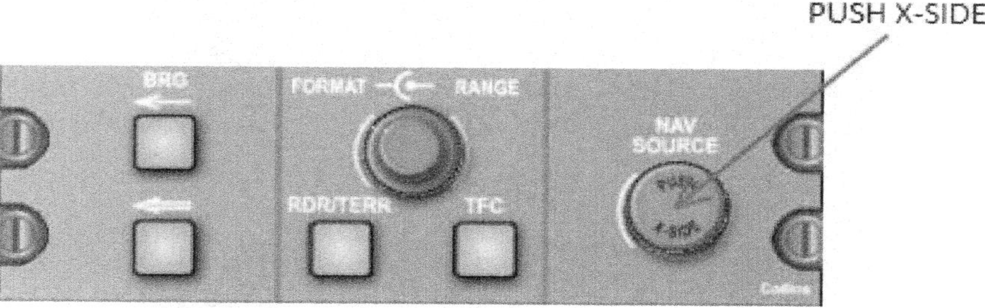

Figure 23 - 42: PUSH X-SIDE Pushbutton

FMS MAP

The FMS MAP displays the track and waypoints as compiled in the flight management system. When in FMS MAP, the actual map heading corresponds to the aircraft heading.

Below the information line on the FMS format is the FROM/TO indications, which give distance and time to the next waypoint.

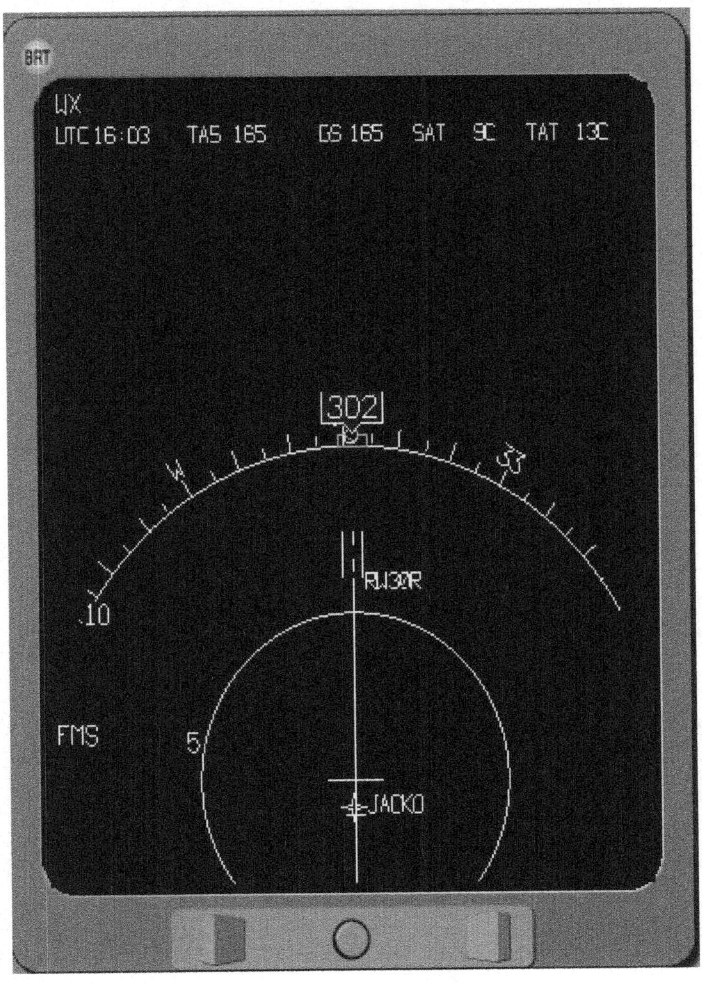

Figure 23 - 43: FMS MAP

FMS PLAN MAP

The FMS PLAN MAP page is used during the initial flight plan setup. This format always shows north at the top of the display. Modified FMS flight plan information can be verified by selecting the up or down scroll keys on the FMS.

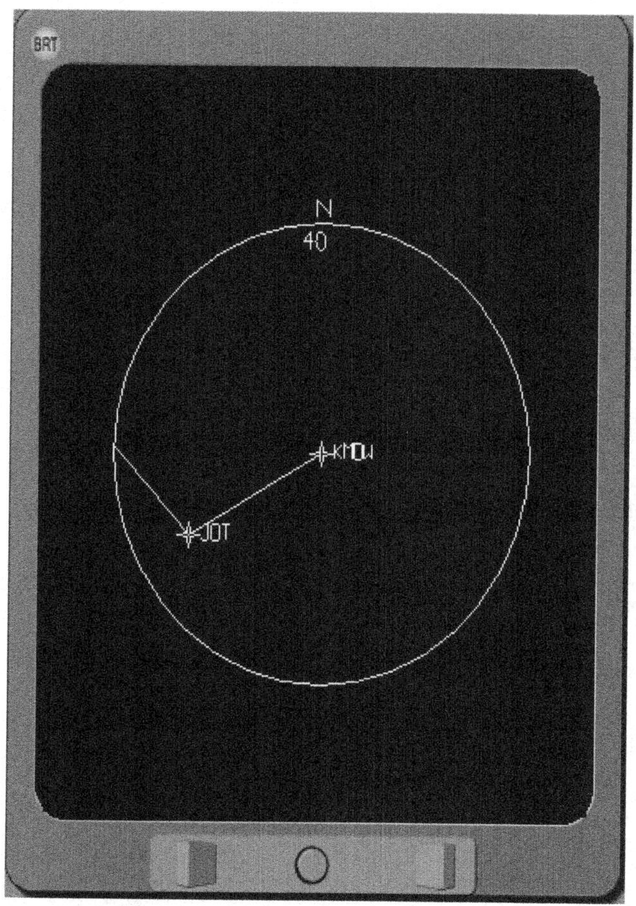

Figure 23 - 44: FMS PLAN MAP

Radar

The RDR selection displays weather radar information on the MFD. A weather radar overlay function is also available for the following MFD formats by selection of the RDR/TERR pushbutton:

- NAV SECTOR
- FMS MAP
- TCAS

The radar/terrain status line operation is displayed on the top line of the radar page. See Navigation chapter, for the following indications:

- Mode
- Antenna stabilization
- Gain
- Antenna tilt
- Ground clutter suppression

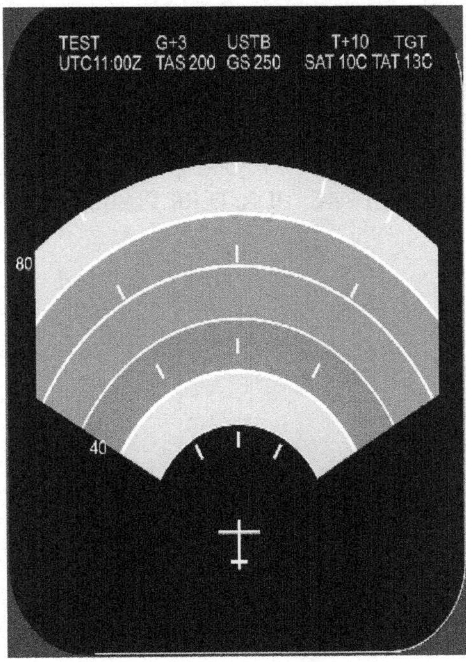

Figure 23 - 45: Radar

RDR/TERR

If the radar is on, pressing the RDR/TERR pushbutton will cause the radar display to be superimposed on the FMS MAP, NAV SECTOR or TCAS pages. A second press of the pushbutton removes the radar information from the display and restores the function TERRAIN.

Enhanced GPWS Terrain Mode

Terrain information is selected by the RDR/TERR pushbutton. Terrain information is displayed on all formats except HSI, FMS PLAN MAP and RADAR.

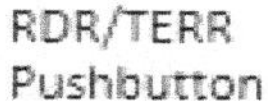

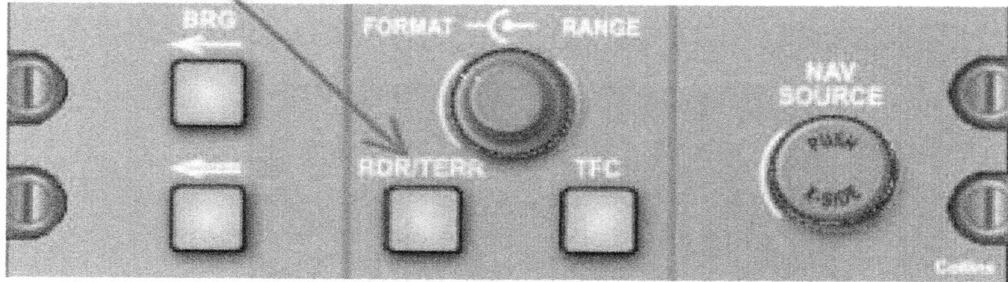

Figure 23 - 46: RDR/TERR Pushbutton

TFC

Pressing the TFC pushbutton on the DCP will display the TCAS traffic page. The TCAS computer identifies the intruding aircraft and categorizes the threat level for display on the TCAS page.

The following traffic symbols are accompanied by altitude data and climb or descent arrows to indicate the vertical speed of the intruding aircraft:

- Resolution advisory (solid red square)
- Traffic advisory (solid yellow circle)
- Proximate traffic (filled cyan diamond)
- Other traffic (white diamond)

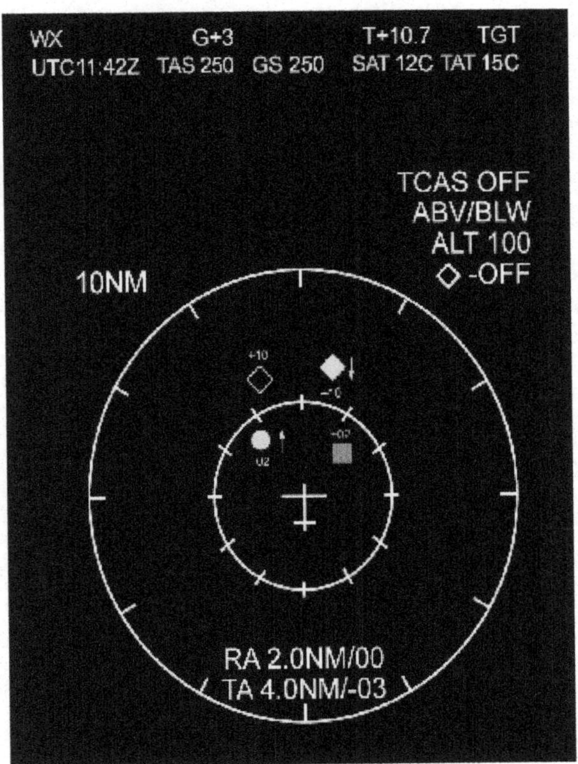

Figure 23 - 47: TCAS Traffic Page

Maintenance Diagnostic Format

The maintenance diagnostic format is used to view data recorded by the maintenance diagnostic computer. When the red-guarded MAINT switch is set to either MFD 1 or MFD 2, the MAINTENANCE MENU page is displayed on the selected MFD. The MAINT switch is located on the circuit breaker panel behind the pilot's seat.

When the maintenance diagnostic switch is selected, the ECP is used to scroll through the MDC data pages. Returning the MAINT switch to the middle OFF position removes the maintenance information from the MFD.

Figure 23 - 48: Maintenance Diagnostic Format

Source Selector Panel

In the event of attitude/heading, air data, EICAS or display control panel (DCP) failure, an alternate source for the disabled system can be selected.

The alternate source is accessed through the source selector panel. The NORM position of each switch is used when both applicable source systems are operational. The '1' or '2' position is used when one system has failed.

Components and Operation

ATT HDG Switch

When NORM is selected, the pilot and copilot electronic flight displays receive data from their individual AHRS/IRS.

When ' 1 ' is selected, the pilot and copilot electronic flight displays receive data from AHRS/IRS 1 only. An amber source message is displayed on the PFD and/or MFD.

When '2' is selected, the pilot and copilot electronic flight displays receive data from AHRS/IRS 2 only. An amber source message is displayed on the PFD and/or MFD.

AIR DATA Switch

When NORM is selected, the pilot and copilot electronic flight displays receive data from their individual air data computers (ADCs).

When '1' is selected, the pilot and copilot electronic flight displays receive data from ADC 1 only. An amber ADC 1 message is displayed on both PFDs.

When '2' is selected, the pilot and copilot electronic flight displays receive data from ADC 2 only. An amber ADC 2 message is displayed on both PFDs.

DSPL CONT Switch

When NORM is selected, the pilot and copilot DCPs control their respective electronic flight displays.

When '1' is selected, the pilot DCP controls the pilot and copilot electronic flight displays. An amber source message is displayed on both PFDs and MFDs.

When '2' is selected, the copilot DCP controls the pilot and copilot electronic flight displays. An amber source message is displayed on both PFDs and MFDs.

EICAS Switch

The EICAS source select is covered in detail in Indicating and Recording System (EICAS) of this manual.

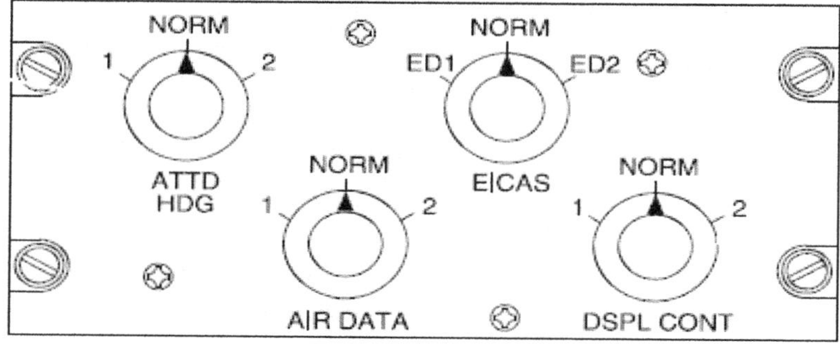

Figure 23 - 49: Source Selector Panel

EFIS Comparison Monitor

Each PFD performs its own software monitoring and continuously compares its information with the data presented on the other PFD. When an EFIS comparison disagreement between the PFDs is detected, a flashing amber data message is presented on the PFD. If after 5 seconds the comparison disagreement is still valid, the flashing is replaced by a steady message and on EICAS, the EFIS COMP MON caution message is displayed.

The messages are displayed as long as the comparison error exists. If the comparison monitor fails, the amber EFIS COMP INOP message is displayed on the primary EICAS page. PFD Comparison Monitor Messages

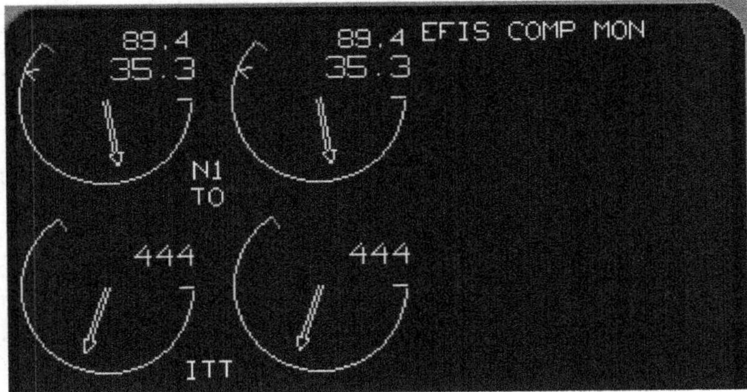

Figure 23 - 50: EFIS COMP MON Caution Message

Heading (HDG)

The HDG message is displayed when a difference o f more than 6 degrees between each AHRS/IRS system is detected.

Roll Attitude (ROL)

The ROL message is displayed when a difference of more than 4 degrees before glide-slope capture and 3 degrees after glide-slope capture is detected.

Pitch Attitude (PIT)

The PIT message is displayed when a difference of more than 4 degrees before glide-slope capture and 3 degrees after glide-slope capture is detected.

Indicated Airspeed (IAS)

The IAS message is displayed when a difference of more than 10 knots is detected.

Altitude (ALT)

The ALT message is displayed when a difference of more than 60 feet at sea level is detected. The allowable difference between PFDs increases with increasing altitude.

Localizer (LOC)

The LOC message is displayed when a difference between the two localizer receivers is detected.

Glide-slope (GS)

The GS message is displayed when a difference between the two glide-slope receivers is detected.

Radio Altimeter (RA)

The RA message is displayed when the radio altimeters are indicating below 1000 feet AGL and a difference is detected.

Display Reversionary Panel

The display reversionary selector switches on the pilot and copilot display reversionary panels (DRP) are used to present a PFD or EICAS format on the associated MFD. The selector can be set to one of three positions:

NORM

The MFD displays navigation, weather radar, EGPWS terrain, TCAS, FMS text data and maintenance diagnosis computer information.

PFD 1 or PFD 2

By turning the knob to the PFD position, the selected PFD information is transferred to the associated MFD.

EICAS

Turning the knob to the EICAS position presents the STATUS page on the MFD. EICAS page selection is then controlled with the EICAS control panel.

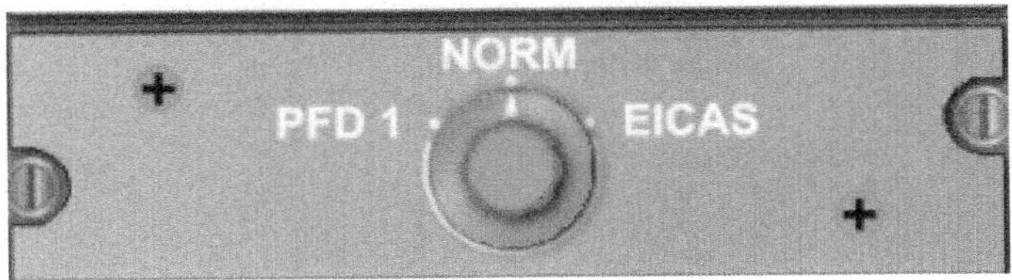

Figure 23 - 51: Display Revisionary Panel

Standby Flight Instruments

A self-contained LCD integrated standby instrument (ISI) that supplies the flight crew with attitude (pitch and roll), air data (altitude, airspeed and Mach), slip/skid, as well as ILS information (localizer and glide-slope).

The ISI is used as an alternative source of flight and navigation data or as a crosscheck t o the electronic flight instrument system.

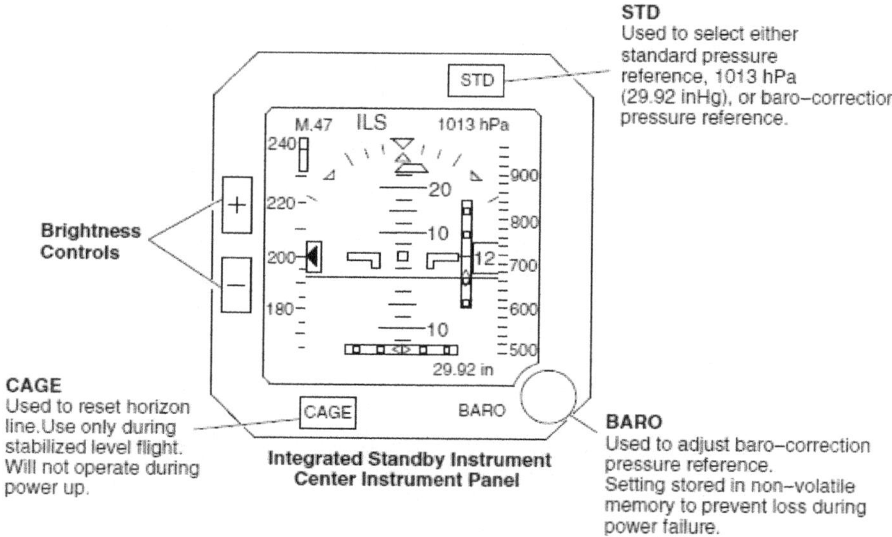

Figure 23 - 52: Integrated Standby Instrument

Components and Operation

Integrated Standby Instrument

The integrated standby instrument (ISI) receives air data information from the standby pitot static system (P3 and S3). The ISI receives electrical power from the 28 vdc battery bus.

Localizer and glide-slope data comes from the VHF NAV 1. The localizer and glide-slope pointers and scales do not come into view until a valid ILS frequency is received. If the localizer or glide-slope data becomes invalid the appropriate red (LOC or GS) flag comes into view and the corresponding pointer and scale are removed.

Aligning Flag

When the ISI is first energized, a n ALIGNING flag i s in view. After approximately 1 minute, the flag is removed from view.

ISI Functions

The brightness control buttons '+' and '-' provide adjustment independent of the instrument panel lighting.

The cage button resets the horizon to 0 when depressed for more than 2 seconds.The standard (STD) pushbutton switches the altimeter display from standard pressure to barometric (BARO) pressure. The rotary BARO knob allows for barometric altimeter adjustment.

Standby Compass

The standby compass is independent and does not interface with other systems. It is a self-contained dry compass which uses eddy current damping to prevent overshooting. A miniature aircraft pointer indicates aircraft heading in relation to magnetic north on a rotating vertical compass card.

A compass correction card, mounted above the instrument, is used to record the values that must be added to or subtracted from the compass indications to correct for the influence of magnetic materials contained in the aircraft and magnetic fields from the avionics systems near the compass. The compass can be illuminated by operating the standby compass switch on the miscellaneous lights panel.

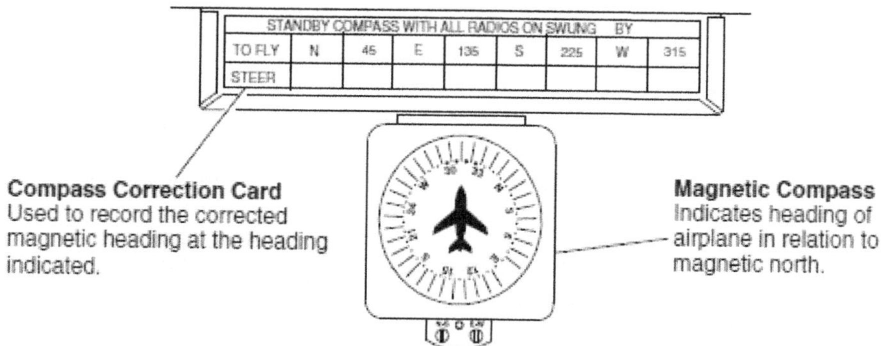

Figure 23 - 53: Standby Compass

Clocks (GPS-Synchronized)

The CRJ airplanes have a digital electronic GPS-synchronized clock installed in each of the pilot and copilot side panels. The GPS clock contains a dual LCD display. The upper six-digit display provides the date (DT), GPS time, internal (INT) UTC time, or local time (LT). The lower four digit display provides the chronometer (CHR) or elapsed time (ET). The clocks are the time base source for the avionics system. A light sensor below the display provides automatic adjustment of display brightness.

Components and Operation

The clocks are synchronized to the GPS input as soon as valid GPS information is received. In the case of invalid GPS data or signal loss, the clocks will operate in internal (INT) mode using the integrated time base of each clock. If there is a valid GPS signal, the clocks do not need to be set, as this will be done automatically at power-up.

The flight crew can disable the GPS signal by entering the time setting mode. The clocks will then ignore the GPS signal until the next primary power reset. The MODE, ET SEL and ET RST pushbuttons are used to set the time and date. To set the clock, push the MODE pushbutton for 2 seconds, then push the MODE pushbutton again to toggle between UTC hours and minutes (when INT is displayed), year, month, and day, (when DT is displayed), and local time hours and minutes (when LT is displayed).

In any of these modes, the ET SEL pushbutton is used to decrease the data and the ET RST pushbutton is used to increase the data. Data changes are in increments of one digit for each press of the ET SEL or ET RST pushbutton. At any time during the time setting process, pressing the MODE pushbutton for a minimum of 2 seconds will exit the time setting mode and restart the clock operation)

Time/Date Display

The time/date display is presented in the upper portion of the clock and displays GPS time, INT time, LT time or date. Time is displayed in hours, minutes, and seconds (HH:MMss). Date is displayed as day/month/year.

ET/CHR Display

The ET/CHR display is presented in the lower portion of the clock and presents elapsed (ET) or chronometer (CHR) time. ET time is displayed in hours and minutes. CHR time is displayed in minutes.

MODE Pushbutton

The MODE pushbutton is used to select the mode of operation (OT, GPS. INT, LT) and is also used in conjunction with the ET SEL and ET RST pushbuttons to set UTC time, date and local time. Holding the MODE pushbutton for more than 2 seconds will enter the manual date/time set function.

CHR Pushbutton

The CHR pushbutton is used to start. stop and reset the chronometer= display. CHR overrides the existing elapsed time display. The first push of the CHR pushbutton starts chronometer. the second push stops chronometer and the third resets chronometer and blanks the display. When in CHR mode, selecting the ET SEL pushbutton will display ET on the lower display although the chronometer will continue to increment in the background.

ET SEL Pushbutton

The ET SEL pushbutton is used to select the elapsed time function when the chronometer function is active. It is also used to decrease the data in one digit increments during manual setting of the clock. Selecting the ET SEL pushbutton before takeoff will display the ET annunciator however no digits will be presented. Once airborne. the digits will appear and begin incrementing. The colon separating the digits will also appear to indicate that the ET counter is incrementing. At touchdown the colon will disappear and the counter will freeze at the elapsed time.Pressing the ET RST pushbutton will reset the elapsed time counter to 0 and blanks the display. The ET RST function is only available on the ground.

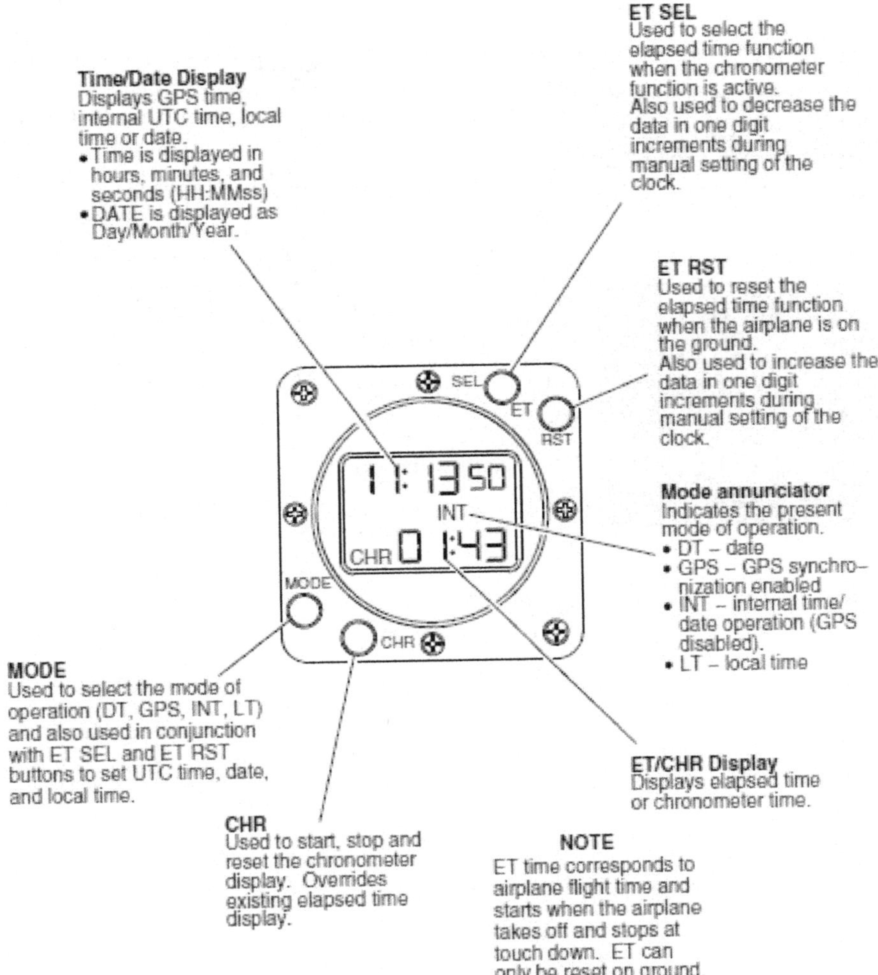

Figure 23 - 54: Digital Clock with GPS Synchronization

EICAS Messages

Master Caution	Cause
EFIS COMP INOP	Comparator information for one or both PFDs is not available
EFIS COMP MON	PFD miscompare has been detected

Status Message	Cause
SPEED REFS INDEP	Captain and First Officer speed bugs (V-speeds) are not synchronized
IRS 1 OVERTEMP	IRS 1 over temperature detected
IRS 2 OVERTEMP	IRS 2 over temperature detected
IRS 1 IN ATT	IRS 1 in attitude mode
IRS 2 IN ATT	IRS 2 in attitude mode

System Schematic

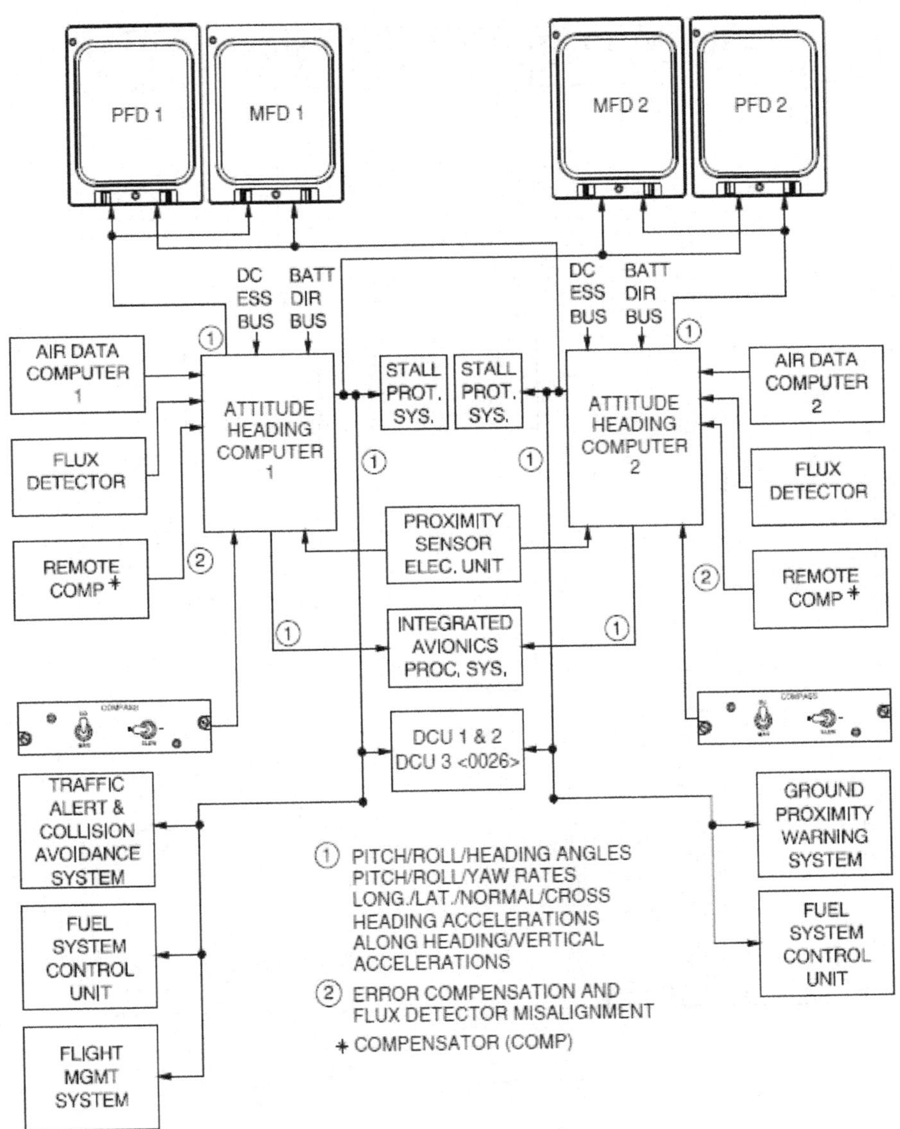

Figure 23 - 55: AHRS Schematic

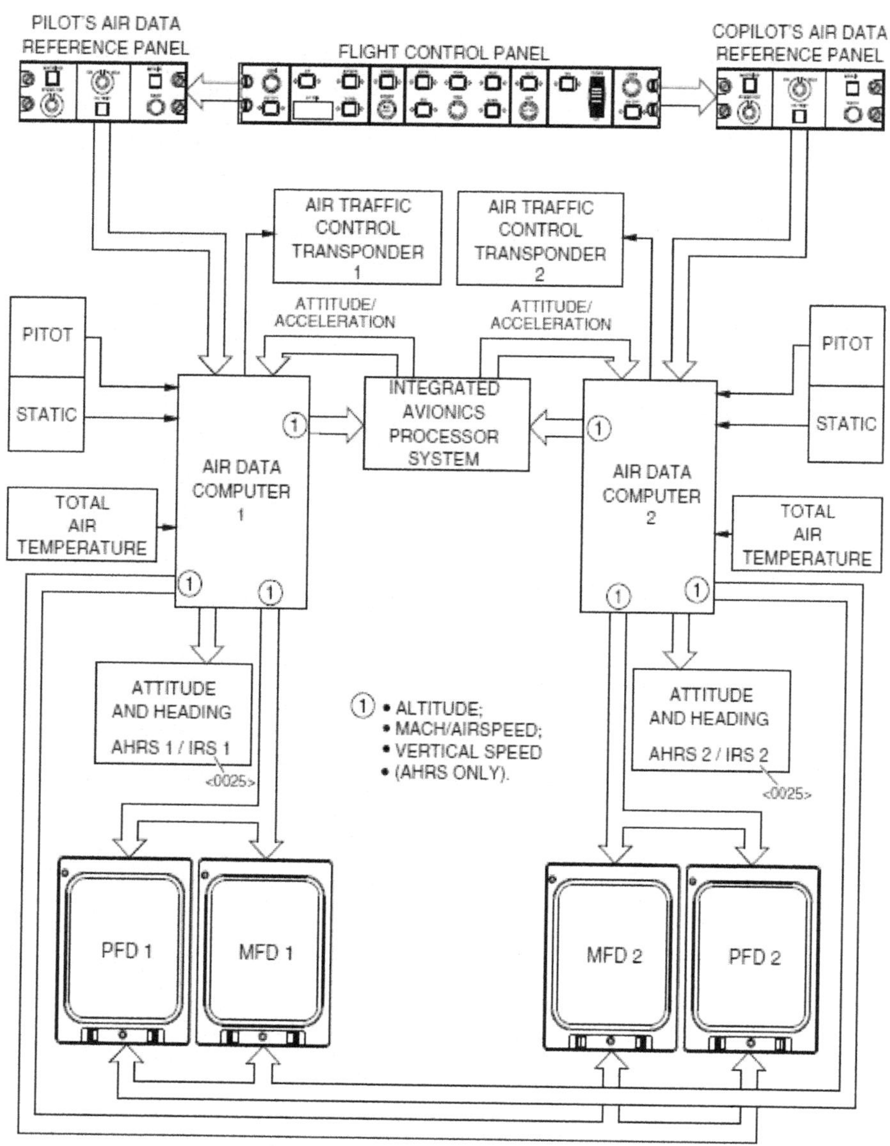

Figure 23 - 56: Air Data System Schematic

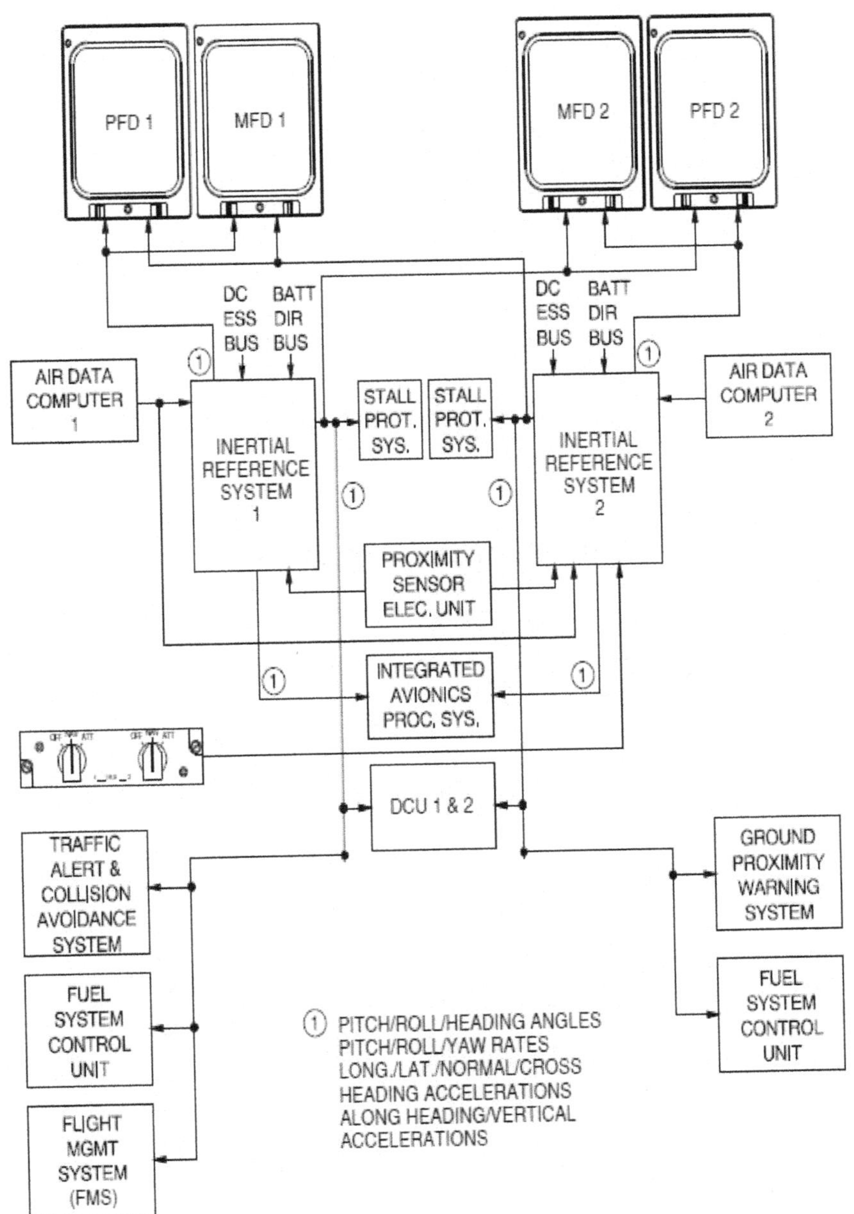

Figure 23 - 57: IRS Schematic

Intentionally Left Blank

Made in the USA
Monee, IL
24 July 2020

36968686R00289